T0190192

Software Testing Automation

Saeed Parsa

Software Testing Automation

Testability Evaluation, Refactoring, Test Data
Generation and Fault Localization

 Springer

Saeed Parsa
School of Computer Engineering
Iran University of Science and Technology
Tehran, Iran

ISBN 978-3-031-22059-3   ISBN 978-3-031-22057-9 (eBook)
https://doi.org/10.1007/978-3-031-22057-9

© The Editor(s) (if applicable) and The Author(s), under exclusive license to Springer Nature
Switzerland AG 2023

This work is subject to copyright. All rights are solely and exclusively licensed by the Publisher, whether
the whole or part of the material is concerned, specifically the rights of translation, reprinting, reuse
of illustrations, recitation, broadcasting, reproduction on microfilms or in any other physical way, and
transmission or information storage and retrieval, electronic adaptation, computer software, or by similar
or dissimilar methodology now known or hereafter developed.

The use of general descriptive names, registered names, trademarks, service marks, etc. in this publication
does not imply, even in the absence of a specific statement, that such names are exempt from the relevant
protective laws and regulations and therefore free for general use.

The publisher, the authors, and the editors are safe to assume that the advice and information in this book
are believed to be true and accurate at the date of publication. Neither the publisher nor the authors or
the editors give a warranty, expressed or implied, with respect to the material contained herein or for any
errors or omissions that may have been made. The publisher remains neutral with regard to jurisdictional
claims in published maps and institutional affiliations.

This Springer imprint is published by the registered company Springer Nature Switzerland AG
The registered company address is: Gewerbestrasse 11, 6330 Cham, Switzerland

Preface

The ability to test software efficiently and thoroughly is a hallmark of high-quality software. For code to be tested thoroughly, it should be testable in the first place. Testers, in a sense, are customers of developers. Developers must undertake an approach that keeps the code clean and efficient. Clean and efficient code is highly testable. Refactoring techniques help develop testable code. As the testability improves, automatic test data generation tools can provide relatively more efficient and effective test data. This book offers a machine learning tool to measure software testability based on which a software development process, Testability-Driven Development (TsDD), in support of test-first approaches is suggested.

Testing is something practical. Software testing is all practice. It relies on knowledge but is more than learning. It is based on programming experience and common sense. This book is all about the design and development of tools for software testing. My intention is to get the reader involved in software testing rather than simply memorizing the concepts. The book has three parts: Software Testability, Fault Localization, and Test data generation.

Part I analyses the test-first development methods, TDD and BDD, and proposes a new method called testability-driven development (TsDD) in support of TDD and BDD. Refactoring is an inevitable part of the TDD and BDD life cycles. The key to developing software tools for automatic refactoring is also covered.

Part II focuses on designing and implementing tools and techniques for automatic fault localization. The essence of this part of the book centers on the design and implementation of a software tool, Diagnoser, to take a program under test and its test suite as input and creates a diagnosis table, representing the statements covered by each test trace and their suspiciousness score.

Part III proposes using software testing as a prominent part of the cyber-physical system software to uncover and model unknown physical system behavior. This part describes the design and implementation of software tools for the automatic generation of test data for a given source code.

I have intentionally begun Chap. 1 of this book with the concept of testability. A straightforward yet fundamental concept overlooked by many programmers. As a software quality factor, testability is the extent to which a software artifact supports its testing. This chapter brings up some critical issues about software preparation for testing. After reading this chapter, the reader will have an idea of what clean code is; why it is fundamental in software development; how to use a compiler generator to develop tools to evaluate the cleanness and testability of software automatically; how to evaluate software architecture; and finally, how to develop tools to automatically re-modularize code.

A good beginning makes a good ending. A good beginning for the testing life cycle is the analysis of technical requirements as the foundation of software development. Requirements form the basis for test case development. Test cases are developed at the requirements analysis stage to verify whether the final software product meets its requirements. During the software development life cycle, test data are generated from the models to examine software structure and behavior.

Chapter 2 briefly describes using Enterprise Architect to define business and technical requirements and generate code from UML models. The chapter proceeds with the practical use of test-driven development (TDD) to complete the generated code. This chapter provides guidelines and step-by-step instructions for installing and using Nunit in Visual Studio and Unitest in Python. In this chapter, the reader will learn how to use requirements specifications and UML models to derive test cases and how to apply TDD.

Chapter 3 describes behavior-driven development (BDD) as an agile methodology to design and develop code based on the acceptance tests satisfying the system requirements defined as user stories. This chapter gives the reader insights into the strengths and weaknesses of different tools support for different programming languages, Java, C#, and Python. BDD supports acceptance testing, while TDD is based on unit tests that test the individual methods in isolation.

Both the TDD and BDD are test-first approaches. Test failures in a test-first approach motivate the developer to design and write code. The test-first philosophy in the first position is to pass the tests with minimal required code and then optimize the code by applying different refactoring techniques—easy peasy. The question is: for what price?

Chapter 4 introduces the new concept of testability-driven development (TsDD) in support of test-first methods. TsDD uses a machine learning model to measure the testability of a given source code without any need to run the code. Frequent refactoring followed by testability measurement helps the developer to write testable code without even running the code. This chapter provides the readers with a code-level description of how to build their own model to measure the coverageability and testability of a given source code.

Chapter 5 describes how to develop software tools to refactor automatically. Compiler generators such as ANTLR and modern compilers such as Roslyn provide programmers with tools to insert, modify, and delete parts of a given source code while walking through the parse tree. In this chapter of the book, the reader will get acquainted with the technique of using a compiler generator to build their own software refactoring tools.

Part II of the book begins with Chap. 6, which describes graph-based coverage criteria to evaluate a test suite in terms of how the paths corresponding to the test cases "cover" a graph abstraction called the control flow graph. The chapter shows the reader how to use the ANTLR compiler generator to develop a software tool to create control flow graphs for a given Java method. Also, it shows how to use the Python AST library to build CFGs for Python functions. This chapter equips the reader with the skill to develop software tools to create CFG for a given source code and to compute prime paths.

Chapter 7 describes the spectrum-based fault localization technique as the most well-known approach to automatically locate faults in a given source code based on the statistics of the program statements in failing or incorrect and passing or correct executions of the code under test. Source code slicing techniques are applied to consider only those statements which directly or indirectly affect the faulty result. This chapter also describes how to develop a software tool to automatically add probes in the code under test to keep track of the execution paths.

Chapter 8 offers the design and implementation of a software tool, Diagnoser, to automatically build a diagnosis matrix for a Python program. A diagnosis matrix is an $n \times m$ matrix, D, in which each entry $D_{i,j}$ indicates whether statement number, S_i, is covered by the test data, T_j. The last column of the matrix shows the suspiciousness score computed for each statement, and the last row represents the test result as failing or passing. Diagnoser takes a Python function and a list of test cases as input and outputs the diagnosis matrix for the function. This chapter also describes how to build software tools to apply and evaluate data flow testing automatically.

Chapter 9 discusses the negative impact of coincidentally correct test cases on the statistics collected for the association of the statements with failing/incorrect and passing/correct runs of the program under test. Based on this hypothesis that sub-paths of a coincidentally correct test appear with a higher probability in the failing runs than the correct/passing ones, a technique to identify coincidentally correct tests is introduced. The execution paths are represented as a sequence of branching conditions or predicates. Each sub-path is a sub-sequence represented as an n-gram. The technique uses cross-entropy to calculate the distance between the n-grams probability distribution in a failing execution path and the others, where each n-gram is a sub-sequence of predicates in an execution path. Code-level description of the software tool to detect coincidentally correct executions in a given program and to locate the suspicious regions of the program is also provided. The readers will get an idea of how to develop their own software tool to detect coincidentally correct executions and locate faults in a given program.

Chapter 10 utilizes graph-based abstraction for boosting coverage-based fault localization. Based on this hypothesis that the root cause of the failure may reside in the discrepancies of a failing execution graph with its nearest neighboring passing graph, the faulty region of code is localized. Such faults are often observable in loops where the fault reveals in a specific iteration of the loop. A combination of Jaccard and cosine vector similarities are used to identify the nearest passing vector to each failing vector. The code-level description of the underlying algorithms of graph-based fault localization enables readers to develop their own graph-based fault localization tools.

Part III of this book begins with Chap. 11, which contains two different meta-heuristic search-based techniques for automatic test data generation. Each test data is a point in the program input space. Test data covering a faulty statement are often very close to each other in the program input space. Therefore, the target of search-based test data generation in the first step should be to generate test data that are very far from each other in the program input space. Once a faulty execution is detected, test data should be generated to determine the domain of inputs covering the faulty execution path and its nearest neighbor to locate the faulty statement. This chapter provides the reader with code-level descriptions of how to write a software tool to locate faults. The reader should learn how to develop their own search-based test data generation tools.

Chapter 12 is centered on software testing to uncover rules underlying unexpected physical system behaviors in cyber-physical systems. Changes in a system's behavior are often at boundaries. Therefore, the concept of behavioral domain coverage to evaluate behavioral domain boundaries is introduced in this chapter. Also, two multiobjective genetic algorithms to generate test data covering the boundaries of behavioral domains are described.

Appropriate test data generation is an essential component of the software testing process. Dynamic-symbolic execution, also known as Concolic execution, is a hybrid method for collecting path constraints and generating test data. The path constraint is given as input to a solver to generate test data, covering the path. Along with the dynamic-symbolic execution of a path, a symbolic execution tree representing all the possible executions through the program is formed gradually. Tree search strategies are conventionally applied to select and flip appropriate branching conditions. Most potential search strategies have their strength and weaknesses. To take advantage of the power of different search strategies and avoid their flaws, the use of meta-strategies, taking advantage of different search strategies is proposed. The question is which combination of search strategies in what order and dependent on what program execution states could be most preferably applied.

Chapter 14 gives the readers insights into using hybrid search strategies and dynamic-symbolic execution to generate test data for white box testing. This chapter focuses on test data generation techniques combining symbolic and concrete (real) executions. The aim is to show the readers how to develop their concolic execution tools and apply a hybrid search method to select the following execution path for concolic execution.

In 1972, Dijkstra made this alarming statement that software testing can show the presence of bugs but never their absence. Following this alarming statement, test adequacy criteria emerged. The question is, how much test data is adequate? Sometimes finding a latent fault in a program is like finding a needle in the sea. It will help execute a path with different input data several times until the fault reveals.

Chapters 11, 12 and 14 introduce a new test adequacy criterion called domain coverage and some algorithms to detect sub-domains of the program input domain that cover a particular path. A key highlight of Chap. 14 is a detailed discussion on developing an algorithm to decide constraint satisfiability. I have proposed a simple algorithm to determine whether a constraint is satisfiable. The algorithm can be applied to non-linear relational expressions, including black-box function calls. The algorithm determines discrete sub-domains of the input domain, satisfying a given path constraint.

In the end, I hope this book will bring light to the dark corners of the software testing industry. No matter how weak the light is, it is still light. Let it shine. "A drop is an ocean only when it is in the ocean; otherwise, a drop is a drop, and an ocean is an ocean."

Tehran, Iran   Saeed Parsa

Acknowledgments

God is *love*, and *love* is God. Thanks God for giving me the *love* and strength to write this book. I would like to express deep gratitude to my students:

- **Morteza Zakeri-Nasrabadi:** My Ph.D. student for helping me by developing software tools and helping with writing chapters four and five of the book about testability and refactoring. These chapters are based on Morteza's thesis and publications with me.
- **Maysan Khazaei:** My M.Sc. student who helped me with writing the fault localization codes described in Chaps. 9 and 10.

Source codes

For more information on the book, copyrighted source code of all examples in the book, lab exercises, and instructor material, refer to the following addresses:

https://figshare.com/authors/saeed_parsa/13893726.

The complete source code of all the software tools described in the book is downloadable from the book's website.

About This Book

As a teacher, I have always tried to make my students understand rather than memorize things. It is easy for a young to memorize something, but it takes time to understand. When students truly understand a term or concept, instead of repeating it verbatim, they can explain it in their own definition, why it is important, and how it relates to other terms and concepts. While memorizing involves understanding the theory behind something, understanding needs a grasp on how the theory plays out in practice.

Software testing is the theory in practice. It relies on knowledge but is more than learning. It is based on programming experience and common sense. To be practical, one needs specific tools to experiment with the concepts and theories of software testing. The lack of such auxiliary software tools has always been a barrier to developing new algorithms and theories for software testing. That is why, in the first place, this book shows the reader how to develop software tools fundamental to software testing. Code-level descriptions of the underlying algorithms are given in the book. A practical understanding of developing the required tools enables the exploration and development of advanced theories and algorithms related to different aspects of software testing. Once the readers learn how to develop the required tools, they can pursue to examine and develop the algorithms and theories described in the book. The source codes are downloadable from the figshare website at:

https://figshare.com/authors/saeed_parsa/13893726

This book is about the design and development of tools for software testing. The book intends to get the reader involved in software testing rather than simply memorizing the concepts. The book has three parts: Software Testability, Fault Localization, and Test data generation.

Part I analyses the test-first development methods, TDD and BDD, and proposes a new method called testability-driven development (TsDD) in support of TDD and BDD. Refactoring is an inevitable part of the TDD and BDD life cycles. The key to developing software tools for automatic refactoring is also covered.

This part of the book focuses on techniques and tools to achieve efficient and effective tests while developing software. The reader will get insight into how to develop machine learning models to measure source code testability. The tool can be used to measure the testability of a given class statically and without any need to run the class.

As the testability of a class improves, the more decidable its testing results will be. Software refactoring tools may improve testability, provided that they improve specific source code metrics affecting testability. The observation is that under certain circumstances, refactoring may degrade testability. For example, if after applying the extract method refactoring, the number of parameters of the extracted method is relatively high, it is observed that the testability degrades.

Therefore, to speed up and improve agile development, a test-first method such as TDD is better off with a cyclical "fail-refactor-pass-refactor" rather than a fail-pass-refactor approach. Testability-Driven Development (TsDD) is a new methodology I introduce in this part of the book. Testability-driven development enables automatic test data generation tools to amplify manually created test suites while ensuring high software testability. It employs two key components to achieve this aim: a testability prediction model and automated refactoring.

I commence this part with an introductory chapter about software testability and its enhancement techniques. Then, describe the most accepted test-first development approaches, including TDD, and BDD, in Chaps. 2 and 3. Chapter 4 introduces a new approach, testability-driven development, in support of TDD and BDD. The part ends with a chapter on automated refactoring, providing the reader with details of how to use a compiler generator to develop software refactoring tools.

Part II focuses on designing and implementing tools and techniques for automatic fault localization. Sometimes finding a fault in a program is like finding a needle in a haystack. However, the main idea behind automatic fault localization is straightforward. The faulty element is the one that is covered by failing test cases more frequently than the passing ones. The difficulty is the lack of some commonly used software tools that researchers and scholars must develop before they can implement and experiment with their new or existing testing algorithms. The software tools include source code instrumentation, slicing, and control flow graph extraction.

This part of the book shows the reader how to use a compiler generator to instrument source code, create control flow graphs, identify prime paths, and slice the source code. On top of these tools, a software tool, Diagnoser, is offered to facilitate experimenting with and developing new fault localization algorithms. Diagnoser accepts a source code and a test suite to test the source code as input. It runs the code with each test case and creates a diagnosis matrix. Each column in the diagnosis represents statements covered by a test case and the test result as failing or passing. The last columns represent the suspiciousness scores computed for each statement. Each row of the diagnosis matrix represents which test cases have covered a particular statement. In addition, the suspiciousness score for each statement using different formulas is shown.

Part III proposes using software testing as a prominent part of the cyber-physical system software to uncover and model unknown physical system behavior. Behaviors are not instantaneous but continuous for a while. They are triggered by specific events but do not change instantaneously. Therefore, precise analysis of unexpected behaviors, especially in cyber-physical systems, requires the domain of inputs triggering the behavior. Unexpected behaviors in a cyber-physical system are sometimes due to unknowns where mathematical models cannot offer judgments. Let us try to cross the unknown by supplying cyber-physical systems with software testing tools to receive test data from the physical system in action and predict unexpected behaviors while extracting the underlying governing physical rules.

This part of the book focuses on developing software tools for test data generation. The reader will get insights into how to develop their own software tools to generate test data for a program under test. Domain coverage adequacy criteria introduced in this part form the basis for evaluating the coverage obtained by the test data.

Contents

Part I
Testability Driven Development

This part of the book focuses on techniques and tools to achieve efficient and effective tests while developing software. The reader will get insight into how to develop machine learning models to measure source code testability. As an inherent property of well-engineered software, testability supports efficient and effective testing and debugging. This part of the book describes code-level details of how to train a machine learning tool to measure the coverageability and testability of a given class based on the class source code attributes. The tool can be used to measure the testability of a given class statically and without any need to run the class.

As the testability of a class improves, the more decidable its testing results will be. Software refactoring tools may improve testability, provided that they improve specific source code metrics affecting testability. The observation is that under certain circumstances, refactoring may degrade testability. For example, if after applying the extract method refactoring, the number of parameters of the extracted method is relatively high, it is observed that the testability degrades.

Therefore, to speed up and improve agile development, a test-first method such as TDD is better off with a cyclical 'fail-refactor-pass-refactor' rather than a fail-pass-refactor approach. Testability-Driven Development (TsDD) is a new methodology I introduce in this part of the book.

Testability-driven development enables automatic test data generation tools to amplify manually created test suites while ensuring high software testability. It employs two key components to achieve this aim: a testability prediction model and automated refactoring.

I commence this part with an introductory chapter about software testability and its enhancement techniques. Then, describe the most accepted test-first development approaches, including TDD, ATDD, and BDD, in Chaps. 2 and 3. Afterward, Chap. 4 introduces the novel testability-driven approach to ensure the development of a testable software product. The part ends with a chapter on automated refactoring, which shows the reader how to use a compiler generator to develop software refactoring tools.

Chapter 1
Software Testability

1.1 Introduction

Software testing should begin with the testability assessment. Testing before assessing code testability is a potentially costly and ineffective approach. Before a piece of code can be tested thoroughly, certain precautions must be taken to ensure the code is in a testable state. Increasing the degree of testability reduces the cost of testing. Testability metrics measure the ease with which a software artifact can be tested. If the software is not testable enough, it should be refactored or rewritten to prepare for the test.

This chapter brings up some critical issues about software preparation for testing. After reading this chapter, the reader will know what clean code is, why it is fundamental in software development, how to develop tools to evaluate the cleanness and testability of software automatically, how to evaluate software architecture, develop tools to automatically re-modularize code, detect patterns in a given source code automatically, and use it as a basis to evaluate the code testability.

1.2 Testability

The question is how to determine the extent to which a software artifact or design is testable. The answer to this question requires making it clear what the main features of testability are. Testable code is clean, modular, and reusable. Evaluation of testability is the subject of Sect. 1.4. The main prerequisite for testability is the cleanness of the code under test.

Clean code is readable, looks like well-written prose, and does not hide bugs. There is no ambiguity and no side effects in clean code. In clean code, functions are small with single responsibilities. They use no global and solely operate on their input parameters and return values, independent of any other function state.

© The Author(s), under exclusive license to Springer Nature Switzerland AG 2023
S. Parsa, *Software Testing Automation*,
https://doi.org/10.1007/978-3-031-22057-9_1

Section 1.3 describes the use of source code analysis tools to develop software tools for automatically evaluating the source code based on clear code principles.

Dependencies between components make it challenging to track and test programs. Dependencies could be lessened by applying modularity principles. The modular code is well structured because it consists of cohesive, loosely coupled, and understandable components/modules. Modular code is reusable and reusable code is testable. Conversely, testable code is reusable. If a piece of code performs as intended, it can be reused elsewhere. The internal details of each module are hidden behind a public interface, making it straightforward to comprehend, test, and refactor independently of the others. Properly encapsulated modular code is more testable than non-modular code. Section 1.5.2 presents a method to automatically re-modularize a given source code based on the modularity principles.

Besides modularity, effective use of certain design patterns such as Factory and Dependency Injection could alleviate dependencies and thereby testability. Factory and Dependency Injection design patterns and their impact on software testability are described in Sect. 1.6.1. In general, appropriate use of design patterns could improve software quality. Therefore, the type and number of design patterns used in a software product could indicate its quality and testability. Section 1.6.2 presents a genetic algorithm to find sub-graphs of a program dependency graph, matching a given pattern class dependency graph.

Testability facilitates testing and defect exposure. Writing software with testability is one of the main elements of writing high-quality software components; otherwise, the code should be refactored to get in a testable state. For instance, consider the following code fragment:

```
if ( GetBladeRotationSpeed() > 4000 )
{
    // Run all the codes for Velocity Governor  .
```

It is needed to get the blade speed in a local variable to facilitate high-level testing and debugging. This way, watching and analyzing the blade speed will be possible. The code is refactored as follows:

```
current_blade_speed = GetBladeRotationSpeed();
if ( current_blade_speed > 4000 /* Rotation per minutes */ )
{
    // Run all the codes for Velocity Governor.
```

Although testability cannot be immediately measured (such as the size of the laptop), it should be considered as the inherent property of a software artifact because it is highly correlated with other keys. There are several metrics to assess the testability of software systems. These metrics are described in the next section.

1.3   Clean Code

The main goal of this book is to provide its readers with solid expertise in the automation of the software testing process. This section is concerned with the use of compiler generators to automatically analyze a program source and highlight the parts that do not comply with clean code principles. In Sect. 1.3.1, a brief description of clean code principles is given. In Sect. 1.3.2, the Microsoft Framework SDK, called Roslyn, is introduced. Section 1.3.3 focuses on using Roslyn to write a program that accepts a C# class or component as its input, evaluates the source code based on the clean code principles, and outputs the evaluation results.

1.3.1   Clean Code Principles

Clean code is a set of rules and principles meant to keep code testable and maintainable [1, 2]. Implementing clean code principles is a fundamental skill that pays off exceptionally well when it's time to refactor or test code. Clean code is easier to understand and test. The foundation is clean software if you see it as part of a home. Clean code is supplied with acceptance and unit tests. Names in clean code are meaningful, and functions are small and have single responsibilities. Clean code is of minimal dependencies to avoid fault propagation through dependencies. The dependencies are explicitly defined and provide clear and minimal APIs. Experienced programmers have embraced the truth that, as the size of the code increases, the number of bugs increases exponentially. While we do our best to eliminate the most critical bugs (through patterns and careful debugging), we never catch them all.

(1)   **Naming conventions**

Clean-code naming conventions suggest using nouns and noun phrases for variables and verbs for functions. An ideal name is self-explanatory without explaining why and for what reason it exists, what it is up to, and how to use it. If a name needs a comment, then it is not suitable.

All the names defined by the programmer in a program are kept at compile time in the symbol table and parse tree. To this end, compiler generators such as CoCo/R [3], supplied with the syntax of C#, and Antler [4], supplied with the syntax of the Java programming language, can be greatly useful. Also, modern compilers such as Roslyn [5], described in the next section, simplify the task of analyzing source code by providing APIs to enable programmers to access the syntax tree built at compile time.

A cyclopedia can be used to find out whether the names are meaningful. Natural language processing software tools such as wordnet and Pos can identify and tag the names as nouns or verbs. Composite names are preferably used to provide the reader with more details. When assigning a composite name, the first word should be a verb or a noun, dependent on whether the name is assigned to a variable or a method. For instance, 'computeTax' is a proper name for a method. Section 1.5.3 describes the use of Roslyn compiler APIs to develop a software tool for evaluating C# source code based on the clean code naming conventions.

Names should be pronounceable and searchable. Single-letter names such as i or j are not searchable. If such names are going to be used in different locations of the program text, they should be replaced with search-friendly names.

(2) **Methods**

Methods should be small, with at most three arguments. If more than three arguments are needed, then the parameters should be wrapped into a class of their own. A method should affect only its own return value. The use of global variables and call-by-reference cause side effects. For instance, it is clear that the instruction:

$$a = f(b, c);$$

returns a value that is copied in 'a'. However, if call by reference were used, it would not be clear how 'b' and 'c' were modified.

A method should have one, and only one, responsibility. The question is what the responsibility of a method is all about. A method's responsibility is typically to do some computations and save the results, display them, or return them to the caller. By using a slicing technique [6], the method could be automatically sliced into a set of cohesive sequences of instructions, where each subset of sequences carries a different responsibility. Slicing is a technique to determine the traces of program statements, directly or indirectly affecting a variable's value at a specific location in the program. For instance, consider the method shown in Fig. 1.1.

Switch statements oppose the single responsibility principle because each case clause in a switch statement may trigger a different responsibility. The suggestion is to avoid using blocks with more than one line in 'if,' 'while,' or 'for.' Statements.

```
1    double computeSumAndProduct(          double ccomputeSum(double no){
2           double no){                         double sum = 0.0;
3        double sum = 0.0;                      for(int i = 1; i < no; ++i)
4        double product = 1.0;                     sum = sum + i;
5        for(int i = 1; i < no; ++i) {          return sum;
6           sum = sum + i;                   }
7           product = product *i;    Responsibility  void ccomputeProduct(double
8        }                                                       no){
9        printf("-product = %f", product);    double product = 1.0;
10       return sum;                           for(int i = 1; i < no; ++i)
11   }                                            sum = sum + i;
12                                             printf(" product = %f",
                                                        product);
                                            }
```

Fig. 1.1 Automatic detection of responsibilities

Also, the use of try/catch blocks is not recommended because they complicate the structure of the code and intermingle normal with error processing. Instead of the try/catch blocks, the use of functions throwing exceptions is recommended. This way, error handling will be separated, making the code readable.

(3) **Classes**

According to SOLID principles [7], a class should have a single responsibility. SOLID defines responsibility as a reason to change because the more responsibilities of a class, the higher the probability of change requests the class will get. For instance, the responsibility of an invoice object should be to hold the attributes and business rules of an invoice. The responsibility of a class is further broken down into some objectives, each being fulfilled by a different method. The methods collaborate to fulfill the class responsibility. That is why the methods of a class with a single responsibility are highly cohesive. Otherwise, the non-cohesive methods should be removed from the class.

In this way, since distinct classes concentrate on distinct functionalities, the test of a certain functionality will be focused on its corresponding class only. A class name should reflect its responsibility. It is also advised to write a comment in about 25 words regarding the class obligation.

1.3.2 *Roslyn Compiler*

Roslyn is a C# compiler with rich code analysis APIs. The APIs provide access to the syntax tree representing the lexical and syntactic structure of the source code.

To install Roslyn, run the visual studio installer. In the installer window, select Modify and click the Individual components tab. Check the box for .Net Compiler Platform SDK in the newly opened window.

- Visual Studio Installer $\Rightarrow$ More $\Rightarrow$ Modify $\Rightarrow$ Individual components $\Rightarrow$ .NET Compiler Platform SDK $\Rightarrow$ Modify

After Roslyn is installed, the following two capabilities will be added to the Visual Studio environment:

1. The syntax tree for any selected part of the program can be visualized via the following option:

$$"View \Rightarrow Other\ Windows \Rightarrow Syntax Visualizer."$$

Syntax Visualizer is a Roslyn SDK tool to display syntax trees.

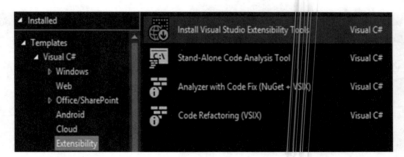

2. In the "new project" window, a new option, Extensibility, enables access to
    Roslyn library functions for the newly developed project.

1.3.3   Using Roslyn to Clean Code

This section presents a detailed step-by-step description of how to develop a C#
program to check a sample class against the following clean code naming and function
principles:

- Chose searchable, unambiguous, and meaningful names,
- Choose verbs to name methods. If a method name consists of more than one word,
  it should begin with a verb. In this case, apply lower case camel forms such as
  mySalary and payTax.

- Chose nouns to name methods. In the case of combined names, the names should begin with a name.
- The number of parameters of a method should not exceed 4.
- The number of lines in a method should not exceed a screen.

The following sample C# class is going to be analyzed by the C# program, described in the continuation:

```
public class CalculateExpression{
        public int AddExpression(int x, int y){
            int result;
            result = x + y;
            return result;
        }
}
```

If the Roslyn compiler is not already installed, use the following commands to install the Microsoft compiler framework SDK:

- Visual Studio Installer ⇒ More ⇒ Modify ⇒ Individual components ⇒.NET Compiler Platform SDK ⇒ Modify

In the visual studio environment, create a console application project called CleanCode. To this aim, the following commands are used:

- File ⇒ New ⇒ Project ⇒ Visual C# ⇒ Windows Desktop ⇒ Console App (.NET Framework) ⇒ Name: CleanCode ⇒ Framework:.NET Framework 4.6.1 ⇒ Ok

As a result, a new project, listed in Fig. 1.2, will be created.

After the project is created, the .Net Compiler Framework should be installed to use the Roslyn APIs. To add Roslyn or, in other words,.Net Compiler Framework SDK to the project, right-click on the project's name in Solution Explorer. Select the Package Manager option. A new window opens. Click on the Browse

Fig. 1.2 The Code to be completed in this section

```
using System;
using System.Collections.Generic;
using System.Linq;
using System.Text;
using System.Threading.Tasks;

namespace CleanCode {
    class Program {
        static void Main(string[] args)
        {
        }
    }
}
```

- Solution Explorer ⇒ References ⇒ Manage NuGet Packages. ⇒ Browse ⇒ Microsoft.CodeAnalysis (Version: 2.0.0) ⇒ Install ⇒ I Accept

It should be noticed that the versions of .NET Compiler Framework and Microsoft.CodeAnalysis should be compatible. For instance, we set the versions of .NET Framework and Microsoft.Code.Analysis to 4.6.1 and 2.0.0, respectively. Now, to generate a parse tree for the sample code, shown in Fig. 1.1, the code in Fig. 1.3 is inserted into the method main in Fig. 1.2.

In the code shown in Fig. 1.3, the CSharpSyntaxTree.ParseText() method is invoked to create a syntax tree for the source code, passed as a string parameter to it. The root of the syntax tree and pointers to its descendants, including class, method, and variable declarations, is accessed by executing the instructions in Fig. 1.4.

The code snippet, in Fig. 1.5, evaluates class, method, and variable names. The method invokes a function isClean() to ensure that a given class or variable name is a noun. Similarly, it evaluates a method name as a verb.

In the above code snippet, the method isClean analyses each class, method, and variable name. The names may be composed of several words. The regular expression in Fig. 1.6 splits the names into distinct words, provided that they are in the Camel case. In the Camel case form, the words in the middle of the name phrase begin with a capital letter.

After splitting composite names into distinct words, each word is looked up in a dictionary by the following method.

According to the clean code naming conventions, class and variable names should be nouns, whereas method names should be verbs. A Part Of Speech Tagger or POS tagger, in short, is a software tool that reads the text in some language and determines whether it is a verb, noun, or adjective. POS trigger may be installed in the visual studio environment by using the following commands.

- Download Stanford POS Tagger full archive with models from the following URL:

```
using Microsoft.CodeAnalysis.CSharp;
var tree = CSharpSyntaxTree.ParseText(@"
    public class CalculateExpression
    {
        public int AddExpression(int x, int y)
        {
            int resultExpression;
            resultExpression = x + y;
            return resultExpression;
        }
    }
    ");
```

Fig. 1.3 Generating pars tree

```
using Microsoft.CodeAnalysis.CSharp.Syntax;
using System.Linq;
var root = (CompilationUnitSyntax)tree.GetRoot();
var classDeclarations = root.DescendantNodes().OfType <
                        ClassDeclarationSyntax>();
var methodDeclarations = root.DescendantNodes().OfType<
                        MethodDeclarationSyntax>();
var variableDeclarator = root.DescendantNodes().OfType<
                        VariableDeclaratorSyntax>();
```

Fig. 1.4 Saving pointers to certan nodes of the syntax tree

```
foreach (var className in classDeclarations){
    isClean(className.Identifier.ToString(), "class", "/NN");
    }
    foreach (var methodName in methodDeclarations) {
    isClean(methodName.Identifier.ToString(), "method", "/VB");
    }
    foreach (var varableName in variableDeclarator) {
    isClean(i.Identifier.ToString(), "variable", "/NN");
    }
```

Fig. 1.5 Evaluation of the method,

```
public static string SplitCamelCase(string input){
    return System.Text.RegularExpressions.Regex.Replace(
        input, "([A-Z])", " $1",
        System.Text.RegularExpressions.RegexOptions.Compiled
        ).Trim();
}
```

Fig. 1.6 A regular expression to split a noun phrase into words

https://sergey-tihon.github.io/Stanford.NLP.NET/StanfordPOSTagger.html.
- Solution Explorer: References ⇒ Manage NuGet Packages… ⇒ Brows ⇒ Stanford.NLP.POSTagger (Version: 3.9.2) ⇒ Install ⇒ I Accept

The method listed in Fig. 1.7 evaluates names as nouns, noun-phrase, or verbs. This method accepts as its input a name and returns the type of the name as noun-phrase, noun or verb as "\NNP," "\NN" or "\VB," respectively.

The method listed in Fig. 1.7 is invoked by the isClean method in Fig. 1.7 to evaluate the names of classes and variables (Fig. 1.8).

The code snippet, listed in Fig. 1.9, counts the number of parameters of the method in the methodDeclarations variable. If the number of parameters of the method exceeds four a message is prompted to notify the user.

```
using java.io;
using java.util;
using edu.stanford.nlp.ling;
using Console = System.Console;
using edu.stanford.nlp.tagger.maxent;
public static string Tag(string x){
    var model = new MaxentTagger(
                        "C://stanford-postagger-full-2016-10-31
                            //models//wsj-0-18-bidirectional-
                            nodistsim.tagger", null, false);
    var sentence = MaxentTagger.tokenizeText(
                            new StringReader(x)).toArray();
    foreach (ArrayList i in sentence){
            var taggedSentence = model.tagSentence(i);
        return SentenceUtils.list string(taggedSentence, false);
    }
    return null;
}
```

Fig. 1.7   Detecting the type of a given name as noun-phrase, noun, or verb

```
public static void isClean(string itemName, string itemEntity, string itemType){
    string text1 = SplitCamelCase(itemName);  // See Figure 1.6
    char Separator = '';
    string[] text2 = text1.Split(Separator);
    for (int j = 0; j < text2.Length; j++) {
        bool isEnglishWord = IsEnglishWord(text2[j]); //See Figure 1.7
        if (isEnglishWord != true) {
            Console.WriteLine("Please check the {0} name {1}.
                                It contains a non-English word.",
                                itemEntity, itemName);
            break;
        }
    }
    if (Tag(text2[0]) != text2[0].ToString() + itemType){
        string w; noune
        if (itemType == „/NN") w = "noun";
        else  w = "verb";
        Console.WriteLine("Please check the {0} name {1}.
                            Its first word is not the {2}.",
                            itemEntity, itemName, w);
    }
}
```

Fig. 1.8   A method to detect the type of the names appearing in a given method

```
foreach (var i in methodDeclarations){
    if (i.ParameterList.Parameters.Count > 4){
        Console.WriteLine("The number of parameters in method named {0}
                           is greater than 4.", i.Identifier);
    }
}
```

Fig. 1.9 A code snippet to count the number of parameters of a method

```
foreach (var i in methodDeclarations){
    var firstLine = tree.GetLineSpan(i.Span).StartLinePosition.Line;
    var lastLine = tree.GetLineSpan(i.Span).EndLinePosition.Line;
    if (firstLine - lastLine + 1 > 24){
        Console.WriteLine("The number of method lines {0} is greater than 24.",
                           i.Identifier);
    }
}
```

Fig. 1.10 A code snippet to count the number of lines of a method body

A method size should not exceed a screen that is 24 lines. The code snippet in Fig. 1.10 counts the number of lines in the method body.

1.4 Testability Metrics

Before testing a software system, it is wise to measure its testability. The main questions to be answered are: which factors affect testability and how to measure and improve testability. Testability is simply the ease with which software can be tested. It is possible to measure testability at both the design and source code levels. The measurement target is often either the number of required test cases or the test complexity. At the design level, the dependencies of the components on each other are the basis for measuring their testability. The fewer the dependencies, the higher the testability of the software components will be. Understandability is also a critical factor for measuring the design testability because the design has to be understood by the developer and tester. Design metrics are the subject of Sect. 1.4.3.

Faults and testability are instinctively correlated. The ease with which errors can be revealed is an aspect of testability. This aspect is the main concern of Sect. 1.4.1. Another aspect of testability concerns the changes to the outputs and inputs of the software components to make their activities both observable and controllable. The main factors affecting software testability are controllability, observability, stability, simplicity, and availability. Traditionally, control-flow and data-flow graphs are used at the source level to determine two testability metrics: controllability and observability. Controllability and observability and their impacts on software testability are the subjects of Sect. 1.4.2.

1.4.1 Object-Oriented System Metrics

Determining which classes of a software system cause it to be relatively harder to test allows for the possibility of resource allocation and planning for modifying the classes to improve their testability. Also, poor testability provides a clue to lousy programming practice. The emphasis is on source-level testability because source code provides a direct rendering of the implementation and is suitable for automatic processing. The main elements of an object-oriented system are methods and classes described in the continuation.

(1) Method Testability

The number of branching conditions in the body of a method affects its testability because the more the branching conditions, the more test cases will be required to cover the alternative branches and paths in the method body. The number of basis paths in the control flow graph of a method, m, is considered as a metric, called Cyclomatic Complexity (CC) [8], to measure the testability of the method:

$$CyclomaticComplexity(m) = E - V + 2^{*}C$$

where,

C   the number of connected components in CFG,
V   the number of vertices in the CFG,
E   the number of edges in the CFG.

For instance, the cyclomatic complexity of the quick-sort algorithm, shown in Fig. 1.11 is:

$$cyclomaticComplexity(quickSort) = 12 - 9 + 2 * 1 = 5$$

The effort required for testing a method can be measured in terms of Cyclomatic Complexity (Fig. 1.11).

Fig. 1.11 Control flow
graph of quik sort

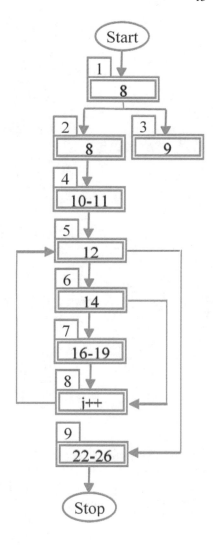

```
1.  /* QuickSort:
2.  arr[] --> Array to be sorted,
3.  low  --> Starting index,
4.  high  --> Ending index */
5.  void quickSort(int array[],
6.  int low, int high)
7.  {
8.    int i, pivot;
9.    if (low >= high)
10.      return;
11.     pivot = array[high] ;
12.     i = (low - 1);
13.     for (int j = low; j <= high- 1; j++)
14.     {
15.       if (arr[j] <= pivot)
16.       { //Swap
17.           i++;
18.           temp = arr[i];
19.           arr[i] = arr[j]);
20.           arr[j] = temp;
21.       }
22.     }
23.     temp = arr[i+1];
24.     arr[i+1] = arr[high]);
25.     arr[high] = temp;
26.     quickSort(array, low, i);
27.     quickSort(arr, i + 2, high);
28.  }
```

Cyclomatic complexity measures the structural complexity of methods [8]. The size complexity can be measured in terms of the number of operators (e.g., $+$, $-$, $*$ and $/$, if, return, break, while), and the number of operands (e.g., Identifiers and constants) [9] as follows:

$$Length = No.Operators + No.Operands$$

$$Vocabulary = No.Unique\ Operators + No.Unique\ Operands$$

$$Volume = Length * \log 2(Vocabulary)$$

$$Difficulty = (No.Unique\ Operators/2) * (No.Operands/No.\ Unique\ Operands)$$

$$Predicted\ No.Bugs\ =\ Volume/3000$$

The above equations were introduced by Halstead [9] in 1977 to measure software complexity. The equations end up with the predicted number of bugs, which indicates fault sensitivity. The fault sensitivity of a method depends on the probability of the method's execution and the probability of the method receiving or propagating fault to others:

$$Fault\ Sensitivity\ =\ Execution\ probability$$
$$*\ Infection\ probability$$
$$*\ Fault\ propagation\ probability$$

The *execution probability* of a method, m, is estimated as the number of times the method is executed divided by the number of the program runs with randomly selected input data. The probability of a method data state getting corrupted by the faults propagated from outside the method gives an estimate for *the infection probability*. The infection probability increases depending on the method's implicit and explicit inputs. Explicit inputs of a method are its class attributes and the accessible fields of the associated classes. Explicit inputs are the parameters of the method. The visibility of a method, m, is estimated as [10]:

$$Visibility(m) = \frac{number\ of\ implicit\ inputs + number\ of\ explicit\ inputs}{number\ of\ implicit\ outputs + number\ of\ exlicit\ outputs}$$

It will be much easier to test a method if there is a one-to-one mapping between its inputs and outputs. The testability of a method, m, depends on its visibility:

$$Testability(m) = K * visibility(m)$$

where K is a count of the variables that may have been modified incorrectly by the method. The value of K is determined by either implicit input parameters, explicit inputs, return values, or exceptions.

A method's testability depends on the number of changes the method makes to map its domain to its range. It is demonstrated [11] that the size of a method in line of code provides an appropriate indication of its testability. The more instructions in a method, the more difficult it will be to test the method.

The domain of a component is its input space. The input space, in general, is an equilateral n-dimensional space, where each dimension of the space represents the values of an input variable. For instance, if there are three inputs, the input space will be three-dimensional, each dimension representing the values of a different input parameter. If a method provides a one-to-one mapping between its input and output spaces, it will be much easier to test. The fault sensitivity of a method can be estimated as the size of its input space or domain divided by the size of its output space or range [12]:

$$Fault\ Sensivity_{a\ Component} = \frac{|Domain_{a\ Component}|}{|Range_{a\ Component}|}$$

If a method provides a many-to-one mapping from its domain to its range, then it will be more fault sensitive because there will be a higher chance of revealing a fault in its inputs. For instance, consider the following method signature:

Boolean a Method(int P1 in range 1..10, int P2 in range 1..10,
int P3 in range 1..10)

The input space, or in other words, the domain of the method, aMethod, is a cube of the size 10*10*10. The fault sensitivity of aMethod is:

$$Fault\ Sensivity_{a\ Component} = \frac{|Domain_{a\ Component}|}{|Range_{a\ Component}|} = \frac{2}{|1000} = 500$$

Domain to Range Ratio (DRR) measure is a simple metric that measures the ratio of the information flowing in a component to the information flowing out of the component, provided that there are no side effects. The testability of a method in terms of its fault sensitivity can be estimated as follows [8]:

$$Testability_{a\ Method} = \frac{Execution\ Rate_{a\ Method}}{Fault\ Sensivity_{a\ Method}}$$

(b)   Class testability

The size of the test suite required to test a class is a measure of the effort required to test the class. Cyclomatic complexity is used to estimate the minimum number of test cases required to execute the basic paths of a method. Therefore, if there are n methods in a class, the minimum number of test cases required to test the class will be equal to:

$$Cyclolmatic\ Complexity(a\ Class)$$

$$= \sum_{i=1}^{n} Cyclomatic\ Complexity(Method_i)$$

A class may inherit methods and attributes through inheritance relation from its superclass. Therefore, the depth of a class in the inheritance tree (DIT) is considered a testability metric. The testability of the methods influences the testability of a class. Another vital metric affecting class testability is the number of methods outside a class invoked at least by a method in that class. Based on these metrics, the testability of a class can be measured as follows [13]:

$$Testability(aClass) = no.Methods\ Of(a\ Class)$$
$$+ Dit(a\ Class) + Coupling(a\ Class, other\ Classed)$$

The depth of a class in the inheritance tree, DIT, highly affects its testability because all the operations and attributes of a superclass are accessible through its subclasses. If a class depends on the others, it will be challenging to test the class in isolation. Conversely, if other classes are dependent on a class, then any changes in the class will be propagated to its dependent classes. Two classes are coupled if one class uses methods or attributes of the other class and vice versa.

The understandability and complexity at the design level are two main quality factors affecting the testability of the classes [14–16]. The understandability factor is estimated in terms of the number of attributes, the number of associations, and the maximum depth of the class in the inheritance tree. The dependencies of a class on other classes aggravate its complexity.

$$Understandability = 1.33515 + 0.12^*no.Associations$$
$$+ 0.0463^*no.Attributes + 0.3405^*MaxDIT$$

$$Complexity = 90.8488 + 10.5849 * Coupling - 102.7527 * Cohesion$$
$$+ 128.0856 * Inheritance$$

$$Testability = -483.65 + 300.92 * Understandability - 0.86 * Complexity$$

Depth in the inheritance tree (DIT), for a class within a generalization hierarchy, is the longest path from the class to the root of the hierarchy.

1.4.2 Observability and Controllability

The primary concern of testability is to have *controllable* input and *observable* output. The *controllability* of a component is the ease with which its input values can be accessed from the system's inputs, while *observability* is how ease outputs of a component can be observed at the system's outputs.

Backward slices connecting inputs of a component to the system inputs determine the component controllability, while observability is measured as the number of forwarding slices from the component outputs to the system outputs.

A backward slice at a location L of a program P for variable V is the set, C, of all the statements S in P that directly or indirectly affect the value of V. The location L and the variable V form the slicing criterion, usually written <L, V> . The forward slice concerning the variable V at location L consists of all the statements, S, in the program P, affected by the value of V at L.

Backward slices for the inputs of a component begin with the inputs and trace back through the sequence of statements affecting directly or indirectly the inputs and end up with the input sources of the program. Conversely, forward slices for the outputs of a component begin with the outputs and follow statements being affected directly or indirectly by the outputs as far as it reaches the outputs of the program.

Controllability concerns the measurement of information available at a component's input, while *observability* quantifies the amount of information propagated from the component output to the system output. In a testability measurement tool called SATAN [17], the testability of a component, M, is computed in terms of its controllability and *observability* for a given *flow*, F, connecting specific inputs to one or several outputs. A *flow* is an information path connecting specific inputs to one or more outputs. Once the set of flows is defined, it is used to select test data covering the *flows*:

$$Testability_F(M) = \bigl(Controllability_F(M);\ Obsevability_F(M)\bigr)$$

$$Controllability_F(M) = \frac{MaxInfoRecived\bigl(Inputs_F;\ Inputs_M\bigr)}{Overall\ Inputs\bigl(Inputs_M\bigr)}$$
$$= \frac{Maximum\ Information\ Recived\ from\ flow\ F\ on\ component\ M\ inputs}{Total\ amount\ of\ information\ that\ can\ be\ receivedbyM}$$

$$Observability_F(M) = \frac{MaxInfoResulted\bigl(Outputs_M;\ Ouputs_F\bigr)}{Overall\ Outputs\bigl(Outputs_M\bigr)}$$
$$= \frac{The\ maximum\ amount\ of\ information\ received\ on\ the\ outputs\ of\ component\ M}{Total\ amount\ of\ information\ quantity\ component\ M\ can\ produce\ on\ its\ outputs}$$

Suppose a component has a single input Inp1 whose domain is [1...100]. Then dependent on the conditions and restrictions imposed by the flow F, the values received by Inp1 may be restricted to the domain [3...40]. In this case, controllability, Cob, of the component is:

$$Cob = (40 - 3 + 1)/(100 - 1 + 1) = 0.38$$

The constraints and branching conditions across a flow impact the information quantity received on component inputs and is a subset of the domain of the component. The input domain of a component is the set of all possible inputs that can be fed into the component. The input domain is determined by the type of inputs and the restrictions imposed on their values.

Built-in test capabilities such as using assertions, stubs, exceptions, and even additional "write" statements can improve observability. Test cases can control inputs and improve controllability. In [18], a specific debugging framework is suggested to improve controllability and observability. The framework improves controllability by letting the tester control the inputs to a function under test. Also, the debugging framework improves the observability by letting the tester to insert observation points to track the changes in the values of the variables.

Considering a functional model, a method like a function provides a mapping from its domain to its range. The domain of a function is the set of possible values of input variables. Similarly, the range of a function is the set of ranges of its outputs. The function's domain, $f(\times 1, \times 2, \ldots, xn)$, is an n-dimensional space, where each dimension, di, represents the values of an input variable xi. The observability of the output values is a precondition for the testability of the function. Observability is supported if each point in the output space corresponds to a single point in the input space. In other words:

$$\forall x, y \in \text{domain}(f) \wedge f(x) \neq f(y) \Rightarrow x \neq y$$

Controllability is the ease with which inputs of a function can be controlled. Again, considering a functional model, controllability can be defined as follows:

$$\forall y \in \text{Range}(f) \exists x \in \text{domain}(f) \cdot f(x) = y$$

Controllability and observability can be improved by adding extra input and output variables so that a one-to-one mapping between the domain and range of the function is obtained.

1.4.3 Dependency Metrics

A software component A is dependent on another component B, if A cannot be compiled or executed without B. Here, by a software component, we mean a function, class, module/component, or software system. Dependencies can be direct, indirect, or cyclic. Dependency graphs are used to represent software component dependencies. A dependency graph G(V, E, L) is a directed graph in which V is the set of software components, E is the set of edges, each representing the dependency of a component on the other ones. Dependency graphs are automatically extracted from call flow graphs. Each node in a call flow graph represents a method, and edges denote method calls. Polymorphic calls and function pointers make it difficult to extract call flow graphs automatically. However, these difficulties are not unsolvable.

A measure for testability is the degree to which a software component such as a function, method, class, component, or System can be tested in isolation and independent of the others. A component's dependencies should be broken before it can be tested in isolation. Fake elements such as stubs and mocks, described in Chap. 2, are used to replace the component on which the component under test is dependent. To measure the impact of eliminating a dependency, d, on the testability, T, of a software component, C, the following relation can be used [19]:

$$Impact(d) = \begin{cases} \left(1 - \frac{T(D/\{d\})}{T(d)}\right), & T(D) \neq 0 \\ 0, & T(D) = 0 \end{cases}$$

where D is the set of dependencies of the component, C. Elimination of dependencies is not for free. Therefore, those dependencies whose impact on the testability is high and their contribution to the system functionality is relatively low are refactored to improve testability.

The testability of component C can be measured as the average time required to compile the components on which C is dependent. Another testability metric is the average component density, ACD:

$$\text{ACD} = \frac{1}{n} \times \sum_{1}^{n} CD_i$$

In the above relation, n is the total number of the dependencies of the components, CDi, of the software under test. The software should be re-modularized if the average dependencies, ACD, are high. Section 1.4.2 it is described how to automatically re-modularize software components.

Cyclic dependencies make it difficult to test components. A metric to measure testability is the number of dependencies that cause cycles in a dependency graph. A cyclic dependency is formed whenever there are direct or indirect dependencies among two or more components. A feedback dependency set consists of the dependencies that cause cycles among dependent components. The number of feedback dependencies is equal to the number of elements in the feedback dependency set:

No.Feedback Dependencies = |Feedback Dependency Set|

Feedback dependencies are back edges in a dependency graph. The following algorithm determines the back edges in a given graph G(v, E).

```
Algorithm DetermineBackedges ( G: Graph )
    VisitedNodes = [ ];
    BackEdges = [ ];
    Stack = [ ];
    Stack.push( G.Root);
    While not Stack.IsEmpty() do
        Vertex = Stack.pop();
        VisitedNodes.Add(Vertex);
        Foreach Child of Vertex do
            If in VisitedNodes(Child )
                Then BackEdges.AddEdge(Child→Vertex )
  +         EndForeach;
    EndWhile
```

1.5 Modularity and Testability

To improve testability, structure the code based on the principles of modularity [20]. Testable code is modular because modular code is cohesive, loosely coupled, and understandable. Modularity principles support testability. The impact of the modularity principles on software testability is described in Sect. 1.4.1. The correlation of testability to modularity can be perceived by seeing that code with redundancy, tight coupling, and weak cohesion is challenging to test. Section 1.4.2 presents an approach to automatically modularizing software.

1.5.1 Modularity Principles

Principles of modularity are widely applied to structure software. Below, these principles and their impacts on software testability are described:

(a) **Modular decomposability**

The modular decomposability principle suggests structuring software systems in cohesive, decoupled, and orthogonal components or modules. Observing the single responsibility principle is a proper way to ensure that an element, such as a method or a class, is orthogonal. Similar to a class, a method should also have a single responsibility. When more than one responsibility is given to a class, these responsibilities often entangle, making it difficult to test the responsibilities separately. Testability requires clarity of responsibilities. Appropriate separation and assignment of responsibilities are some of the critical skills in object-oriented design. There are nine general responsibility assignment patterns called GRASP [21]. GRASP is described in Sect. 1.4.

When a class has a relatively large number of methods or complex methods with more than one responsibility, it violates the decomposition principle because of inadequate abstraction. This violation is called insufficient modularity smell [22].

(b) **Modular continuity**

The modular continuity principle suggests structuring software architecture so that a slight change in the problem specification triggers a change of just one module or a small number of modules. This principle supports testability by minimizing change side effects because the modular continuity criterion is satisfied if changes in a module necessitate no or slight changes to the other modules. In large software systems, the most challenging part of making changes is not the change itself but making the change in a way that does not regress any other necessary behavior. Appropriate, fast tests that can provide basic assurances are an enormous productivity improvement.

Interfaces are used to decouple dependencies. Direct dependencies are created between classes when a method of a class calls the methods of another class, a class extends another class, or aggregates objects of another class.

(c) **Modular understandability**

According to the modular understandability criterion, each module, or in other words, the unit of software, should be understandable without the need for knowing the others. This principle leads to the heuristic that a control structure related to a particular functionality is preferably assigned to one class rather than distributed over several classes. The distribution of control structures over several classes makes testing and understanding its related functionality difficult.

(d) **Modular protection**

The modular protection principle suggests architectures in which the run time error in a module confines to that module or, at worst, propagates to a few neighboring modules. In effect, processing error side-effects should be minimized. This feature enhances the testability because the bug localization process will be confounded to the module where the fault is observed.

(5) **Modular composability**

"A method satisfies Modular Composability if it favors the production of software elements which may then be freely combined to produce new systems, possibly in an environment quite different from the one in which they were initially developed."

A module, or in other words, a component, tends to be a standalone, independent unit of code. On the other hand, testability is the degree to which a software component can be tested in isolation. According to Bertrand Meyer, classes are the primary form of modules.

1.5.2 *Automatic Re-modularization*

Testability is the ease with which a piece of code can be tested efficiently and effectively. As described above, before beginning with the actual testing of a piece of code, it is beneficial to enhance its testability as much as possible. The first thing to do is to ensure that the code is structured based on modularity principles. If not, it should be re-modularized.

Modularity is in close correlation with testability. By virtue of low coupling, a modular component is easier for being tested in isolation. The fewer the dependencies, the less it takes to write tests for a module. Likewise, module tests with fewer external dependencies also run faster-at least marginally, because there are fewer things to do for each test.

A software system can be automatically re-modularized to divide its functionality into units. Figure 1.12 shows the steps to be taken to re-modularize a software system and build component and package diagrams representing the modularized architecture in a case-tool environment such as Rational Rose.

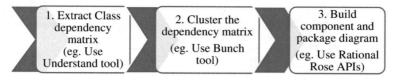

Fig. 1.12 Software modularization process

i. Extract class dependency graph

The class dependency graph should be derived from its text in the first step of reconstructing a program structure. A class, A, is dependent on another class, B, if A has an attribute of type B, invokes a method from B, or is a subclass of B. The class dependency graph is a weighted graph in which the total number of dependencies between every two classes is attached to the edge connecting their corresponding nodes.

Understand is a source code analysis tool. It can derive a class dependency graph from a given project source files or any directory, including source files. For this purpose, as shown in Fig. 1.13, select Reports, Dependency, Class Dependencies, and Export Matrix CSV from the main menu bar in the Understand environment.

Understand, will begin to analyze the source files and outputs an Excel file, including the adjacency matrix of the class dependency graph. Figure 1.14 shows the dependency matrix derived from the Peach open source. The first row and first column of the matrix contain the names of the classes. Each entry of the adjacency matrix is a number indicating the degree of correlation of the corresponding pair of classes. This number is the weight of the edge connecting the class nodes in the class dependency graph. For each inheritance relationship or invocation between the two classes, one is added to the weight of the edge, connecting their corresponding

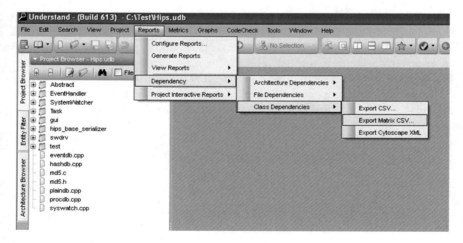

Fig. 1.13 Generating class dependency matrix in the understand environment

```
   HipsClassDependency.csv
1    Dependent Class,CNetRMSettings,CNetRulesManager,cEHSysWatchWin,cEHSysWatch,cAu
2    CNetRMSettings,,,,,,,,,,,,,,,,,,,4,,,,,,,,,,,,,,,,,,,,,,,,,,
3    CNetRulesManager,21,,,,,,,,,,,,,,,,,21,6,,,,,,,,,,,,,,,,,,,,,,,
4    cEHSysWatchWin,,,,,1,,,,,,,,,,,,,,,,,,,,,,,,,,,,,,,,,,,
5    cEHSysWatch,,,,,,,,,,,,,,,,,,1,,,,,1,,,,,,,,,,,,,,,,,,,,,
6    cAutoLockerCS,,,,,,,,,,,,,,,,,,,,5,,,,,,,,,,,,,,,,,,,,,,
7    cSystemWatcherData,,,,,,,,,,,,,,,,,,1,,,,,1,,4,2,,,,,,,,,,,,,,,,,,
8    CHipsRuleManager,,,,,,,,,,,3,,,,,,6,,,,,,,,,,,9,15,28,6,22,,,,,,,,,,,,,,
9    cFileAL,,,,,,,,,,,,,,,,,,,,,,,,,,,,,,,,,8,,,,,,,,,,,,,
10   CHipsManager,,,,,,,,27,,,,,7,,,14,22,,,,,,,,,,,,,,14,,,,22,,,,,,,,,,,,
11   cSystemWatcherDrv,,,,,1,7,,,,,,,,,,,,,,,,2,,,,,,,,,,,38,39,,,,,,,,,,,
12   CHipsLocalCash,,,,,,,,,,,,,,,,,,,,,,,,,,,,,,,,,,,,,,2,2,,,,,,,,,
13   cHashDB,,,,,,,,,,,,,,,,,,,,,,,,,,,,,,,,,,,,,8,14,,,,,,
14   cPlainDB,,,,,,,,,,,,,,,,,,,,,,,,,,,,,,,,,,,,,,6,,,,,
15   CHipsTask,1,16,,,,,,7,,,59,,,1,,,,,,,,,,,,,,,,,39,,,,,,,,,,,,,,,
16   CallBackHipsStruct,,,,,,,,,,,1,,,,,,1,,,,,,,,,,,,,,,1,,,,,,,,,,,,,,,
17   cSystemWatcher,,,,,,,,,,,,,,,,,,,,,,,,,,,,,,,,,,,,,,,,,,,7,1,11,3,1
18
```

Fig. 1.14 Class dependency matrix, derived from the Peach open source

nodes. The point to note here is that if a call is made within a loop, the Understand software will not consider the loop iterations.

ii.   Cluster the dependency matrix

In the second stage, the adjacency matrix, representing the correlations between pairs of the classes, is clustered. To this end, a clustering environment such as Bunch can be used. The Bunch tool accepts a text file with three columns. The first and the second columns include the class names, and the third column indicates their dependencies. Therefore, the adjacency matrix generated by Understand should be transformed into the format acceptable by Bunch.

The "Basic" tab of the Bunch main window accepts the name of the class dependency file, the name, and format of the output file, and whether hierarchical or user-managed clustering is to be performed. Also, under the "action" label in the main window, it is possible to introduce your clustering algorithm by selecting the "user-driven clustering algorithm" option. The main window of Bunch is shown in Fig. 1.15.

Under the "clustering option" tab, select Basic Modularity Quality (BasicMQ) as the clustering objective function. BasicMQ evaluates the difference between cohesion and coupling:

$$BsicMQ = AverageCohesion - AverageCoupling$$

$$AerageCohesion = \frac{1}{No.Clusters} \sum_{i=1}^{No.Clusters} Cohesion_i$$

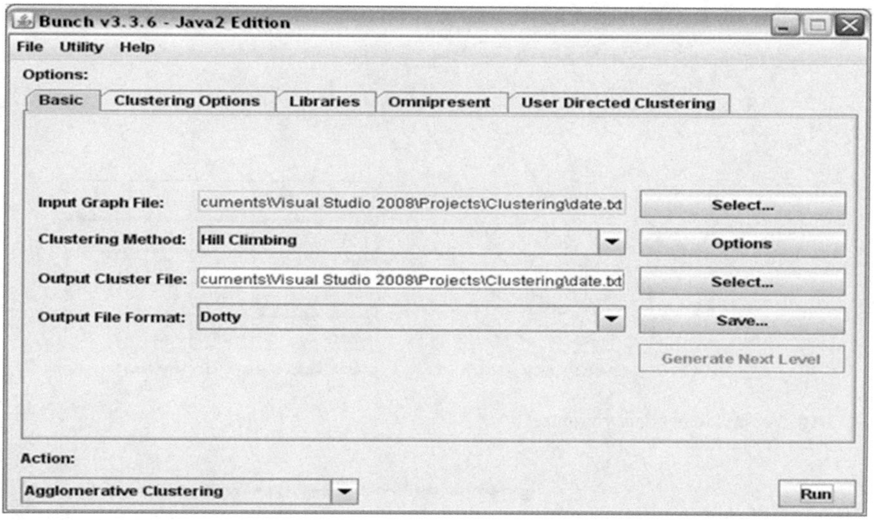

Fig. 1.15 The main window of bunch

$$AverageCoupling = \frac{1}{No.Clusters * (No.Clusters - 1)} \sum_{i,j=1}^{No.Clusters} Coupling_{i,j}$$

$Cohesion_i$ = the Total number of class dependencies in the ith cluster.

$Coupling_{i,jj}$ = Total number of dependencies between the ith and j^{th} clusters

The class dependency graph in Fig. 1.16a illustrates the class dependency model of the four classes a, b, c, and d.

The hill-climbing algorithm in the Bunch environment has been applied to partition the graph into two clusters, shown in Fig. 1.17a. The output file generated by the bunch software is in the Dotty format, a popular format for displaying graph structures. Graphviz 2.28 software was used to view this file, as shown in the graph shown in Fig. 1.17b.

iii. Build component and package diagrams

Component diagrams illustrate components and their dependencies. A component is a software system module whose behavior is defined by its interfaces. Considering modularity principles, spreading classes among components will profoundly enhance the testability of the software artifact. Components may be grouped into packages.

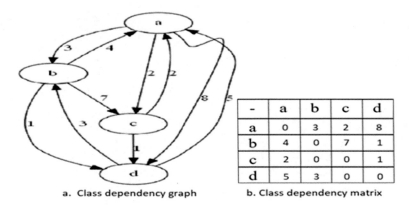

a. Class dependency graph   b. Class dependency matrix

-	a	b	c	d
a	0	3	2	8
b	4	0	7	1
c	2	0	0	1
d	5	3	0	0

Fig. 1.16 A class dependency graph

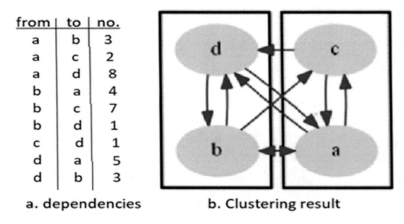

from	to	no.
a	b	3
a	c	2
a	d	8
b	a	4
b	c	7
b	d	1
c	d	1
d	a	5
d	b	3

a. dependencies   b. Clustering result

Fig. 1.17 Dependency graph partitioned into two clusters

As a part of the software testing classwork, my students are supposed to develop a program, ModelExtractor, to accept the class dependency matrix and its clustering files as its input and output a .mdl file, including all the required commands to display class, component and package diagrams, derived from the class dependency and clustering files. ModelExtractor considers each cluster as a software component in the component view of the software architecture. To improve accessibility, for each component, an interface class consisting of objects of the type of the public classes in the component is also automatically built. ModelExtractor creates a .mdl file, including all necessary commands to display and modify the derived class, component, and package diagrams in the Rational-Rose case-tool environment.

Rational Rose supplies a set of APIs, Rose Extensibility Interface (REI), to allow application programs to display and manipulate diagrams and other Rose model elements. REI provides more than 130 COM objects. There are two methods in VC++ to use REI classes: one method is to use ClassWizerd to locate the class to be imported from Rose's type library, rationalrose.tlb; another method is to use # import to import COM objects as follows:

#import "c:\\Program Files\\Rational\\Rose\\rationalrose.tlb"

As shown in Fig. 1.18, the rationalrose.tlb file can also be added to the projects from the:

Project → Add Reference → Browse

in the C# environment in the visual studio. The path to this file is often:

C:ProgramFiles\Rational\Rose\rationalrose.tlb.

Fig. 1.18 Adding rationalrose.tlb to projects

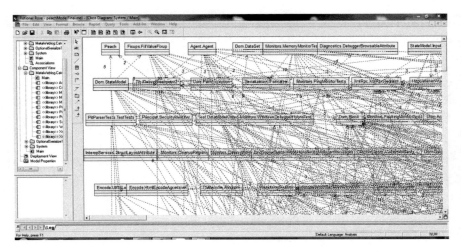

Fig. 1.19 Class dependency graph for the Peach open source in the Rose environment

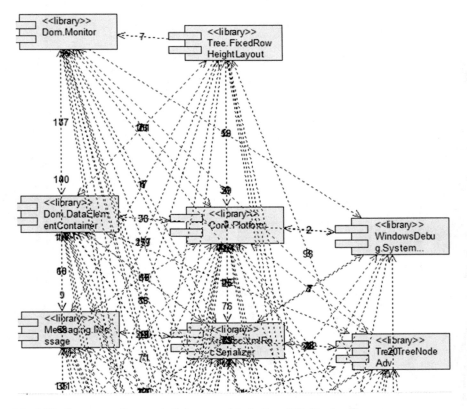

Fig. 1.20 Component diagram, built by re-modularization of the Peach software

To get access to the rationalrose.tlb library resources in C#, use the following command:

using RationalRose;

The inputs to the C# program are:

(1) The class dependency matrix, derived from the given source code, OpenFileDialog DependencyFiles = new OpenFileDialog();
(2) The program modules provided by clustering the dependency matrix, OpenFileDialog ClusterFiles = new OpenFileDialog();

The output is a file named Models.mdl:

- RoseApplication rapp = new RationalRose.RoseApplication();
- RoseModel rmdl = rapp.NewModel();

Models.mdl is in a format acceptable by Rational Rose. It includes all the commands to display package, component and class diagrams for the given source file. For instance, in Fig. 1.19, the class diagram for the open source, Peach, is shown.

In addition to the class diagram, ModelExtractor, inserts a command in Models.mdl to display component diagrams. For instance, a component diagram for the Peach class diagram, saved in Models.mdl, is shown in Fig. 1.20.

ModelExtractor, also partitions the components into packages. Figure 1.21 shows the package diagram for the Peach software.

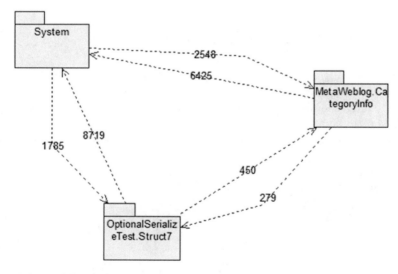

Fig. 1.21 Package diagram, built for the Peach software

1.6   Design Patterns

The ratio of design patterns applied in a software design provides a basis for evaluating software testability because patterns are time-tested and proven solutions to common design problems. The main issue is the rationalization of the use of such patterns in the design of a software system. Sometimes, using a small number of patterns could be beneficial, while high usage could be fatal. However, using design patterns as time-tested solutions to known problems makes it easier to understand and, subsequently, simpler to test the software. Some examples of design patterns and their impact on testability are given in Sect. 1.6.1. The exciting part is to develop a software tool to automatically distinguish and extract any design pattern used in a given source code. There are two questions concerning design patterns. Firstly, which patterns affect testability? Secondly, how to measure testability based on the percentage of code covered by the patterns?

Class diagrams represent the structural properties of design patterns. A class diagram is a graph whose nodes address classes, and edges represent the relationships among the classes. The detection of a specific design pattern in the source code is then carried out by graph isomorphism to identify sub-graphs similar to a template graph in a big graph, representing the class diagram of the source code. Section 1.6.2 presents a practical approach to the inexact matching of patterns in a given source code.

1.6.1   Design Patterns and Testability

The designer's choice of design patterns can influence the testability and testing methodology employed by the tester. Some design patterns ease the construction of test fixtures and test case design. For instance, Dependency Injection, Dependency Lookup, Sub-classed Singleton, and Factory patterns improve testability by solving the problem of tight coupling via interfaces. In support of isolated testing, a

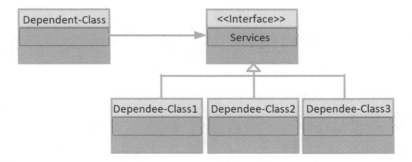

Fig. 1.22   Using interface as a mediator between dependent class and concrete services

good programming practice is to avoid direct dependencies. As shown in Fig. 1.22, dependencies are broken down by connecting the dependent objects and components through an interface to the concrete services.

Each subclass is a kind of its superclass, and a superclass is the essence of its subclasses. In real life, there are certain situations where one does not want to bother about the kinds; therefore, they ask an expert to decide on selecting a kind amongst the existing ones. Delegating decision-making tasks on selecting and creating instances of an appropriate type amongst the existing ones is a recommended programming practice provided by the Factory pattern. The Factory pattern delegates the details of the selection and creation of instances of sub-classes implementing the same interface to factory methods. In continuation, the Factory and Dependency Injection patterns are described to provide an insight into the concept of design patterns.

(i) **Factory pattern**

The Factory pattern promotes loose coupling by delegating the selection and creation of instances of different classes holding the various implementation of the same method to a factory method. For instance, in Fig. 1.23, airShipment, landShipment, and stubClass are different kinds of shipments. Selecting and creating instances of these subclasses is assigned to the shipmentFactory() class.

The freightInvoice class determines the type of shipment based on the orders it receives at runtime. It asks the shipmentFactory class to create appropriate instances of the subclasses of the shipment class. The shipmentFactory class generates an object of concrete classes, airShipmet, landShipmet and stubClass, based on the information received from shipmentFactory. The stubClass provides dummy methods, to allow isolated unit testing of the createInvoice method. Principally, a stub is a dummy method, used as a test fake to substitute the real type. The stub substitution prevents tests from extending beyond the boundaries of "unit." Unit testing is the subject of Chap. 2. The C# code for the class diagram in Fig. 1.23 is given below:

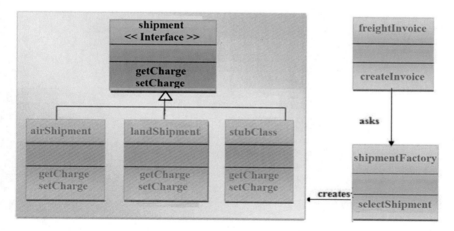

Fig. 1.23 Factory-method creates instances of the classes

```
namespace factoryPattern
{
interface shipment
{ // The interface class includes two methods:
   double getCharge();
   double setCharge();
}

public class airShipment : shipment
{ // airShipment implements the interface class

   public double rate;

   public double getCharge();
   {
      //looks for the latest air delivery charges.
      rate = lookupTheLatestAirDeliveryRate();
      return rate;
   }
    public double setCharge();
    {
     rate = getCharge();
     return rate;
    }
}
public class landShipment : shipment
{
   public double rate;

   public double getCharge();
   {
      //looks for the latest land delivery charges.
      rate = lookupTheLatestLandDeliveryRate();
      return rate;
   }
    public double setCharge();
    {
     rate = getCharge();
     return rate;
    }
}

public class stubClass : shipment
{ /* Stubs are used to replace concrete methods
   With dummy ones.*/

   public double getCharge();
   {
     return 1;
   }
    public double setCharge();
    {
     return 1;
    }
}}
```

```
namespace factoryPattern
{
public enum shipmentKinds { byAir, byLand, dummy, null };
public class shipmentFactory
{
  public shipmentKinds getKind()
    {
    //determines the type of shipment
    …
    return shipmentKind;
    }
  public shipment selectShipment(shipmentKinds kind);
    {
    if (kind == shipmentKinds.null)kind = getKind();
    if (kind == shipmentKinds.byLand)
      return new landShipment();
        else if (kind == shipmentKinds.byAir)
      return new airShipment();
        else if (kind == shipmentKinds.dummy)
      return new stubClass();
    else printErrorMessage();
    }
}}
```

```
namespace factoryPattern

{public class freightInvoice
  {
    Public void createInvoice()
      {
      /* To provide  loose  coupling, the decision about the
        type of delivery is deligated to the shipmentFactory
        class. */
      shipmentFactory theShipmentFactory = new shipmentFactory();
      shipment theShipment;
      theShipment = theShipmentFactory.selectShipment();
      double rate = theShipment.getCharge();
      …
      }
}}
```

(ii) Dependency injection pattern

According to the clean code principles, a method should focus on fulfilling its assigned responsibility rather than creating objects required to accomplish its responsibility. In general, the inversion of control principle, one of the SOLID principles [23], suggests delegating additional responsibilities, such as creating dependent objects to achieve loose coupling. That is where dependency injection comes into play by providing the required objects. The dependency injection pattern restricts dependencies to interfaces. Objects of the concrete classes implementing an interface

Fig. 1.24 Dependency
Injection pattern classes

are created and passed as instances of the interface class to the dependent classes. The dependency injection pattern consists of four main classes, shown in Fig. 1.24.

The injector class injects instances of service classes to the client class. There are three types of dependency injection:

- *Constructor* injection: In this type of injection, the injector passes instances of the services to the client class constructor when creating objects of the client class.
- *Setter* Injection/Property Injection: In this type of injection, instead of the constructor, instances of the services are passed by the injector to the setter method by the client class.
- *Interface* Injection: This type of injection is similar to the Setter Injection method, except that with this type, an interface class, including the signature of the setter method, is defined. Both the injector and the client class should implement the interface.

A sample class diagram for the dependency injection pattern is shown in Fig. 1.25.

In the class diagram shown in Fig. 1.25, the Client class depends on three services, Service1, Service2, and service3. Interface classes break direct dependencies. Instead of the client class, an Injector class creates the objects and injects them into the Client as instances of the interface classes. The three methods mentioned above for implementing the injector class are described in the following.

1. **Constructor injection method**

The constructor method requires the Client class to receive instances of the interface classes as parameters of its constructor:

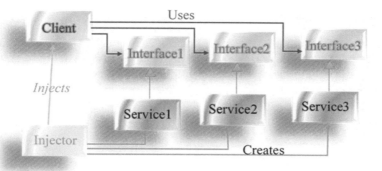

Fig. 1.25 The dependency injection pattern class diagram

```
namespace setterDependencyInjection
{
  public class Injector
    {
      public Injector(Client theClient)
        {
          Service1 theService1 = new Service1();
          Service2 theService2 = new Service2();
          Service3 theService3 = new Service3();
          // theClient = new Client(Service1, Service2, Service3);
          theClient.setService(theService1, theService2, theService3);
        }
    }

namespace setterDependencyInjection
{
    public class Client
{
  private Interface1 theService1;
  private Interface2 theService2;
  private Interface3 theService3;

  void setService(Interface1 theObject1, Interface2 theObject2, Interface3 theObject3)
    {
      theService1 = theObject1;
      theService2 = theObject2;
      theService3 = theObject3;
    }

// All the methods within this client may use the services:
//     theService1, theService2 and theService3.
...
...
```

2. Setter Injection method

A significant difficulty with the Constructor Injection method is that every time new service objects are required, a new instance of the client class has to be created. To avoid this, instead of the constructor, a different type of method called setter is used to do the same job as the constructor in Constructor Injection.

```
namespace setterDependencyInjection
{
  public class Injector
    {
     public Injector(Client theClient)
       {
         Service1 theService1 = new Service1();
         Service2 theService2 = new Service2();
         Service3 theService3 = new Service3();
         // theClient = new Client(Service1, Service2, Service3);
         theClient.setService(theService1, theService2, theService3);
       }
    }

namespace setterDependencyInjection
{
    public class Client
      {
       private Interface1 theService1;
       private Interface2 theService2;
       private Interface3 theService3;

       void setService(Interface1 theObject1, Interface2 theObject2, Interface3 theObject3)
         {
           theService1 = theObject1;
           theService2 = theObject2;
           theService3 = theObject3;
         }

    // All the methods within this client may use the services:
    //     theService1, theService2 and theService3.
    ...
    ...
```

3. Interface Injection method

The Interface Injection method provides an interface class, including the setter method. Both the Injector and Client classes should implement this interface.

```
namespace interfaceDependencyInjection
{
  public class serviceSetter
  {
    public void setService(Client theClient);
  }
}
```

```
namespace interfaceDependencyInjection
{
  public class Injector : serviceSetter
  {
    public Injector(Client theClient)
    {
      Service1 theService1 = new Service1();
      Service2 theService2 = new Service2();
      Service3 theService3 = new Service3();
      // theClient = new Client(Service1, Service2, Service3);
      theClient.setService(theService1, theService2, theService3);
    }
  }
}
```

```
namespace inerfaceDependencyInjection
{
  public class Client: serviceSetter
  {
    private Interface1 theService1;
    private Interface2 theService2;
    private Interface3 theService3;

    void setService(Interface1 theObject1, Interface2 theObject2,
                    Interface3 theObject3)
    {
      theService1 = theObject1;
      theService2 = theObject2;
      theService3 = theObject3;
    }

    // All the methods within this client may use the services:
    //    theService1, theService2 and theService3.
    ...
    ...
}}
```

1.7 Summary

Before testing a program, one should ensure it is clean and testable. The clean code is intrinsically testable. If the code is not clean, it should be refactored. Compiler generators such as ANTLR and COCO can be used to develop software tools to evaluate the program under test quality based on clean code principles. Also, a new generation of compilers, Roslyn, provides APIs to access syntax trees and refactor source codes. However, despite its complete support for the analysis and refactoring of source code, Roslyn operates on C# source codes. Compiler generators such as ANLR and COCO are more flexible than Roslyn. C++, Python, and JAVA grammars are available for ANTLR and COCO. Section 1.3.2 describes how to use Roslyn to develop a software tool for evaluating the quality of C# programs based on clean code principles.

Dependencies make it challenging to trace and test software components in isolation. Faults propagate across the dependent components. Dependencies oppose testability. Automatic re-modularization of the software architecture may lessen the dependencies by providing high cohesion between elements of components and low coupling between the components. Section 1.4.2 describes how to develop a software tool to automatically re-modularize the architecture of programs simply by clustering their class dependency graphs based on the BasicMq criterion. As described in Sect. 1.4.2, the interesting point is that components provided by CASE tools such as Rational Rose simplify the development of tools to build files to display and edit package, component, and class diagrams of the re-modularized software.

Specific design patterns such as Dependency Injection and Factory design patterns break direct dependencies and improve testability. Design patterns provide readability and clarity for programs because patterns are well-known solutions to common problems. Therefore, the automatic extraction of design patterns could help the reader understand the program's semantics. Section 1.4.2 proposes a genetic algorithm for automatically detecting design patterns in source code.

1.8 Exercises

1. What are the testability metrics? Name the reasons why these metrics are useful.
2. Readability and independence of software components from one another are the two main characteristics of software affecting its testability. Describe why design patterns could improve readability. List some design patterns to reduce dependencies.
3. Which software attributes and entity the cyclomatic complexity measure. Why cyclomatic complexity of more than 10 is problematic.
4. Write a C# program that inputs a C# method, uses Roslyn to extract the methods control flow graph, and outputs the method's cyclomatic complexity.

5. A good design should provide high module cohesion and low coupling. Briefly describe what you understand this assertion to mean.
6. The relation presented in Sect. 1.4.1 for measuring the design understandability is derived from the data set in [12] by applying linear regression. Use the same data set to make a more accurate model by applying non-linear regression or other machine learning techniques.
7. Experienced programmers use their own programming skills to develop quality software. Suggest a graph mining method to extract unknown design patterns from source codes for specific systems such as stock control, accounting, banking, manufacturing, or ECommerce.
8. According to the modularity principles, components as building blocks of software architecture should be loosely coupled and tightly cohesive. Suggest some patterns further to reduce the dependencies of the components on each other.
9. Use Roslyn to develop a program to input a C# project, determine the number of dependencies between the program methods and classes, and outputs the results as a matrix representing the dependencies between methods and another matrix representing the dependencies between the classes.
10. Describe how the dependency matrices described in exercise 5 could be applied to measure a class's understandability, complexity, and testability. Extend the program in exercise 5 to determine the testability.
11. Write an algorithm in pseudo-code that uses APIs provided by Roslyn to predict the number of faults in a given source code. Implement the algorithm.
12. Develop a C# windows form application that takes a.sln solution and analysis the program based on the clean code principles. The analysis result should be displayed as shown below:

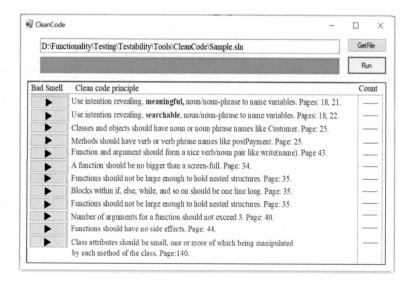

The count column should display the number of items found in the program during the analysis of the program source code based on the principle listed in the "clean code principle" column. Click the action button in the "Bad smell" column to get a list of the items found while analyzing the code. Page numbers refer to the Clean Code book by Robert C. Martin.

Due to specific faults in Roslyn, you have to create a "console application" project and then convert it to "windows form"; otherwise, Roslyn can not analyze ".sln" files. To convert a console application into a windows form, refer to the following link:

https://gregs-blog.com/2008/02/09/how-to-convert-a-console-app-into-a-windows-app-in-c-part-two/.

References

1. Martin, R.C.: Clean Architecture: A Craftsman's Guide to Software Structure and Design. Prentice-Hall (2012)
2. Martin, R.C.: Clean Code: A Handbook of Agile Software Craftsmanship. Prentice-Hall (2009)
3. Mössenböck, H.: The Compiler Generator Coco/R User Manual. University of Linz, Hanspeter Mössenböck (2010)
4. Stuff, M.: Antlr Tutorial—Hello World. http://meri-stuff.blogspot.com/2011/08/antlr-tutorial-hello-word.html
5. Vasani, M.: Roslyn Cookbook. Packt Publishing, Verlag (2017)
6. Silva, J.: A vocabulary of program slicing-based techniques. ACM Comput. Surv. **44**(1), 1–41 (2012)
7. https://scotch.io/bar-talk/s-o-l-i-d-the-first-five-principles-of-object-oriented-design
8. Watson, McCabe, T.: Structured testing: A software testing methodology using the cyclomatic complexity metric. T. NIST Special Publication 500–235, U.S. Department of Commerce/National Institute of Standards and Technology, Washington, D.C. (1996)
9. Halstead (1977)
10. McGregor, J., Srinivas, S.: A measure of testing effort. In: Object-Oriented Technologies Conference, USENIX Association, pp. 129–142 (1996)
11. Bruntink, M.: Testability of Object-Oriented Systems: A Metrics-Based Approach. University VAN Amsterdam, Software Improvement Group (2003)
12. Voas, J.M., Miller, K.W., Payne, J.W.: An Empirical Comparison of a Dynamic Software Testability Metric to Static Cyclomatic Complexity (1993)
13. Badri, M., Kout, A., Toure, F.: An empirical analysis of a testability model for object-oriented programs. ACM SIGSOFT Softw. Eng. Notes **36**(4), 1 (2011)
14. Nazir, M.: An empirical validation of complexity quantification model. Int. J. Adv. Res. Comput. Sci. Softw. Eng. **3**(1) (2013)
15. Nazir, M., Mustafa, K.: An empirical validation of testability estimation model. Int. J. Adv. Res. Comput. Sci. Softw. Eng. **3**(9), 1298–1301 (2013)
16. McCabe, T.J., Butler, C.W.: Design complexity measurement and testing. Commun. ACM **32**(12), 1415–1425 (1989)
17. Nguyen, T.B., Delaunay, M., Robach, C.: Testability analysis of data-flow software. Electron. Notes Theoret. Comput. Sci. **116**, 213–225 (2005)
18. Guo, Q., Derrick, J., Walkinshaw, N.: Applying testability transformations to achieve structural coverage of Erlang programs. Testing of Software and Communication Systems, pp. 81–96 (2009)
19. Jungmayr, S.: Testability measurements and software dependencies. In: 12th Workshop on Software Measurement (2002)

20. Meyer, B.: Object-oriented software construction, 2nd edn. Prentice-Hall, I S B N: 0-136-29155-4 (cit. on pp. 55, 57) (1997)
21. Noorullah, R.M.: Grasp and GOF patterns in solving design problems. Int. J. Eng. Technol. **1**(3) (2011)
22. Understand User Guide and Reference Manual: Scientific Toolworks Inc. (2018)
23. Martin, R.C.: Design Principles and Design Patterns. http://www.objectmentor.com/resources/articles/Principles_and_Patterns.pdf. Accessed 18 June 2015

Chapter 2
Unit Testing and Test-Driven Development (TDD)

2.1 Introduction

Testing typically begins with the units of the software under test. Integration tests can begin when the unit tests pass and the results appear convincing. In software testing, a unit refers to the smallest possible part of a program that can be tested in isolation. However, a program unit such as a method/function often collaborates with others to fulfill a requirement. Therefore, Mock and Stub solutions are used to isolate methods and allow them to be run and tested independently. Before a program can be tested thoroughly, it should ensure that all its building blocks and units perform as required. Unit testing is the cornerstone of Test-Driven Development, TDD. TDD is an Agile development technique. It is a test-first approach because it begins with writing tests for a program component before writing the code. This chapter teaches the reader how to apply unit testing in practice and use it for agile development.

Unit testing

© The Author(s), under exclusive license to Springer Nature Switzerland AG 2023
S. Parsa, *Software Testing Automation*,
https://doi.org/10.1007/978-3-031-22057-9_2

2.2   Unit Testing

Unit testing is a significant ingredient of TDD. A unit is the smallest possible program part that can be logically isolated and tested. A unit's instance is a method, function, procedure, or subroutine. The first phase of testing is unit testing.

Let us start this discussion with a motivating example, the one shown in Fig. 2.1. The C# class Point gets the point coordinates and simplifies building and operating on geometric shapes such as triangles and rectangles as different connections between points. The class has some public methods we want to unit test in the visual studio environment.

The question is which framework is the most suitable for generating unit tests for C#. In C#, there are three Frameworks used for testing, which are:

- NUnit
- MsTest
- XUnit.NET

I prefer the NUnit testing Framework due to its ease of use. Below, I describe how to use NUnit to generate test classes, including test methods. The generated test methods can be modified for parameterized testing. Parametrized testing extends the closed, traditional unit tests by allowing parameters to form test methods. As described in Sect. 2.2.1, NUnit generates test methods and allows it to pass test data and its expected results to the generated methods as parameters. Section 2.2.2 describes how to pass a collection of test data and their expected results, kept in a file, to the test method.

2.2.1   Generating a Test Class with NUnit

Nunit automatically generates a test class to unit test the class under test. In Fig. 2.1, there are two methods, "GetDistance" and "rotate," in the class point. The GetDistance(p, q) method measures the distance between the two points, p, and q, and the rotate(xc, yc, alpha) method rotates the point (x,y) around the point (xc, yc) for alpha degrees. To test the method GetDistance, right-click on its name in the code editor window and a pop-up menu, shown in Fig. 2.2. Select the "Create unit test" option from the menu.

The following window pops up by selecting the "Create Unit Test" option. Click OK to accept the defaults to create unit tests or change the values used to create and name the unit test project and the unit tests.

By clicking on the OK push button in Fig. 2.3, the class PointTest, including a method, GetDistanceTest(), is generated as follows:

In Fig. 2.4, TestFixture is an NUnit class level attribute indicating the class includes test methods. I modify the code that is added by default to the unit test method, GetDistanceTest() as follows:

```csharp
using System;
using System.Collections.Generic;
using System.Linq;
using System.Text;
using System.Threading.Tasks;
namespace Geometry
{
    public class Point : GraphicalObject
    {
        public double x, y;  //coordinates of the Point

        //1- The class constructor
        public Point(double xCoord, double yCoord)
            { x = xCoord; y = yCoord; }
        //2- Computes the distance between two points p and q
        public static double dist(Point p, Point q)
        {
            double res = 0;
            if (p.x == q.x) { res = Math.Abs(p.y - q.y); return res; }
            if (p.y == q.y) { res = Math.Abs(p.x - q.x); return res; }
            double dx = p.x - q.x; double dy = p.y - q.y;
            res = Math.Sqrt(dx * dx + dy * dy);
            return res;
        }

        //3- rotates (x,y) around the point (xc, yc) for the angel degree
        public void rotate(int xc, int yc, double angel)
        {
            //point ComparedWith = new point(xc, yc);
            if (xc == x && yc == y) return;
            else
            {
                double cosa = Math.Cos(angel); double sina = Math.Sin(angel);
                double dx = x - xc; double dy = y - yc;
                x = (int)Math.Round(cosa * dx - sina * dy + xc);
                y = (int)Math.Round(sina * dx + cosa * dy + yc);
            }
        }
    }
```

Fig. 2.1 A sample class, point

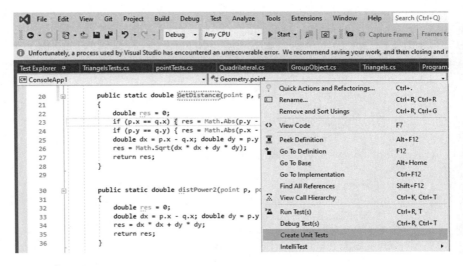

Fig. 2.2 Visual studio.Net IDE for C#

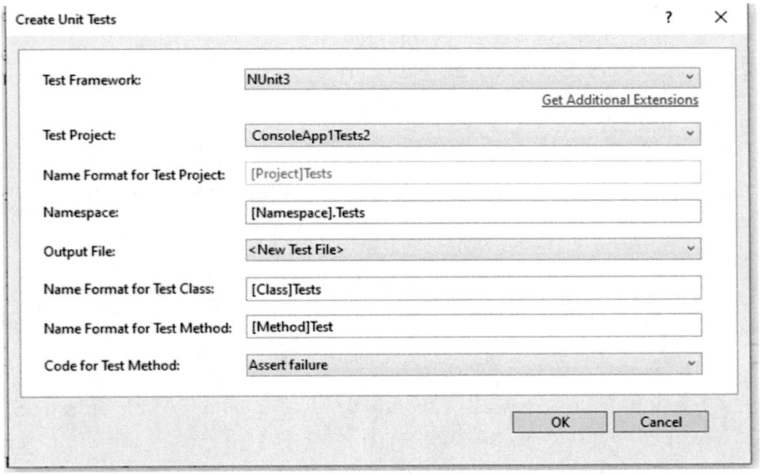

Fig. 2.3 The window used to name the test project, class, and methods

Fig. 2.4 Test class,
PointTests, generated by
NUnit

```
using NUnit.Framework;
using Geometry;
using System;
using System.Collections.Generic;
using System.Linq;
using System.Text;
using System.Threading.Tasks;
using System.Reflection;
using System.IO;
using System.Collections;
namespace Geometry.Tests
{ [TestFixture()]
    public class PointTests
    {
        [Test()]
        public void GetDistanceTest()
        {
            Assert.Fail();
        }
    }
}
```

```
[TestFixture]
public class PointTests
{
    [Test()]
    public void GetDistanceTest()
    {
        Point p = new Point(1, 10);
        point q = new Point(1, 20);
        Assert.That(Point.GetDistance(1, 2), Is.EqualTo(3));
    }
}
```

In Fig. 2.4, nine test cases are defined immediately before the GetDistanceTest
method. The method is run with each of the test cases. Each time the test method
executes it invokes the GetDistance method. The return value is compared with the
ExpectedResult by the NUnit test runner.

Parameterized testing is passing values into a test method through the method
parameters. In this example, we have used nine TestCase attributes on the same
method, GetDistanceTest(), with different parameters. In Fig. 2.5, there are nine test
cases comprising test data and their expected return values.

The parameters to the TestCase attribute should match up to the test method arguments, and the values are injected when the test is executed. The method GetDistanceTest executes nine times each time with one of the test cases. To run the test

```
namespace Geometry.Tests
{ [TestFixture()]
  public class PointTests
  {
      [Test()]

      [TestCase(1, 1, 2, 1, ExpectedResult = 4)]
      [TestCase(2, 5, 1, 9, ExpectedResult = 4)]
      [TestCase(1, 0, 1, 1, ExpectedResult = 2)]
      [TestCase(0, 0, 2, 0, ExpectedResult = 2)]
      [TestCase(1, 1, 8, 1, ExpectedResult = 7)]
      [TestCase(0, 2, 1, 3, ExpectedResult = 5)]
      [TestCase(3, 2, 3, 5, ExpectedResult = 3)]
      [TestCase(1, 2, 1, 5, ExpectedResult = 4)]
      [TestCase(3, 5, 4, 5, ExpectedResult = 1)]
      public void GetDistanceTest(double x1, double y1, double x2,
                                  double y2)
      {
          Point p = new Point(x1, y1);
          Point q = new Point(x2, y2);
          var distance = Point.GetDistance(p, q);
          return distance;
      }
  }
}
```

Fig. 2.5 Passing parameters to GetDistanceTest() to test the method GetDistance()

cases and see the status, move to Test Explorer in Visual Studio, where you find all the Unit Test Cases list. To move to Test Explorer, choose Test > Test Explorer from the top menu bar (or press Ctrl + E, T). Run unit tests by clicking "Run All" (or press Ctrl + R, V). The output is shown in Fig. 2.6.

Test Explorer displays the results in groups of Passed Tests, Failed Tests, Skipped Tests, and Not Ran Tests as you run your tests. The details pane at the side of the Test Explorer displays a summary of the test run. Figure 2.7 describes the tabs of the menu bar at the top of the test explorer window.

To view the details of an individual test, select the test. TestCaseAttribute supports some additional named parameters, which may be used as follows:

2.2.2   Reading Test Cases from a File

This section describes using Nunit to supply the test method with a test suite kept in a text file. A test suite is a logical grouping of all the tests for a given system. It

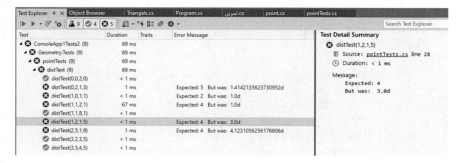

Fig. 2.6 Test results for the GetDistanceTest() method

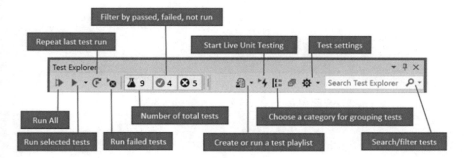

Fig. 2.7 The test explorer menu bar tabs description

consists of one or more test cases. Formally, a test case consists of an input (data) and its expected results or test oracles. A test oracle determines whether the software is executed correctly for a test case. We define a test oracle to contain two essential parts: oracle information representing expected output; and an oracle procedure that compares the oracle information with the actual output.

In Fig. 2.8, test cases are taken from a test suite in a text file, D:\TestCases.txt. The test suite contains the following test cases:

    0,0,0,1,1
    0,0,2,0,2
    0,0,3,4,5
    0,0,0,0,0
    0,2,1,3,5

As a test runner, NUnit provides an environment where test suites can be loaded, executed, and verified [1]. To run the method GetDistanceTest with the above test cases, select "Test > Run All" from the menu bar. The output is shown in Fig. 2.9.

```csharp
namespace Geometry.Tests
{  [TestFixture()]
   public class PointTests
   {
       [Test(), TestCaseSource(typeof(MyTestCases), "TestCasesFromFile")]]

       public void GetDistanceTest(Point p, Point q)
       {
          var distance = Point.GetDistance(p, q);
         return distance;
       }
    }

   public class MyTestCases
   {
            public static IEnumerable TestCasesFromFile
       {
          get
          {
            using (StreamReader file = new StreamReader(
                                       @"D:\TestCases.txt"))
            {
               string ln;
               while ((ln = file.ReadLine()) != null)
               {
                  string[] values = ln.Split(',');
                  Point p = new Point(Convert.ToDouble(values[0]),
                  Convert.ToDouble(values[1]));
                  Point q = new Point(Convert.ToDouble(values[2]),
                  Convert.ToDouble(values[3]));
                  yield return new TestCaseData(p, q).Returns(
                                    Convert.ToDouble(values[4]));
               }
            }
          }
       }
   }
}
```

Fig. 2.8 Getting test cases from a text file

2.2.3 *Testing in Isolation*

Unit testing is preferably performed in isolation and independent of the intervention of the other units [2]. Otherwise, in the case of the test failure, it is unclear whether

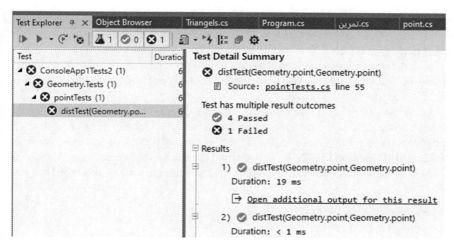

Fig. 2.9 Results of running GetDistanceTest() with the test suite

the root cause of the failure resides in the unit under test or the other units it depends on. Dependency inversion is a valuable principle for mitigating dependencies. It is implemented using the dependency injection design pattern. The pattern creates dependent objects outside of a class and provides those objects to the class. When using dependency injection, the creation and binding of the dependent objects are moved outside of the class that depends on them.

2.2.3.1 Dependency Injection

The dependency injection pattern suggests that if a class depends on an object to work for it, it does not create the object. Instead, pass the required object to the class either via its constructor or setter. Let us take an example of a Triangle class.

The class Triangel, listed in Fig. 2.10, is dependent on the class Point. For instance, the method "GetType" invokes the method Point.GetDistance(). Therefore, in the case of any test failure, it is not clear whether the Point.GetDistance(), ShapeType(), or any other methods invoked by GetType() are faulty.

Class interfaces are often used to remove dependencies. An interface class is a particular type of class that only defines function/method signatures. Other classes implementing an interface must implement all functions defined by the interface class [1].

In the first step, an interface, PointInterface, including signatures for the public methods of the class Point, is created. As shown in Fig. 2.11, the class Point implements the interface class PointInterface.

In the second step, as shown in Fig. 2.12, Triangle's dependent class is modified to depend on PointInterface rather than Point. In this respect, any occurrence of the name Point is replaced with the PointInterface. The class Triangle does not create instances of Point anymore; instead, it is modified to accept the instances via its constructor or preferably via a setter method.

To test the method, right-click on its name and select the "Create Unit Tests" option as described above. Then complete the GetTypeTest() method as shown in Fig. 2.13.

As shown in Fig. 2.12, the class Triangle no longer depends on the class Point. Instead, it depends on the Interface class PointInterface. Decoupling a class from its dependencies, we can simply unit test the class in isolation. For instance, consider the GetType() method Fig. 2.13. This method determines the type of the Triangle as Scalene, Isosceles, Equilateral, and Right-angled. It invokes the method GetDistance

```
namespace Geometry {
  public enum shapeTypes {
                          Isosceles, Equilateral, Right, Ordinary, Rectangle,
                          Square, Parallelogram, Diamond }

  public class Triangel  {
    public Point p1, p2, p3;
    public Triangel(double x1, double y1, double x2, double y2
                  double x3, double y3)  {
      p1 = new Point (x1, y1); p2 = new Point (x2, y2);
      p3 = new Point (x3,y3);
      if (Point.isCoLinear(p1, p2, p3))
        dump();
    }

    public void rotate(int xc, int yc, double angle) {
      p1.rotate(xc, yc, angle); p2.rotate(xc, yc, angle);
       p3.rotate(xc, yc, angle);
    }

    public shapeTypes GetType() {
      double a = Point.GetDistance (p1, p2);
      double b = Point. GetDistance (p1, p3);
      double c = Point. GetDistance (p2, p3);
      if (isIsosceles(a, b, c)) return shapeTypes.Isosceles;
      if (isEquilateral(a, b, c)) return shapeTypes.Equilateral;
      if (isRight(a, b, c)) return shapeTypes.Right;
      return shapeTypes.Ordinary;
    }
  }}
```

Fig. 2.10 The class triangle is dependent on the class Point, listed in Fig. 2.1

```
using System;
using System.Collections.Generic;
using System.Linq;
using System.Text;
using System.Threading.Tasks;
namespace Geometry
{
    public interface PointInerface
        {
            double GetDistance(PointInerface p);
            void rotate(double xc, double yc, double angle);
             bool IsCoLinear(PointInerface p, PointInerface q);
        }
    public class Point : PointInerface
    {
        public double x, y;  //coordinates of the Point

        //1- The class constructor
        public Point(double xCoord, double yCoord)
        { x = xCoord; y = yCoord; }

        //2- Computes the distance between two points p and q
        public double GetDistance(PointInterface p1)
        {
            double res = 0;
                Point p = (Point)p1; Point q = (Point)this;
            if (p.x == q.x) { res = Math.Abs(p.y - q.y); return res; }
            if (p.y == q.y) { res = Math.Abs(p.x - q.x); return res; }
            double dx = p.x - q.x; double dy = p.y - q.y;
            res = Math.Sqrt(dx * dx + dy * dy);
            return res;
        }

        //3- rotates (x,y) around the point (xc, yc) for the angel degree
        public void rotate(double xc, double yc, double angel)
        {
            //point ComparedWith = new point(xc, yc);
            if (xc == x && yc == y) return;
            double cosa = Math.Cos(angel); double sina = Math.Sin(angel);
            double dx = x - xc; double dy = y - yc;
             x = Math.Round(cosa * dx - sina * dy + xc);
             y = Math.Round(sina * dx + cosa * dy + yc);
        }
        public bool isCoLinear(PointInterface p, PointInterface q)
        {
```

Fig. 2.11   Class Point implements the interface PointInterface

```
                  double tmp = GetDistance (p) - (GetDistance (q)
                                          + p.GetDistance (q));
                  if (Math.Abs(tmp) <= 0) return true; else return false;
               }
           }
```

Fig. 2.11 (continued)

```
namespace Geometry
{
    public enum shapeTypes {Isosceles, Equilateral, RightAngled, Scalene,
                            Rectangel, Square, Parallelogram, Diamond };
    public class Triangle
    {
        public PointInterface p1, p2, p3;
        public Triangle(PointInterface po1, PointInterface po2,
                        PointInterface po3)
        {
            p1 = po1; p2 = po2; p3 = po3;
            if (po1.isCoLinear(po2, po3))
                dump();
        }
        // Rotates a triangle around the point (xc, yc) for angle degree
        public void rotate(double xc, double yc, double angle)
        {
            p1.rotate(xc, yc, angle); p2.rotate(xc, yc, angle);
            p3.rotate(xc, yc, angle);
        }
        public shapeTypes GetType()
        {
            double a = p1.GetDistance (p2);
            double b = p2.GetDistance (p3);
            double c = p3.GetDistance (p1);
            if (isIsosceles(a, b, c)) return shapeTypes.Isosceles;
            if (isEquilateral(a, b, c)) return shapeTypes.Equilateral;
            if (isRight(a, b, c)) return shapeTypes.RightAngled; //I know!!!!
            return shapeTypes. Scalene;
        }}}
```

Fig. 2.12 The class triangle is dependent on the class Point in Fig. 2.11

```
namespace Geometry.Tests{
  [TestFixture()]
  public class TriangleTests {
    [Test()]
    public void GetTypeTest() {
      PointInterface p1  = new point(2, 0);
      PointInterface p2 = new point(0, 1);
      PointInterface p3 = new point(2, 1);
      Triangle MyTriangle = new Triangle(p1, p2, p3);
      Assert.That( MyTriangel.GetType() == shapeTypes.RightAngeled);
    }
  }
}
```

Fig. 2.13 Testing the method GetType() in the Triangle class.

Fig. 2.14 The class
StubPoint is a stub class,
partially implementing
PointInterface.

```
public class StubPoint: PointInterface
{
    public double side {set, get};
    double GetDistance(double From)
        { return side; }
    bool isColinear(double p,  double q)
        {return false; }
}
```

from the class Point. The stub class, StubPoint, in Fig. 2.14, is used to replace the
dependency of Tringle on the Point class. the class StubPoint in Fig. 2.14 facilitates
the test of the GetType method in isolation,

Using the class Stub instead of the Point, we can ensure that if there is a bug in
the execution of the GetType method and the bug is not caused by the GetDistance
method of the Point class. To test the method, right-click on its name and select
the "Create Unit Tests" option described above. Then complete the GetTypeTest()
method as follows:

Testing a unit is difficult when it is dependent on other units. Methods that receive
input from the methods they call are data-dependent on the called methods. To run
tests in isolation, we must break down such dependencies somehow. As described
above, we refactor the class under the test to a dependency injection style to remove
the dependencies. Once the unit under test is refactored, stubs are created to get
around external dependencies.

A method stub or simply stub is an implementation of an interface used by a
method to receive data from the outside. It is a small piece of code that, during
testing, takes the place of another component. The advantage of using a stub is that
it yields expected outcomes, making writing the test simpler, and even if the other
components do not work yet, tests can be continued.

```
namespace Geometry.Tests{
  [TestFixture()]
  public class TriangleTests {
    [Test()]
    public void GetTypeTest(){
       PointInterface side_1  = new StupPoint(1.0);
       PointInterface side_2  = new StupPoint(2.0);
       PointInterface side_3  = new StupPoint(Math.Sqrt(5.0));

       Triangle MyTriangle = new Triangle(side_1, side_2, side_3);
       Assert.That( MyTriangel.GetType() == shapeTypes.RightAngeled);
    }
  }
}
```

Fig. 2.15   The class Triangle is tested with the objects of type StubPoint

For instance, in Fig. 2.15, the method GetType() of the class Triangle is the one that is supposed to be tested. This class uses another class, Point, to get the length of its sides. Nevertheless, Point returns different results every time its methods are invoked, making it hard to test Triangle. Therefore, it is substituted with a different class, StubPoint, while testing the GetType() method.

One of the awkward things we run into is the various names for stubs, such as mocks, fakes, and dummies that people use to stub out parts of a system for testing. The generic term used for different types of stubs is Test Double.

A Test Double refers to any instance where you replace a production object with one used for testing. A Mock is a kind of Test double that stands in for another class in a test. A mock returns a predefined value when calling its method or attribute. A stub is Like a mock class, except that it does not provide the ability to verify that methods have been called/not called. Shims is a test double used to replace calls to assemblies whose source code is unavailable and you cannot modify.

2.2.3.2   Stubs and Mocks in Visual Studio

Microsoft Fakes is a framework that enables tests in isolation by replacing code dependencies with stubs and shims [3]. Microsoft Fakes highly supports testing in isolation by offering tools and techniques to implement test doubles rapidly.

As described in the previous section, you could write the stubs as classes in the usual way. However, you may speed up writing test doubles such as shims and stubs by using Microsoft Fakes. Microsoft Fakes provides a relatively more dynamic way to create a stub for every test.

After creating the test class, TriangleTest, for the class Triangle, as shown in Fig. 2.16, right-click the project name in the project explorer, and select the "Add

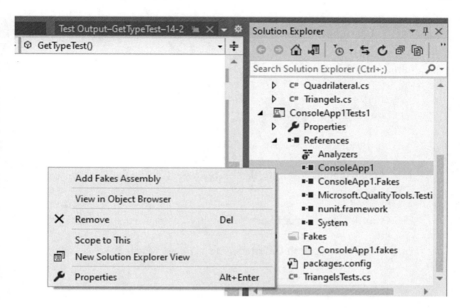

Fig. 2.16 Right-Click on the project name, ConsoleApp1, and select "Add fakes Assembly"

Fake Assembly" option. This generates a new assembly that must be referenced to create the shim objects (Fig. 2.16).

Now, modify the GetTypeTest() method, listed in Fig. 2.15, as shown in Fig. 2.17.

Below, in Fig. 2.18, the GetType method is listed. GetType invokes the GetDistance methods three times. The method GetDistancePointerInterface(), in Fig. 2.17, stubs the GetDistance() method. As shown in Fig. 2.17, the first time the GetDistance method is invoked by GetType(), the value of "a" will be 1. In the second and third invocations the GetDistance() method that is stubbed by GetDistancePointerInterface() returns the values 2, and Sqrt(5), respectively. These values returned by the GetDistancePointerInterface() method are the lengths of the three sides of MyTriangle.

MyTriangle is a RightAngeled triangle. The results of running the GetTypeTest(), as shown in Fig. 2.19, confirm that MyTriangle is RightAngeled.

2.3 Test-Driven Development (TDD)

Test Drive Development (TDD) focuses on generating test data to ensure the software under development solely satisfies the business requirements defined as features. It suggests generating test data to unit-test the satisfaction of the current working requirement before even writing a single line of code for a feature defined as a user story. User stories describe the System's functionality, termed requirements in Agile.

```
namespace Geometry.Tests
{
  [TestFixture()]
  public class TriangleTests
  {
    [Test()]
    public void GetTypeTest()
    {
      PointInterface Side1 = new Geometry.Fakes.StubPointInterface()
      {
       GetDisttancetPointInterface = (p_1) => { return 1; }
      };

      PointInterface Side2 = new Geometry.Fakes.StubPointInterface()
      {
        GetDisttancePointInterface = (p_1) => { return 2; }
      };

      PointInterface Side3 = new Geometry.Fakes.StubPointInterface()
      {
        GetDisttancePointInterface = (p_1) => { return Math.Sqrt(5); }
      };

      Triangle MyTriangle = new Triangle(side_1, side_2, side_3);
      Assert.That( MyTriangel.GetType() == shapeTypes.RightAngeled);
    }
  }
}
```

Fig. 2.17   Using Microsoft Fakes to write a stub

```
public shapeTypes GetType()
{
    double a = p1.GetDistance (p2);
    double b = p2.GetDistance (p3);
    double c = p3.GetDistance (p1);
    if (isIsosceles(a, b, c)) return shapeTypes.Isosceles;
    if (isEquilateral(a, b, c)) return shapeTypes.Equilateral;
    if (isRight(a, b, c)) return shapeTypes.RightAngled; //I know!!!!
    return shapeTypes. Scalene;
}
```

Fig. 2.18   GetType() in class Triangle invokes GetDistance() three times

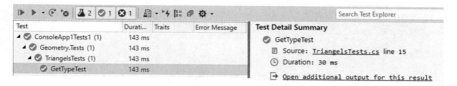

Fig. 2.19 The results of testing GetType() in isolation

The question is how to align requirements and tests. The answer is to use acceptance criteria (scenario) defined for each user story to collect test cases, ensuring the satisfaction of the requirements. For instance, consider the following user story:

User story:

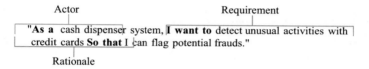

Acceptance criteria:

- When a customer makes a single transaction above $2000, mark the transaction as suspicious and display the amount.
- When a customer makes multiple transactions with an accumulative amount above $4500 during 6 min or fewer intervals, mark the transaction as suspicious and display the total amount.
- When a customer makes multiple transactions with an accumulative amount above $4000 in a single period of activity with an hour inactivity gap, mark the transaction as suspicious and display the total amount.

The acceptance criteria are then analyzed to extract test cases, used as a basis for designing one or more classes satisfying the corresponding requirement. Some examples of test cases are shown in Table 2.1:

Table 2.1 Test cases derived from the acceptance criteria

Test ID	Transaction amount	Time intervals	1-h gap	Mark as suspicious	Display
1.0.10	>$20,000	–	–	✓	✓
11.0.20	<= $20,000	–	–	–	–
21–30	N transactions with total amount > $4500	<=6 min	–	✓	✓
31–40	N transactions with total amount >= $4500	>6 min	–	–	–
41–50	N transactions with total amount > $4000	–	✓	✓	✓

In general, the following template can be used to provide a comprehensive description of a test case:

Test Case:

Test case ID: a unique number or identifier
Objective: a summary of the test objective
Pre-requisites: preconditions for using the test case
Test Procedure: Step-by-step procedure for execution.
Expected and Actual Results.

A good test plan covers both happy and unhappy paths. Happy paths are described by positive test cases, whereas unhappy path scenarios are composed of negative test cases. As shown in Table 2.1, both the positive and negative test cases should be created.

- Happy (positive) and unhappy (negative) path testing.

The developers' mistake is only to test happy paths or what users should do when using an application while ignoring unhappy testing paths or the many ways to break the software. By doing so, they may not be able to cope with errors properly. When writing acceptance criteria/tests for a given user story, an unhappy path scenario may be like this:

Scenario: User leaves the webpage before the upload finishes
Given that a user begins uploading a document
When the user closes the webpage
Then the upload should be disregarded.

Stakeholders are concerned with user behavior, while product owners describe the happy path between business requirements and intended user behavior. The followings are examples of happy paths:

The customer creates a new account
The customer logs in with his/her new credentials
The customer retrieves information about his/her account

However, what if an unintended behavior occurs, like when users enter invalid inputs? When designing or developing software systems, such scenarios need to be considered. Following are some examples of unhappy paths:

The customer creates a new account without completing the required details
The customer logs in with invalid credentials
The customer attempts to retrieve information about the other accounts

As another example, consider a login and registration system. Before starting a red-green-blue (TDD) cycle to develop the System, features covering happy and unhappy paths should be specified. For instance, consider the "Managing user registration" feature shown in Fig. 2.20.

In Fig. 2.20, information is inserted into the scenario as text offset by delimiters consisting of three double-quote marks """" on lines by themselves. Multiple strings,

Feature: Managing user registration
   As a software developer
   I want to be able to manage registration
   So that a user can sign up
  Background:
   Given users details as follow:

```
|Id.    | Name-Family | Email address  | password   |
| 100   |  Kevin-Lili | Lili@yahoo.com | Salford2000 |
| 200   |  Lu-Suzuki  | Lu@yahoo.com   | Tokyo2001   |
```

   And I set header "Content-Type" with value "application/json"
\# Happy path
Scenario: Can register with valid data
  When I send a "POST" request to "/register" with body:
    """

```
{
  "email": "Tom@gmail.com",
  "username": "Tom",
  "plainPassword": {
  "first": "savage666",
  "second": "123Infinity"
  }
  }
  """
```

Then the response code should be 201
And the response should include "The user is registered successfully."

  \# Unhappy path
  When the user successfully logs in with username: "Tom",
      and password: "savage666"
  And I send a "GET" request to "/profile/2"
  And the response should contain json:
    """

```
{
  "id": "222",
  "username": "Tom",
  "email": "Tom@Gmail.com"
}
"""
```

  Scenario: Cannot register with your current username
   When I submit a "POST" request to "/register" with body:
    """

```
{
  "email": "Lili@a-different-domain.com",
  "username": "Kevin-Lili",
  "plainPassword": {
```

Fig. 2.20 A feature including happy and unhappy scenarios

```
            "first": "York2020",
            "second": "Shefiled1012"
        }
    }
    """
```

Then the response code should be 400
And the response should include "Something is wrong."

Fig. 2.20   (continued)

known as the so-called DocString, can be inserted into a scenario If the information specified in the scenario does not fit on a single line.

The Test-Driven Development (TDD) approach uses the test cases derived from the user stories to specify and validate what the code is supposed to do before writing the code.

2.3.1   TDD Life Cycle for Object-Oriented Software

TDD relies on software requirements being converted to test cases before the software is fully developed [4]. As shown in Fig. 2.21, the project life cycle begins with selecting a user story to implement and generating test cases from the acceptance criteria defined for the selected user story. An example of test case generation from the acceptance criteria for a user story is given above.

After generating the test cases, classes with empty methods are developed to satisfy requirements. For instance, in Fig. 2.22, a class diagram depicting group objects, including geometric shapes, is illustrated (Fig. 2.22).

In the next step, the class diagram is converted into source code. For instance, below, the class point is shown. The class should be added to a C# solution.

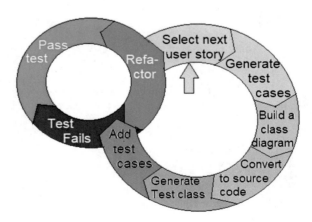

Fig. 2.21   TDD cycle

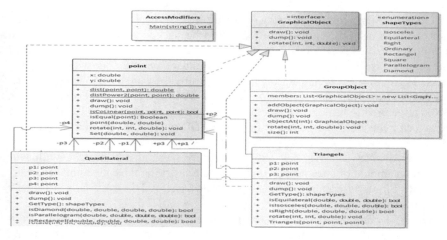

Fig. 2.22 Group objects class diagram

```
///////////////////////////////////////////////////////
//  point.cs
//  Implementation of the Class point
//  Generated by Enterprise Architect
//  Created on:     09-Apr-2022 5:44:14 PM
//  Original author: Asus
///////////////////////////////////////////////////////

using System;
using System.Collections.Generic;
using System.Text;
using System.IO;
using Geometry;

namespace Geometry {
  public class point : GraphicalObject {
      public double x;
      public double y;
      public point(){  }
      ~point(){          }
      public point(double xCoord, double yCoord){}
      public static double dist(point p, point q){return 0;}
      public static bool isCoLinear(point p, point q, point r)
                                {return false; }
      public void rotate(int xc, int yc, double angel){  }
   }//end point
}//end namespace Geometry
```

Once the empty classes are ready, a unit testing tool should be invoked to generate unit tests. For instance, the test class for testing the class point is shown in 0. The unit testing process using NUnit is described in Sect. 2.2.1. Once the unit tests pass, and the code is correct. The developer may clean up and refactor the code with all the tests passed to remove any deficiencies. For instance, after passing the unit tests, the class may suffer from God class [5], the long method [5], or so many other smells that need to be refactored to remove the smells. Refactoring is the subject of Chap. 6. After refactoring the class and getting it right, the TDD process is repeated by selecting the following user story and trying to get it right and efficient.

To sum it up, as shown in Fig. 2.23, TDD involves writing a failing test, writing the minimum amount of code needed to fix it, and refactoring the code to make it more readable. Similarly, each requirement gets its own test, and the cycle fails ->passes-> refactors continue (Fig. 2.23).

Once selecting the following user story in the TDD life cycle, the question is on which basis we should design classes to fulfill the user story and satisfy its acceptance criteria. The answer is to apply the single responsibility principle because it ensures that the class is responsible for only one requirement, and any changes to the class are rooted in the requirements defined in the user story.

Fig. 2.23   TDD life cycle

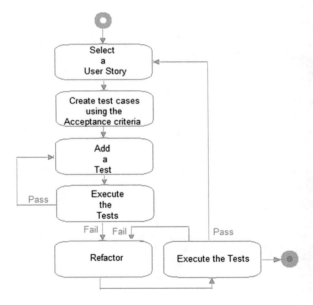

2.3.2   Single Responsibility Principle

According to the Single Responsibility Principle (SRP), each class should be responsible for only one actor [6]. An actor is a role. Therefore, to follow the single responsibility principle, a class should not satisfy user stories [7] with more than one actor. For instance, consider the following user stories defined for an event logger class:

1. As a software developer, I want access to event log data so that I can use it in my application code.
2. As a software developer, I want to destroy the event log when my application software stops running.
3. As a test engineer, I want to get events logged as they occur on the test system so that I can use them when something goes wrong
4. As a test engineer, I want event logs to be readable so that I can visually analyze them
5. As a test engineer, I want the event logs to be renewed every day so that log files do not grow unbounded
6. As a database administrator, I want to be able to store event logs to a database so that they can persist globally
7. As a database administrator, I want to retrieve events from the database so that the software can import them.

The user stories identify three roles/actors, software developer, test engineer, and database administrator. Therefore, to follow the SRP, at least three different classes, each responsible for another actor, should be developed.

Also, according to SRP, there should not be more than one reason for a class to change. The reason for to change is the change in the requirement. Therefore, a class should satisfy only a single requirement. A class meeting only a single requirement is cohesive and thereby testable.

Requirements are placed at the leaves of the goal model. Therefore, to follow the SRP, each software component/unit's responsibility should be to achieve a single goal in the goal model. In this way, the modular structure of the software will be the same as the goal hierarchy. For instance, consider the goal model shown in Fig. 2.24.

The modular architecture corresponding to the goal model in Fig. 2.24 is shown in Fig. 2.25.

Hierarchical architecture in one to one correspondence with the goal model. Each package (functional unit) is supposed to realize a different goal scenario.

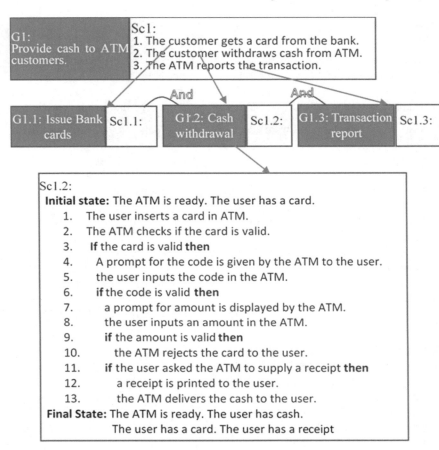

G1:
Provide cash to ATM
customers.

Sc1:
1. The customer gets a card from the bank.
2. The customer withdraws cash from ATM.
3. The ATM reports the transaction.

And   And

G1.1: Issue Bank   Sc1.1:   G1.2: Cash   Sc1.2:   G1.3: Transaction   Sc1.3:
cards   withdrawal   report

Sc1.2:
Initial state: The ATM is ready. The user has a card.
  1.   The user inserts a card in ATM.
  2.   The ATM checks if the card is valid.
  3.   **If** the card is valid **then**
  4.   A prompt for the code is given by the ATM to the user.
  5.   the user inputs the code in the ATM.
  6.   **if** the code is valid **then**
  7.   a prompt for amount is displayed by the ATM.
  8.   the user inputs an amount in the ATM.
  9.   **if** the amount is valid **then**
 10.   the ATM rejects the card to the user.
 11.   **if** the user asked the ATM to supply a receipt **then**
 12.   a receipt is printed to the user.
 13.   the ATM delivers the cash to the user.
Final State: The ATM is ready. The user has cash.
              The user has a card. The user has a receipt

Fig. 2.24   An actor goal-scenario model []

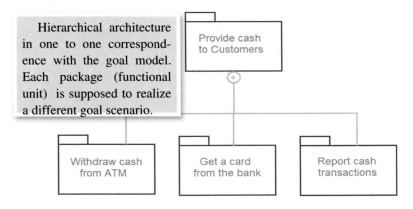

Hierarchical architecture in one to one correspondence with the goal model. Each package (functional unit) is supposed to realize a different goal scenario.

Provide cash
to Customers

Withdraw cash   Get a card   Report cash
from ATM   from the bank   transactions

Fig. 2.25   The hierarchical architecture is derived from the goal hierarchy in Fig. 2.24

2.4 Unit Testing in Python

TDD focuses on component-level testing, code quality, and fast feedback. Component level testing begins with unit testing as the lowest level of testing. TDD is the core practice of agile methods, and unit testing is the core of TDD as a test first method.

The question is that what makes unit testing so much worthwhile? Suppose a car's front lights do not turn on. Then the alternatives maybe the broken bulbs, dead batteries, broken alternators, or a failing computer. Suppose the battery is brand new and the computer is all right. Therefore, bulbs and alternators should be tested as the priority. If there is no problem with these two, the wires connecting these parts to the lights should be tested. Testing each part on its own is unit testing, while testing the connections is similar to an integration test. Modern electric cars use a form of unit testing to test different parts of the car.

Test runners orchestrate tests and present the results to users. There are many test runners for Python, four of which are relatively more popular:

1. Nose or nose2
2. Unittest
3. Pytest
4. MonkeyPatch versus Mock

2.4.1 Nose

Nose is a Python-based unit test framework. Nose2 is a successor of Nose based on the Python unit test; hence it is referred to as extend unit test [8]. It is, in essence, unit test with the plugin designed to make testing simple and more straightforward. Consider the following simple Python program:

```
import nose
def add(a, b):
    return a + b
def test_add_integers():
    assert add(5, 3) == 8
def test_add_floats():
    assert add(1.5, 2.5) == 4
if __name__ == '__main__':
    nose.run()
```

The test result for the simple 'add' function using the Noze test runner is as follows:

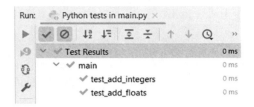

2.4.2   Unit Test

Unittest is the standard test runner built into the Python standard library. Consider the following string_handler class in Python:

```
class string_handler(object):
    """Perform basic string operations"""
    def __init__(self):
        pass
    def add_str(self, *str_):
        s = ''
        for i in str_:
            s += i
        return s
    def subtract_str(self, from_ : str, str_: str):
        s = from_.replace(str_, '')
        return s
```

The class string_handler has two methods, add_str, and subtract_str. The add_str method concatenates two or more strings, while the subtract_str removes any occurrences of a substring in a target string. In order to unit test a class, 'c', a class named "ctest" should be created. Similarly, to test a method, 'm', in the class 'c,' a method named "mtest" is added to "ctest."

In the PyCharm environment, to create a unit test for a function, the below steps are followed:

1. Right-click on the function body
2. Click on the Go-To options
3. Click test
4. Create new test
5. Choose a directory to save the test file
6. Enter the test file name
7. Enter test class name
8. Click OK

Each test class may include a setUp() and a tearDown() method, to be executed respectively before and after each method run in the test class.

```python
class string_handlerTest(unittest.TestCase):
    """Tests performed on the Calculator class"""

    def setUp(self):
        """
        Set up our test: The method is automatically invoked
        before each test.
        """
        self.calc = string_handler()

    def test_add_str(self):
        self.assertEqual(self.calc.add_str("ABC", "DE", "F"), "ABCDEF")
        self.assertEqual(self.calc.add_str("I am", " who ", "I am"), "I am who I am")

    def test_subtract_str(self):
        try:
            self.assertEqual(self.calc.subtract_str("xyzw", "yz"), "xw")
        except:
            print("Error -1:", self.calc.subtract_str("xyzw", "yz"), "xw")
        try:
            self.assertEqual(self.calc.subtract_str("ABCD", "BC"), "AD")
        except:
            print("Error -2:", self.calc.subtract_str("ABCD", "BC"), "AD")

    def tearDown(self) -> None:
        pass
```

The above class string_handlerTest is a short script to test two string methods, add_str, and subtract_str. It inherits from a class called unittest.TestCase to create test cases. Several assert methods in the TestCase class can be used to report and check for failures. The assert methods assertEqual(a, b), assertNotEqual(a, b), assertIn(a, b), and assertNotIn(a, b) respectively control whether a = = b, a ! = b, a in b, and a not in b. The TestCase class also provides four methods to run the test, gather information about the test, check conditions, and report failures.

```python
def main():
    sys.path.insert(0, os.path.dirname(__file__))
    unittest.main()
    sys.exit(0)
if __name__ == '__main__':
    main()
```

The unittest.main() function looks for all the method names starting with test_ and executes them. Also, any Python file whose name starts with test_ in the project directory or any one of its subdirectories will be executed. The nose framework extends unittest with a set of plugins to make testing easier. The test result for the string handler test class using the unittest test runner is shown below:

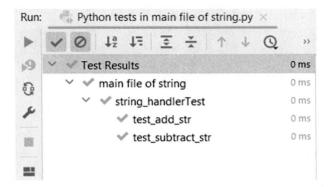

2.4.3   PyTest

PpyTest [10] supports the execution of Unittest test cases [9]. The advantage of
PyTest over Unittest comes from writing sequences of test cases in the test class
before each test method to run the method with the test cases. For instance, below,
the test runner invokes the test_add_str function five times, each time with one of
the five test cases in the test_suite.

```python
import pytest
from string_handler import string_handler

test_suite = [(["AB","CD","EF"],"ABCDEF"),
              (["qw","er","ty"],"qwerty"),
              (["so","m"],"som"),
              (["x","y","z"],"xyz"),
              (["st","ri","ng","dd"],"stringdd")]

@pytest.mark.parametrize("test_input, expected" ,test_suite)
def test_add_str(test_input, expected):
    returned_result = string_handler().add_str(*test_input)
    assert returned_result == expected

test_suite = [(["ABCDEF","EF"],"ABCD"),
              (["qwerty","qwe"],"rty"),
              (["iran","ran"], "i"),
              (["xyzw","yz"], "xw"),
              (["ABCD","CD"], "AB"),]

@pytest.mark.parametrize("test_input,expected",test_suite)
def test_subtract_str(test_input, expected):
    returned_result = string_handler().subtract_str(test_input[0],test_input[1])
    assert returned_result == expected
```

The result of the test for more inputs using test Pytest is as follows:

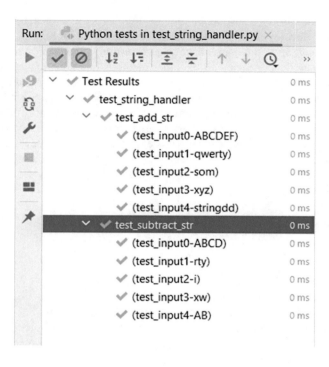

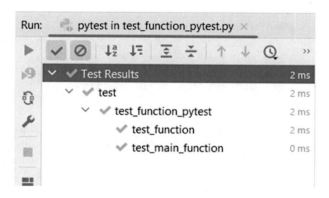

2.5   Summary

The most basic form of testing is unit testing. Unit testing is the foundation of test-driven development, and the heart of agile methodologies. Tests should cover one thing at a time and have one assertion. However, when testing a method/function, dependencies on the other methods/functions make it difficult to test the method/function in isolation. The dependencies are controlled by using the injector pattern. Fake objects are used to remove the dependencies. Stubs and mocks, also known as test doubles, simplify unit testing by providing dummy objects. Stubs hold predefined data to answer calls. Mocks are helpful if you have a dependency on an external system whose source code is inaccessible. Test-driven development habit is red, green, and blue. Try to get the code correct by writing the simplest possible code. Once the code is correct, improve its quality by applying various automated and manual refactoring techniques.

2.6   Exercises

(1) Compare and contrast NUnit versus JUnit and Xunit.
(2) Draw the class diagram for an online shopping system and follow the TDD procedure described in this chapter to implement the System.
(3) Use a top-down TDD style, starting from the user story to implement a hospital reception system. Below is the activity chart of the hospital reception system depicted in the Enterprise Architect (EA) environment:

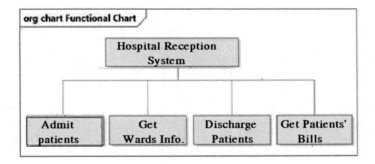

Requirements model and test cases associated with each requirement, modeled in EA, is as follows:

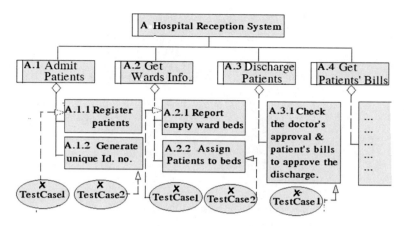

The test cases associated with each requirement in the EA environment are as follows:

☒ A. Hospital Reception System
  ☒ A.1. Admit Patients
    ☒ A.1.1 Register patients
      ○ TestCase1:
        provide correct national insurance number
        provide incorrect national insurance number
        provide certified medical doctor name
        provide uncertified medical doctor name
        provide correct address
        provide incorrect address
    ☒ A.1.2 Generate unique Id. No.
      ○ TestCase2:
        Provide correct unique id.
        Provide incorrect id.
  ☒ A.2. Get Wards Info.

    ☒ A.2.1 Report empty ward beds
      o TestCase3:
        Assign than one bed to a patient
        Assign a bed to a patient when there are empty beds
        Assign a bed to a patient when there is no empty beds
    ☒ A.2.2 Assign patients to bed
  ☒ A.3. Discharge Patients
    ☒ A.3.1 Check the doctor's approval & patient's bills to approve the
             Discharge.
  ☒ A4. Get Patients' Bills.

Below is a class in C# that Enterprise Architect has generated for the TestCase1, derived from the "A.1.1 register patients" requirement:

```
/////////////////////////////////////////////////////////
//  TestCase_A_1_1.cs
//  Implementation of the UseCase TestCase_A_1_1.cs
//  Generated by Enterprise Architect
//  Created on:     13-Mar-2022 12:11:01 PM
//  Original author:Parsa
/////////////////////////////////////////////////////////

using System;
using System.Collections.Generic;
using System.Text;
using System.IO;

namespace Requirments.Requirements {
  /// <summary>
  /// Successful implementation of patients' registration
  /// <ul>
  ///        <li>provide correct national insurance number</li>
  ///        <li>provide incorrect national insurance number</li>
  ///        <li>provide certified medical doctor name</li>
  ///        <li>provide uncertified medical doctor name</li>
  ///        <li>provide correct address</li>
  ///        <li>provide incorrect address</li>
  /// </ul>
  /// </summary>
  public class TestCase_A_1_1 {
    public TestCase_A_1_1 {
    }
    ~ TestCase_A_1_1
    }
  }//end C.1:TestRegister
}//end namespace Requirements
```

(4) User stories often indirectly imply some authentication and authorization requirement defined in "As a" part of the user story template. The "I want to" part of the user story template describes a functional requirement. The third part, "so that," explains the consequences from which test cases can be derived. Generate test cases for the user story described in Sect. 2.3.

(5) Depict the following class diagram in the Enterprise Architect environment and then apply the TDD process described in this chapter to generate code for at least two classes in the diagram.

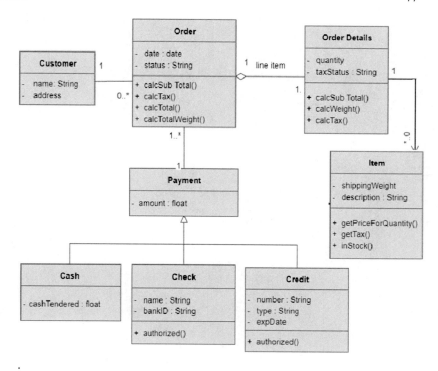

References

1. Hamilton, B.: NUnit Pocket Reference: Up and Running with NUnit. Pocket Reference, O'Reilly Media; 1st edn, Kindle Edition (2004)
2. Hunt, A., Thomas, D., Hargett, M.: Pragmatic Unit Testing in C# with NUnit, 2nd edn. The Pragmatic Programmers Publisher (2007)
3. Schaub, W.-P.: Better Unit Testing with Microsoft Fakes v1 eBook. Microsoft (2013)
4. Ramonyai, J.: Programming Practices & Methodologies-Clean Code, Reuse, Continuous Delivery, Integration, Domain Design, Open Source, Extreme Pair Programming, Polymorphism, TDD, Test-Driven Development, Kindle Edition. Amazon (2021)
5. Suryanarayana, G., Samarthyam, G., Sharma, T.: Refactoring for Software Design Smells Managing Technical Debt. Elsevier (2015)
6. Martin, R.C.: Clean Architecture. Pearson Education (2018)
7. Cohn, M.: User Stories Applied for Agile Software Development. Addison Wesley (2009)
8. Beck, K.: Test Driven Development: By Example. Addison Wesley (2003)
9. Gift, N., Deza, A.: Testing in Python: Robust Testing for Professionals. Amazon (2018)
10. Okken, B.: Python Testing with PyTest. The Pragmatic Bookshelf (2017)

Chapter 3
Acceptance Testing and Behavior Driven Development (BDD)

3.1 Introduction

Throughout this chapter, the reader will learn how to specify requirements in natural language so they can be directly and automatically converted into executable specifications. Requirements form the basis of the software development life cycle. Software projects begin with requirements gathering and end when all the requirements are satisfied by the final software product. Acceptance tests verify the satisfaction of the requirements. Success in the acceptance tests depends on how well the requirements are treated and met while developing the software. A reasonable requirement states something *necessary*, *verifiable*, and *attainable*.

To ensure a requirement is something *necessary*, document it as a user story expressed by the end-user. User stories are the subject of Sect. 3.3.

The *verifiability* is supported by defining acceptance criteria to attain a required feature together with the user story. Acceptance criteria are the subject of Sect. 3.3.1. Acceptance criteria are defined together with examples to clarify the expectations of the final software product and provide test cases to examine what the product is supposed to do while developing the software. Specification by examples is the subject of Sect. 3.3.2.

Finally, the *attainability* of a requirement is assured by converting it to executable specifications, further supported by the detailed specifications in a TDD manner. The executable specification is automatically obtainable when specifying a user story and its acceptance scenarios as a feature of the expected software in Gherkin's domain-specific human-readable language. The Gherkin language is described in Sect. 3.3.6.

© The Author(s), under exclusive license to Springer Nature Switzerland AG 2023   79
S. Parsa, *Software Testing Automation*,
https://doi.org/10.1007/978-3-031-22057-9_3

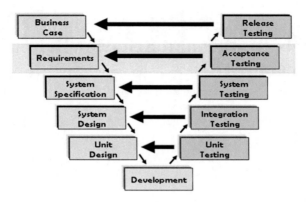

Fig. 3.1 Different phases in a V-model of the SDLC

Behavioral Driven Development, or BDD, is a new methodology that uses the Gherkin language to provide executable specifications. BDD methodology and executable specifications are the subjects of Sect. 3.3.

3.2   Acceptance Testing

Acceptance tests verify that the software product meets all the agreed requirements before being delivered to the customer [1, 2]. It is a "black box" testing method involving end-users. User acceptance testing corresponds to the requirements phase of the V-model software development life cycle, in Fig. 3.1. V-model is an SDLC model where the process executes sequentially in a V-shape. As shown in 00, each step of the software development life cycle in the V-model is associated with a testing phase. User acceptance testing formally proceeds in a user environment. Acceptance testing verifies whether the software system is ready for delivery to the customer.

3.3   User Stories

People like stories. Good storytellers capture the imagination of their audience. However, not everybody is a good storyteller. It is an art.

So, how can we expect a business owner to tell us a good user story? It might help if user stories are short and follow the following template:

As a <role/actor>, I want to <requirement/goal> so that <benefits/reasons>.

As an example, consider the following user story:

As a coach

I want to register my team using the swimming portal

so that I can add my team details and register for the competition.

In addition, to help with the acceptance tests, acceptance criteria are added to the definition of user stories [3].

3.3.1 Acceptance Criteria

The business owner, as a knowledge and information provider, should discuss and approve the acceptance criteria when writing user stories. Acceptance criteria allow you to confirm when the application performs as expected, i.e., the user story is completed. By approving the acceptance criteria, we can ensure that there will be no ambiguity and doubt to prove the project's performance according to the user's expectations during the final delivery of the project. In addition, the developers will know which criteria they must meet. Acceptance criteria synchronize the development team with the client. As an example, consider the following user story:

User Story: Doctor views calendar

As a doctor, I want to coordinate my patient appointments with my existing available time slots **so that** I can find free times to do my personal affairs.

Acceptance criteria: Given that the doctor is signed in and has scheduled appointments, the doctor should have enough time to do personal affairs.

The question is how much time is enough for doing the housework. Typically, specification by example is the recommended practice.

3.3.2   Specification by Example

We love practical examples. Examples allow us to communicate better and have a clear picture in mind. Examples that are easy to visualize help us define goals and requirements without ambiguity. Examples are one of the most potent ways to disambiguate because they provide the context in which a requirement is understood. To avoid ambiguities, try to include one or more examples of using features that meet a requirement.

Specification by example (SBE) encourages defining and illustrating software requirements based on realistic examples rather than abstract statements [4, 5]. It is also known as example-driven development, Test-Driven Requirements (TDR), executable requirements, acceptance test-driven development (ATDD or A-TDD), and Agile Acceptance Testing.

Examples are helpful not just for clarifying requirements but also for designing test cases. Figure 3.2 shows the relationship between requirements, examples, and tests. During requirement engineering, use examples to elaborate requirements and transform these into tests.

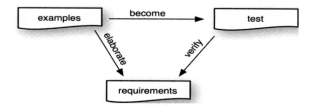

Fig. 3.2 Tests are derived from examples explaining requirements

3.3.3 Acceptance Criteria/Tests

Product Owners and Agility testers should collaborate in writing the acceptance criteria for their user stories. Testers attempt to design test cases that exemplify how the customer expects the software system to work. Acceptance criteria/tests are an essential part of a user story to determine the conditions under which the requirement defined by the user story is satisfied [6]. For instance, consider the following user story:

User Story
  # As a [role]
    As an online customer,
  # Requirement
    I want to see the total price of the items I add to my shopping cart
  # So that I [can get some benefits or consequences]
    So that I can keep account of what I am going to purchase
  # Exemplified test scenario
  Acceptance criteria/tests
    # Given [input | precondition]
      Given that the sum of the prices of the items in my shopping cart is $150
      And the total allowed amount for shopping is $666
    # When [acction | triger]
      When I add a new item priced at $100
      And an item priced at $316
    # Then [output | consequences]
      Then the total amount exceeding $666 should be notified

Test cases can be derived from the acceptance criteria or, in other words, test scenarios written for a user story. To have more test data, you may repeat the scenario or the acceptance test with different values [7]:

Feature: Controlling expenses
  As a current account holder
  I want to be able to limit my expenses for weekdays
  So that I can control my expenditures
  Scenario 1: Unacceptable Expenses
      # Given [input | precondition]
      Given that my weekly expenses are limited to "$1000,"
      And the deadline is not expired.
      # When [acction | triger]
      When I attempt to spend "$1000,"
       # Then [output | consequences]
      Then I should receive the notification "Failed."

  Scenario 2: Acceptable Expenses
      # Given [input | precondition]
      Given that my weekly expenses are limited to "$1000,"
      And the deadline is not expired.
      # When [acction | triger]
      When I attempt to spend "$900,"
       # Then [output | consequences]
      Then I should receive the notification "Passed."

3.3.4   Background Section

The Given step in the two scenarios in the above feature definition is the same. Whenever scenarios of a feature file share common steps, instead of duplicating the steps, a Background [7] section is used, as shown below:

Feature: Controlling expenses
  As a current account holder
  I want to be able to limit my expenses for weekdays
  So that I can control my expenditures
  Background:
        # Given [input | precondition]
        Given that my weekly expenses are limited to "$1000,"
        And the deadline is not expired.
  Scenario 1: Unacceptable Expenses
        # When [acction | triger]
        When I attempt to spend "$1000,"
         # Then [output | consequences]
        Then I should receive the notification "Failed."

  Scenario 2: Acceptable Expenses
        # When [acction | triger]
        When I attempt to spend "$900,"
        # Then [output | consequences]
        Then I should receive the notification "Passed."

The background scenario is copied in the enclosing feature definition at the start of each scenario section.

3.3.5   Example Table

The above feature repeats the same scenario twice with different data values. Instead of repeating the same scenario for different test data, we can use a scenario outline replacing example variables/keywords with the value from a data table. For instance, the above scenarios are replaced with an outline scenario that repeats twice with two different values for the data in an example table.

Feature: Controlling expenses
  As a current account holder
  I want to be able to limit my expenses for weekdays
  So that I can control my expenditures
  Outline Scenario: Accepted and unacceptable expenses
        # Given [input | precondition]
        Given that my weekly expenses are limited to "<allowed>."
        And the deadline is not expired
        # When [acction | triger]
        When I attempt to spend "<expenses>"
        # Then [output | consequences]
        Then I should receive the notification "<status>."
  Examples:
        | allowed | expenses | status |
        | 1000 | 1000 | Failed |
        | 1000 | 999 | Passed |

In the above feature definition, The three input parameters, allowed, expenses, and status, receive their values from the Examples table.

TDD and BDD both recommend writing tests first before writing code. As for BDD, this means acceptance tests, followed by unit tests. BDD captures the required behavior in User Stories from which acceptance tests are extracted. BDD builds upon TDD to encourage collaboration between developers and stakeholders by creating tests that are easy to understand by both of them. This is due to using a non-technical language, Gherkin, to define acceptance tests that the developers execute. Gherkin uses feature files with scenarios written in natural language that is aimed at testing software.

3.3.6 Gherkin Language

The basis of a high-quality software development process is the requirements. The software testing will be limited to fault detection if the requirements are not considered. Generally, a defect occurs when something does not work as expected, while a fault occurs when something does not work correctly [4].

Behavior-Driven Development, or BDD, is about requirements-driven or acceptance tests. It begins with a focus on features that deliver business values. A feature is a functionality that helps users and other stakeholders achieve a specific business goal. BDD uses a ubiquitous language, Gherkin, to define user stories, as shown in Fig. 3.3.

Feature: [Title (one line describing the feature or story)]

\# 1. Description of the feature or narrative of the story
Narrative:
    As a [role]
    I want [feature: something, requirement]
    So that [benefit: achieve some business goal]

\# 2. The background is executed once before each scenario
Background:
    Given [some condition]
    And [one more thing]

\# 3. Acceptance Criteria: (presented as Scenarios)
Scenario 1: Title
    Given [context]
      And [some more context]
    When [event]
    Then [outcome]
      And [another outcome]...
    For instance,
Scenario 2: ...

\# 4. Templates with placeholders require a table.
Scenario Outline:
    Given I have <something>
      And I also have <number> <thing>
    Examples:
     | something | number | thing |
     | ... | ... | ... |
     | ... | ... | ... |

Fig. 3.3 Features template

For Example, consider an online shopping feature to add and remove products from a shopping basket. The feature includes one background and two scenario steps.

Feature: Adding-Removing-items

As a store owner,
I want to give a discount as customers add more items to their baskets.
So that I can encourage the customers to buy more items.

Background:
 Given that the discount rate is %10
 And the minimum discountable amount for the discount to apply is $200
 And a customer logs into the System

Scenario: Adding items to my basket
 When adding "Tea shirts" at the price of $100 to the basket
 And adding "sunglasses" at the price of "$135" to the basket
 Then the discount will be "$23.5"

Scenario: Removing items from the basket
 When adding "Mobile" at the price of "$600" to the basket
 And adding "Cookies" at the price of "$126" to the basket
 And removing "Mobile" at the price of "$600" from the basket
 Then the total should be "$0.0"

The Background steps get duplicated across the scenarios in the feature, as follows:

 Scenario: Adding items to my basket
 Given that the discount rate is %10
 And the minimum discountable amount for the discount to apply is $200
 And a customer logs into the System
 When adding "Tea shirts" at the price of $100 to the basket
 And a customer logs into the System
 And adding "sunglasses" at the price of "$135" to the basket
 Then the discount will be "$23.5"

 Scenario: Removing items from the basket basket
 Given that the discount rate is %10
 And the minimum discountable amount for the discount to apply is $200
 And a customer logs into the System
 When adding "Mobile" at the price of "$600" to the basket
 And adding "Cookies" at the price of "$126" to the basket
 And removing "Mobile" at the price of "$600" from the basket
 Then the total should be "$126."

The two scenarios are the same but with different data. We can rewrite them as a scenario outline, including input parameters taking their values from tables.

Feature: Adding-Removing-items
As a store owner,
I want to give a discount as customers add more items to their basket
So that I can encourage the customers to buy more items

Scenario Outline: Add and Removing items from the basket
Given that the discount rate is <discount-rate>
And the minimum discountable amount <min-amount>
And a customer logs into the System
When adding and removing items from the basket

Item	Price
Tea Shirt	$100
Sunglasses	$135
Cookies	$126
Sunglasses	$600
Cookies	-$126

Then the total should be <total>
And the discount should be <discount>
Examples:

discount-rate	min-amount	total	discount
%10	$200	$835	$83.5

The above feature includes a data table with two columns and an examples table with four columns. The tables have a header row so that the compiler can match the header columns to the placeholders in the scenario outline steps. The examples table appears at the end of the scenario outline. For each row of the examples table, the whole scenario is repeated.

3.4 Behavior-Driven Development (BDD)

A good beginning makes a good ending. A software development process ends when acceptance tests succeed. The good beginning is where the behavior-driven development process brings together the tester, business owner, and developer to devise concrete examples of acceptance criteria in a user story. These examples are described using a domain-specific language, like Gherkin, and put into a feature file. The feature is converted into an executable specification where developers can then write an actual executable test. When it comes to examining the behavior by running the acceptance tests, TDD is used to implement and get the code ready for the acceptance tests.

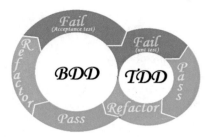

Fig. 3.4 BDD complements TDD

In fact, as shown in Fig. 3.4, BDD complements TDD by putting acceptance tests first. BDD cycle starts with selecting a feature [8]. The feature is automatically translated into executable code. Afterward, the executable code, including the acceptance tests for the selected feature, is executed. BDD goes through red in this stage because the code is not complete yet. A TDD development approach is used to complete the code.

TDD uses the art of iteration and incremental to get the code right. As part of the iterative development process, test cases are developed before any code is written, and code is constantly tested against the test cases through the "red-green-refactor loop" until the whole unit is complete. TDD developers create test cases of the methods and functions, allowing for incremental development and tests of the units. As a result, TDD development cycles are often very short.

TDD focuses on getting a piece of functionality right, while BDD foresees the end when the software product should pass all the acceptance tests in the presence of the customer's representatives. Poor handling of user requirements causes severe damages later. That is why Behavior-Driven Development emphasizes executable requirements.

BDD is a software development process to bridge the gap between user requirements and the software feature. To begin with, a working example of the BDD process to develop an application software is described in the next section.

3.4.1 Case Study: A Medical Prescription Software System

As an end-user, the doctor wants to get a patient's records, create new patients, and modify or remove an existing one. Also, the doctor should be capable of writing new electrical prescriptions or modifying existing ones. When writing a new prescription, in the first position depending on the patient's disease, a list of the drugs should be suggested by the system. The doctor should also have access to the complete database of the drugs. After selecting the drug, the doctor determines the dose of the prescribed drug and approves the drug to be added to the final prescription. Finally, the final prescription is prepared for the patient after the drugs are added. After the doctor's approval, the prescription will be printed and stored in the database. The dosage for each drug in the prescriptions should be modifiable. The doctor should be capable of reviewing the patient's records, making a copy of an existing prescription, changing the dosage of the drugs, adding some new drugs, or removing some existing ones.

As another end-user, the patient wants to have access to their prescriptions. The prescriptions should be accessible to the patient through the medical center website. The patients may log into the website using their national security number. Also, the patients should be able to make a report about their disease and treatment.

Dose checking is a safety precaution to ensure the doctor has not prescribed a dangerously small or large dose. Using the formulary id for the generic drug name, look up the formulary and retrieve the recommended maximum and minimum dose. Check the prescribed dose against the minimum and maximum. If outside the range, issue an error message saying that the dose is too high or too low. If within the range, enable the 'Confirm' button.

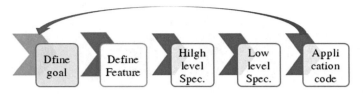

Fig. 3.5 The BDD process

3.4.2 Defining Goals

As shown in Fig. 3.5, the BDD process starts with defining the goals. The medical prescription software aims to avoid overdose toxicity by helping doctors determine drug dosage. The doctors should be capable of getting the details of the drugs from a valid database and adding it to their prescriptions. Once the prescription is completed, the doctor should be able to transfer it to the patient's mobile number.

Goals identify, describe, and correlate requirements. Agile development suggests presenting requirements as user stories. User stories talk about requirements and do not define requirements as functionalities. Goals identify, describe, and correlate requirements. Agile development suggests presenting requirements as user stories. *User stories* talk about requirements and do not define requirements as functionalities. User stories describe the *features* the customer wants to see implemented in the software product.

3.4.3 Defining Features

The BDD software development process continues with mapping business needs to the software product features that satisfy the needs and provide value.

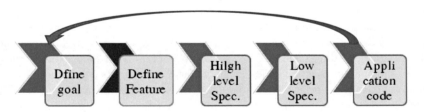

A feature describes what the end users expect from the software product and how they will be satisfied with the final product. The feature in Fig. 3.6 describes the doctor's requirements for issuing medical prescriptions.

saved in Features\Feaure1.feature

Feature: Issuing medical prescription
  As a doctor,
  I want to have access to the database of drugs
  So that I can provide a complete description of the drugs I prescribe
  Scenario Outline: sign in to the application
    Given that the doctor is signed in
    And the doctor issues a prescription for reimbursed medications
    When the doctor clicks on the drugs button and selects a database
    And the doctor selects a <drug>
    Then the <drug>, <dosage>, <usage>, and a brief <description>
          are inserted in the prescription
    And the prescription is sent to the patient's <mobile>
  Examples:

drug	dosage	usage	description	mobile
Fluoxetine	60 mg	Every day moon	Mental depression	09421000
Guanabenz	8 mg a day	Same times	Blood pressure	09221666
Aspirin	75 mg per day	Same times	Reduces pain	09226666
Caffein	400 mg per day	Same tines	Drowsiness, fatigue	09421666
Tramadol	200 to 400 mg	Every four hours	Severe pains	06221666

Fig. 3.6 Issue medical prescription feature saved in features\Feature1.feature file

Another feature concerning the over dosage toxicity avoidance goal is to control the drug dosage as doctors write prescriptions. The feature is shown in Fig. 3.7.

The last feature of the medical prescription software tool is to provide the doctor with the patient's medical recode. The feature is presented in Fig. 3.8.

Features are written in a text file with a *feature* extension, and the file should be saved in a directory named "features" within the project root directory. Figure 3.9 illustrates the directory hierarchy for the medical prescription project in the *PyCharm* environment.

The Python module called *behave* [9] allows backing Gherkin requirements with concrete tests. Behave by default expects Gherkin files to be placed in a folder called features. It was initially a command-line tool. Pip, the standard Python package installation tool, can be used to install Behave:

$ pip install behave

\# saved in Features\Feaure2.feature
Feature: Automatic control of drug dosage
  As a doctor,
  I want to have my prescribed drug dosage be controlled
  So that I can avoid any risk of causing huge health problems for patients
Scenario Outline:
  Given that the doctor is signed in
  And the doctor is writing a prescription
  When the doctor selects a <drug>
  And enters the drug dosage
  Then the <min_dosage>, and <max_dosage>for <weeks> is controlled
      with the database
  Examples:

drug	Min_dosage	Max_dosage	weeks
Priftin	50 mg	900 mg	12 weeks
Pantoprazole	40 mg	120 mg	28 weeks
Cephalexin	250 mg	333 mg	14 days every 6 hours
Penicillin V	125 mg	250 mg	10 days every 6 hours
Valium	2 mg	10 mg	4 times daily 6 month

Fig. 3.7 A feature to control the drug dosage prescribed by the doctor

Fig. 3.8 A feature to view patients' medical records

\# saved in Features\Feaure3.feature
Feature: View records
  As a doctor
  I want to be able to view my patient's medical records
  So that I can analyze their medical status
Scenario Outline: Doctor asks for records
  Given the <username> is signed in
  When the user asks for the records of a <patient>
  Then the user sees the patient records
  Examples:
    | username | patient |
    | @daviddoc | @jamesp |
    | @jamesp | @jamesp |

3.4.4 Executable Specifications

As shown below, the third task in the BDD process is to convert a feature scenario into an executable specification:

Fig. 3.9 A typical directory
structure for the behave
software development life
cycle

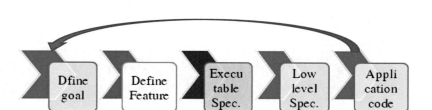

An executable specification provides a formal description for acceptance criteria, illustrating and verifying how the application delivers a specific system requirement. It consists of methods/functions, each implementing a step, Given, When, Then, and of a feature scenario representing the acceptance criteria.

BDD follows an outside-in approach to implement the scenario steps. In fact, the whole software product roots at step functions, satisfying the acceptance criteria. The programmer begins with implementing functions considering the feature scenario steps. Typically, a scenario has three steps: given, when, and then. All the steps defined in the scenarios of a feature should be saved in a Python file in the "features/steps" directory.

As shown in Fig. 3.10, behave allows for generating step definitions skeleton from within the feature file editor as follows:

(a)   Right-click on the Scenario and click on "Show Context Actions."
(b)   Select "Create all step definitions" to convert a BDD feature file into a step. It is also possible to create step definitions manually.

As shown in Fig. 3.10, there are two options to create a definition for one step or all the selected scenario steps. After selecting the "create all step definition" option, the window in Fig. 3.11 will be displayed to enter the step file.

(1)   Enter the step file name (the step file must have the same name as the feature file).
(2)   Select the step file type.

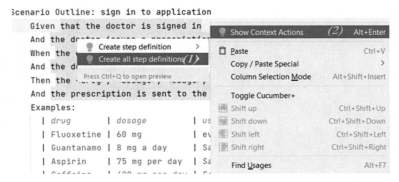

Fig. 3.10 Mapping scenario steps into Python functions

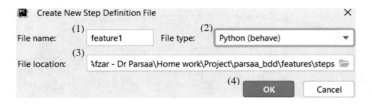

Fig. 3.11 Naming step definition files

(3) By default, Behave creates a directory, Step, inside the feature directory and saves the step file, e.g., feature_name.py, in the directory.

(4) Click on OK.

As a result, a new file called feature1.py is created and saved in the Feature\Step directory in the main project folder, shown in Fig. 3.12.

Step functions are identified using step decorators. All step implementations should typically start with the import line. Figure 3.13 represents the Python module. Feature1.py, generated by 'behave' from feature1.feature:

The code in Fig. 3.13 is automatically generated from the feature1 scenario steps. Now, to satisfy the acceptance criteria defined by the scenario steps, the developer

Fig. 3.12 The features directory includes all the .feature files and a step directory

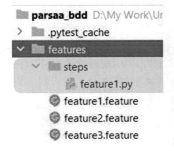

```
#feature1.py:
from behave import *
use_step_matcher("re")

@given("that the doctor is signed in")
def step_impl(context):
    raise NotImplementedError(u'STEP: Given that the doctor is signed in')

@step("the doctor issues a prescription for reimbursed medications")
def step_impl(context):
    raise NotImplementedError(u'STEP:
        And the doctor issues a prescription for reimbursed medications')

@when("the doctor clicks on the drugs button and selects a database")
def step_impl(context):
    raise NotImplementedError(u'STEP:
        When the doctor clicks on the drugs button and selects a database')

@step("the doctor selects a (?P<drug>.+)")
def step_impl(context, drug):
    raise NotImplementedError(u'STEP: And the doctor selects a <drug>')

@then(
    "the  (?P<drug>.+), (?P<dosage>.+), (?P<usage>.+),
        and a brief (?P<description>.+) are inserted in the prescription")
def step_impl(context, drug, dosage, usage, description):
    raise NotImplementedError(
        u'STEP: Then the  <drug>, <dosage>, <usage>,
            and a brief <description> are inserted in the prescription')

@step("the prescription is sent to the patient's (?P<mobile>.+)")
def step_impl(context, mobile):
    raise NotImplementedError(u'STEP:
        And the prescription is sent to the patient\'s <mobile>')
```

Fig. 3.13 Step functions generated by 'behave' from the featre1 scenario

should define classes and methods to complete the code for each step. As shown in Fig. 3.14, three four, Doctor, Drugs, prescription, and DrugsDb, are introduced to complete the step functions generated by 'behave.' In the following code, a context object is a global dictionary that allows for sharing variables between the steps.

In the given step, the doctor should be signed in. The class doctor is designed to hold any attributes and methods concerned with the doctor as an object. The sign_in operation should be delegated to the login class. In the "when" step, the doctor chooses a drug to issue a prescription. Two classes, Drug and Prescription, are required to implement the "when" step. In the "then" step, the prescription should be sent to the patient's mobile number. Three classes called 'mobile, 'patient,' and

```
from behave import *
use_step_matcher("re")
from Log_in import Log_in
from Drug import Drug

@given("that the doctor is signed in")
def step_impl(context):
    log_object = Log_in()
    if(log_object.sign_in() == True):
        print("---- ---- ---- ---- ---- ---- ----\n")
        print("---- Doctor signed in ----")


@step("the doctor issues a prescription for reimbursed medications")
def step_impl(context):
    print("---- Doctor wants issues a prescription ----")

@when("the doctor clicks on the drugs button and selects a database")
def step_impl(context):
    drug_object = Drug()
    print("Drugs List =" , end = " ")
    for medicine in drug_object.drug_list:
        print(medicine[0] , end = " ,")

@step("the doctor selects a (?P<drug>.+)")
def step_impl(context, drug):
    drug_object = Drug()
    prescription = []
    for drug_obj in drug_object.drug_list:
        if drug == drug_obj[0]:
            prescription = drug_obj
            context.prescription = prescription
    print("\nSelected Drug = {}".format(prescription[0]))

@then("the (?P<drug>.+), (?P<dosage>.+), (?P<usage>.+), and a brief
        (?P<description>.+) are inserted in the prescription")
def step_impl(context, drug, dosage, usage, description):
    assert context.prescription[0] == drug
    print("Drug Name = {} , Dosage = {} , Usage = {} ,
            Description = {}".format(drug, dosage, usage, description))
    print('prescription issued successfully.')

@step("the prescription is sent to the patient's (?P<mobile>.+)")
def step_impl(context, mobile):
    print('prescription send to {} successfully.'.format(mobile))
```

Fig. 3.14 Step functions are further completed to satisfy the system requirements

Fig. 3.15 Test results for the feature1 scenario

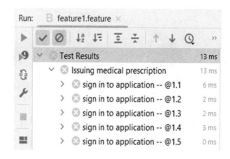

Fig. 3.16 The BDD life cycle starts with the 'red' step

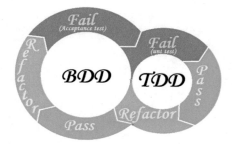

prescription are needed to implement this step. It is observed that the complexity of project development is broken down into designing classes and methods to implement the 'given,' 'when,' and 'then' steps.

To start the test in the Pycharm environment, click the green play button on the gutter, shown in Fig. 3.15. Since the classes are not completed yet, the test fails, and the following messages are shown.

As shown in Fig. 3.15, the third task of the BDD process is to ensure the acceptance tests fail because the designated classes are not implemented yet. This is the red stage because all the tests fail. As shown in Fig. 3.16, the TDD process cycle starts in the next stage.

3.4.5 Low-Level Specification

The fourth task in the BDD process is to provide low-level specifications. The low-level specification is reasoning about what the actual code should do prior to writing the code. The TDD process is followed to write the code.

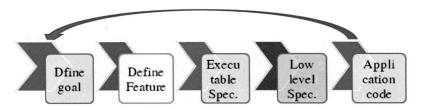

As described above, when trying to write acceptance tests, classes and methods are defined to implement the 'given,' 'when,' and then steps, respectively. In the case of object-oriented software, low-level specification starts with creating class diagrams. It is worth mentioning that in the case of web design, client and server pages are two stereotypes of classes. Figure 3.17 represents the class diagram to implement the methods and classes used in the scenario steps shown in Fig. 3.14.

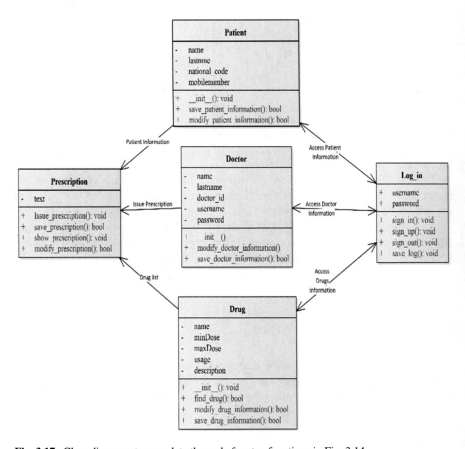

Fig. 3.17 Class diagrams to complete the code for step functions in Fig. 3.14

The acceptance criteria let the tester know when particular step functions are completed. This helps the tester create test data to unit test the classes in the class diagram to implement the scenario steps. The class diagram shown in Fig. 3.17 has been depicted in the Enterprise Architect environment.

Enterprise Architect's Code Template Framework provides flexible forward engineering of the UML class diagrams into source code. Generating code from the model is known as forwarding engineering. Enterprise Architect (EA) enables creating a source code equivalent to the Class or Interface element in many different programming languages such as C, C++, C#, Delphi, Java, PHP, and Python. The source code generated includes Class definitions, variables, and function stubs for each attribute and method in the class diagram. As shown in Fig. 3.18 the following steps should be taken to generate code from a class diagram in EA:

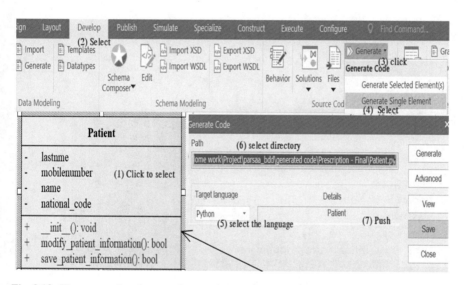

Fig. 3.18 The steps to be taken to select any language rather than C#

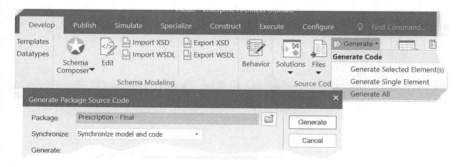

Fig. 3.19 Generate Python code for the class diagram

Step 1:

1. In the class diagram, click on one of the classes,
2. Click on the "Generate" code button and select the following option:

"Generate a single element."

Step2:

1. **Choose the path to save code in that directory**,
2. Select the source code language,
3. Save and close (Fig. 3.18).

After saving the configuration, in the next step Develop tab is selected from the main menu bar. After a short delay, the Develop menu bar pops up. Click on the

```
###############################################################
#
# Drug.py
# Python implementation of the Class Drug
# Generated by Enterprise Architect
# Created on:      01-Aug-2022 12:18:14 PM
# Original author: ----
#
###############################################################
import Log_in
import Prescription
class Drug:
    m_Log_in = Log_in()
    m_Prescription = Prescription()
    def __init__(self, name : String, minDose : int, maxDose : int,
                 usage : String, description : String) -> void:
    pass

    def find_drug(name : string) -> bool:
    pass

    def modify_drug_information(name : string, minDose : int, maxDose : int,
                    usage : string, description : string) -> bool:
    pass

    def save_drug_information() -> bool:
    pass
```

Fig. 3.20 Java source code as a result of forward-engineering the class diagram in Fig. 3.17

Config tab a menu with three options, shown in Fig. 3.19, will be displayed. Select the generate all option (Fig. 3.19).

Step3:

1. Click on the Develop tab.
2. Click on generate,
3. Select the 'generate all' option,
4. **Click Generate**.

Figure 3.20 represents the Java class generated for the Prescription class. In the next step, before writing code for the methods of the prescription class test, test cases for unit testing of the class should be prepared manually.

As part of BDD, every unit test is a low-level specification illustrating how a class or component works. It is important to use tests immediately if one follows a genuine TDD approach. Tests are indeed harder to add in retrospect. However, implementing code to satisfy the acceptance criteria naturally helps developers design and document the application code in the context of delivering high-level features. TDD is a software development approach in which tests are written before writing the code. Figure 3.21 represents the test case and test function for unit testing the find_drug method of the Drug class in Fig. 3.20. The PyTest library is used for writing readable tests.

The code development for the function find_drug starts with testing the empty function. Apparently, this is a failing red test because the body of the "find_drug" method is empty. Figure 3.22 illustrates the test results generated by Pytest.

The red step in TDD follows the red step in the BDD process. Test first development begins with an abstraction of intent as test cases to repeatedly validate the functionality inside a testing framework. The test is needed to turn red before any production code can be written to make it turn green (Fig. 3.23).

```
import pytest
from Drug import Drug

test_suite = [
    (["Fluoxetine"], True),
    (["Aspirin"], True),
    (["Caffein"], True),
    (["Tramadol"], True),
    (["Guanabenz"], True),
    (["Lorazpom"], False),
    (["Antibiotic"], False)
]

@pytest.mark.parametrize("test_input,expected", test_suite)
def test_find_drug(test_input,expected):
    drug_object = Drug()
    find_result = drug_object.find_drug(test_input[0])
    assert find_result == expected
```

Fig. 3.21 Test function for the find_drug method

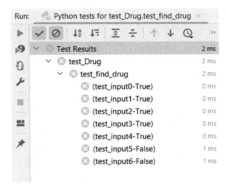

Fig. 3.22 Failing test results

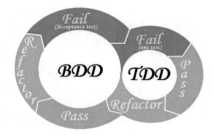

Fig. 3.23 The TDD process starts after the BDD tests fail

3.4.6 Creating Test Functions

The following steps are taken to automatically generate test functions for each method of the class under test:

1. Right-click on the function body (e.g., find_drug),
2. Select the 'Go To' option,
3. Select the 'Test' option,
4. Click on Create New Test,
5. Enter the target directory address to save the test file,
6. Enter test file name (e.g., find_drug.py),
7. Enter test class name,
8. Click OK.

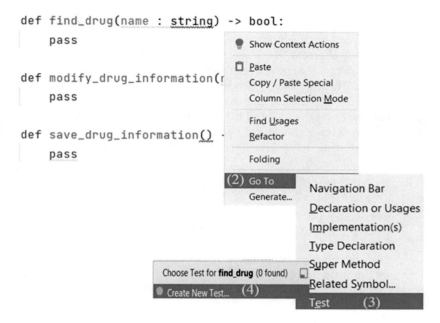

3.4.7 Application Code

The last step in the BDD process is to write the application code. The TDD approach is followed to write and design the application code while trying to make the test pass. This is usually a three-step red/green/blue process, where red indicates a failed test, green is a passing one, and blue means refactor [10].

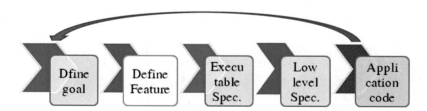

The three main activities of TDD are coding, designing, and testing. The code is not expected to be perfect from the beginning. At first, the minimum code containing only the functionality necessary for the test to succeed is developed. The written code has to be only designed to get the test pass; there should be no code for which no test has been designed.

Once all the tests pass, the code is cleaned and refactored without changing the semantics. Even after refactoring, the code must be tested to ensure that the code's semantics is not altered. The completed code for the drug class is given in Fig. 3.24.

After completing the body of the Drug class, the methods are tested. This time all the tests pass. The test result is shown in (Fig. 3.25).

As shown in Fig. 3.25, all the tests pass. This is called the green phase. During the green step, everything works well but not necessarily optimally. In the green phase, everything works as it should, but not necessarily to its full potential. The question is whether automatic refactoring improves every aspect of the code!

```python
class Drug:
    def __init__(self):
        global drug_list
        f = open("D:\\My Work\\Uni iust\\Term 2\\3 - Mohandesi Narm Afzar –
                  Dr Parsaa\\Home work\\Project\\parsaa_bdd\\drugs_list.txt", "r")
        drug_list = f.read().splitlines()
        for i in range(len(drug_list)):
            drug_list[i] = drug_list[i].split(',')
        f.close()
        self.drug_list = drug_list

    def find_drug(self,drug_name):
        for drug in drug_list:
            if drug_name == drug[0]:
                return True
        return False

    def modify_drug_information(self, name, minDose, maxDose,
                                usage, description):
        if(self.find_drug(name)):
            for drug in drug_list:
                if name == drug[0]:
                    drug[1] = minDose
                    drug[2] = maxDose
                    drug[3] = usage
                    drug[4] = description
                    return True
        else:
            return False

    def save_drug_information(self):
        pass
```

Fig. 3.24 The completed code for the Drug class

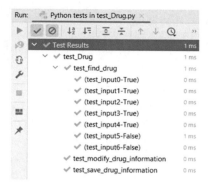

Fig. 3.25   Unit test results for the class Drug

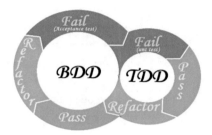

Fig. 3.26   The green phase is when all the unit tests pass

As shown in Fig. 3.26, the next step is to optimize code using refactorings. Refactoring is the subject of Chap. 6. It may take several cycles to complete and get the code ready for acceptance testing. Figure 3.27 shows the results of running feature1 step functions with behave.

Figure 3.28 shows the details of the acceptance test results provided by the 'behave' test framework.

As shown above, finally, all the acceptance tests for the first feature are passed. As shown below, this stage seems to be green. However, there are some questions raised in mind:

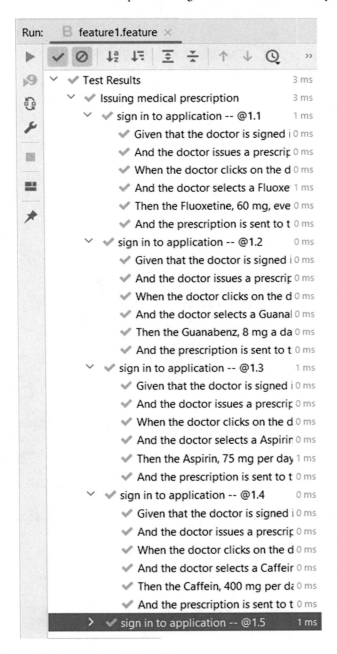

Fig. 3.27 Feature1.feature acceptance test results

Fig. 3.28 Report folder creation

```
---- ---- ---- ---- ---- ---- ----

---- Doctor signed in ----
---- Doctor wants issues a prescription ----
Drugs List = Fluoxetine ,Aspirin ,Caffein ,Tramadol ,Guanabenz ,
Selected Drug = Fluoxetine
Drug Name = Fluoxetine , Dosage = 60 mg , Usage = every day morning , De-
scription = Mental depression
prescription made successfully.
prescription send to 09421000 successfully.
---- ---- ---- ---- ---- ---- ----

---- Doctor signed in ----
---- Doctor wants issues a prescription ----
Drugs List = Fluoxetine ,Aspirin ,Caffein ,Tramadol ,Guanabenz ,
Selected Drug = Guanabenz
Drug Name = Guanabenz , Dosage = 8 mg a day , Usage = Same times Blood ,
Description = pressure
prescription made successfully.
prescription send to 09221666 successfully.
---- ---- ---- ---- ---- ---- ----
```

---- Doctor signed in ----
---- Doctor wants issues a prescription ----
Drugs List = Fluoxetine ,Aspirin ,Caffein ,Tramadol ,Guanabenz ,
Selected Drug = Aspirin
Drug Name = Aspirin , Dosage = 75 mg per day , Usage = Same times , De-
scription = Reduces pain
prescription made successfully.
prescription send to 09226666 successfully.
---- ---- ---- ---- ---- ---- ----

---- Doctor signed in ----
---- Doctor wants issues a prescription ----
Drugs List = Fluoxetine ,Aspirin ,Caffein ,Tramadol ,Guanabenz ,
Selected Drug = Caffein
Drug Name = Caffein , Dosage = 400 mg per day , Usage = Same tines , De-
scription = Drowsiness, fatigue
prescription made successfully.
prescription send to 09421666 successfully.
---- ---- ---- ---- ---- ---- ----

---- Doctor signed in ----
---- Doctor wants issues a prescription ----
Drugs List = Fluoxetine ,Aspirin ,Caffein ,Tramadol ,Guanabenz ,
Selected Drug = Tramadol
Drug Name = Tramadol , Dosage = 200 to 400 mg , Usage = Every four hours ,
Description = Severe pains
prescription made successfully.
prescription send to 06221666 successfully.

1. Are the tests adequate?
2. Are there no more bugs in the code?
3. Which refactorings should be applied to improve the code quality?
4. Does refactoring always improve the code testability?

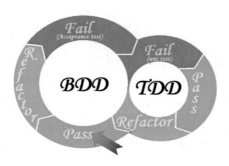

I am going to answer the above questions in the following chapters.

3.4.8 Behave Reports

Report generation is one of the most critical steps toward the test automation framework. At the end of the execution, one cannot rely on the console output; instead, there should be a detailed report. The report should provide typical data such as the number of tests (i.e., the number of passed, failed, and skipped tests), test coverage, and test results. Behave does not produce an in-built report, but it can output in multiple formats, and third-party tools can also help to generate sophisticated reports. All the available formatters employed by 'behave' are displayed with the command:

behave --format help

As a result, the screen, shown in Table 3.1, will appear, showing all the tools Behave provides.

Some of the common Behave reports are:

- Allure Report.
- Output JSON Report.
- JUnit Report

The following command creates JUnit reports:

behave --junit

Table 3.1 List of available formatters

Available formatters	
Json	JSON dump of test run
Json.Pretty	JSON dump of test run (human readable)
Null	Provides formatter that does not output anything
Plain	Very basic formatter with maximum compatibility
Pretty	Standard colorized pretty formatter
Progress	Shows dotted progress for each executed scenario
Progress2	Shows dotted progress for each step of a scenario
Progress3	Shows detailed progress for each step of a scenario
Rerun	Emits scenario file locations of failing scenarios
Sphinx.steps	Generates sphinx-based documentation for step definitions
Steps	Shows step definition (step implementation)
Steps-catalog	Shows non-technical documentation for step definitions
Step.doc	Shows documentation for step definitions
Steps-stage	Shows how step definitions are used by steps
Tags	Shows tags (and how often they are used)
Tags.location	Shows tags and the location where they are used

The reports folder, shown in Fig. 3.28, is used to keep the reports generated for each feature.

In Fig. 3.28, feature1 and feature2, and feature3 are the names of the feature files. A folder called reports includes an.xml files named TESTS-<feature file name> .xml, for each feature. The following command generates reports and saves them in a specific folder, my_reports:

behave --junit --junit-directory my_reports

To create a JSON output in the console, run the command:

behave -f json

The following command creates a JSON output in a more readable format:

behave -f json.pretty

The following command generates and saves reports in a specific folder, my_reports.json:

behave –f json.pretty –o my_reports.json

3.4.9 Behave Hooks

Behave setup and teardown functions are implemented in a file called the environment.py, within the same directory containing the steps folder. The setup functions include—browser open, database connection, configurations, and so on. The teardown functions include browser closure, database connection termination, reversing changes, etc. The **environment.py** file contains the following functions:

- before_feature (context, feature)—Executes prior every feature.
- before_scenario (context, scenario)—Executes prior every scenario.
- before_step (context, step)—Executes prior to every step.
- before_tag (context, tag)—Executes prior every tag.
- before_all (context)—Executes prior everything.
- after_feature (context, feature)—Executes post every feature.
- after_scenario (context, scenario)—Executes post every scenario.
- after_step (context, step)—Executes post every step.
- after_tag (context, tag)—Executes post every tag.
- after_all (context)—Executes post everything.

The above functions are used as hooks in Behave. The directory structure should be as follows:

```
>   .pytest_cache
∨   features
    >   steps
        environment.py
        feature1.feature
        feature2.feature
        feature3.feature
>   reports
    Database.py
    Doctor.py
    Drug.py
    drugs_list.txt
    TestClass.py
External Libraries
Scratches and Consoles
```

The following code fragment shows how to use the hook functions:

```python
# before all
def before_all(context):
    print('Before all executed')
# before every scenario
def before_scenario(scenario, context):
    print('Before scenario executed')
# after every feature
def after_feature(scenario, context):
    print('After feature executed')
# after all
def after_all(context):
    print('After all executed')
```

The output obtained after running the feature files is as follows:

```
Before all executed
Before scenario executed
---- ---- ---- ---- ---- ---- ----

---- Doctor signed in ----
---- Doctor wants issues a prescription ----
Drugs List = Fluoxetine ,Aspirin ,Caffein ,Tramadol ,Guanabenz ,
Selected Drug = Fluoxetine
Drug Name = Fluoxetine , Dosage = 60 mg , Usage = every day mornir
prescription made successfully.
prescription send to 09421000 successfully.
Before scenario executed
---- ---- ---- ---- ---- ---- ----
```

```
---- Doctor signed in ----
---- Doctor wants issues a prescription ----
Drugs List = Fluoxetine ,Aspirin ,Caffein ,Tramadol ,Guanabenz ,
Selected Drug = Guanabenz
Drug Name = Guanabenz , Dosage = 8 mg a day , Usage = Same times E
prescription made successfully.
prescription send to 09221666 successfully.
Before scenario executed
---- ---- ---- ---- ---- ---- ----
```

3.5   BDD Process

BDD is a software development process that leads to a shared understanding of requirements between the product owner and the Agile Teams. BDD puts acceptance tests in front as a result of which the developer knows from the very first day which tests the final product should pass. Streamlining development, reducing rework, and increasing flow are the consequences of such an understanding of the objectives. Figure 3.29 summarizes the BDD software development process.

As shown in Fig. 3.29, the BDD practice starts by identifying business goals. Goal models capture system goals and their hierarchical decomposition into lower-level goals and objectives. The system requirements appear at the bottom of the goal hierarchy. Requirements defined in natural language may be ambiguous and misunderstood. The Gherkin language helps the analyst define requirements unambiguity and quickly transform them into automated acceptance tests.

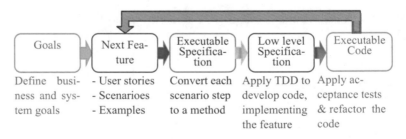

Fig. 3.29 The BDD software development process

3.5.1 Example

Fuel cards are alternative payment solutions businesses provide drivers exclusively for motoring expenses and fuel purchases. The use of fuel cards helps businesses manage their fleet effectively, simplify expenses, and fuel their vehicles cost-effectively. Below is a feature required by an imaginary fuel cards company:

Feature
As a fuel card company owner
I want to have only valid transactions get approved
So that I can prevent misuse of the fuel card,
Scenario: Only items that are necessary for the car are allowed
Given the car drives on petrol
When Jhon pays for the <item>
Then the transaction is <outcome>

Item	Outcome
Petrol	Approved
Motor oil	Approved
Windscreen wiper	Approved
Tea	not approved
Carwash	not approved
Petrol and Tea	not approved

After a feature is defined, the scenario steps, specifying the acceptance criteria, are translated into the following step definitions:

are translated into the following step definitions:

```
@When("When john pays for the \"([^\"]*)\" ")
    public void WhenJhonPays( String items ) {
            transaction.checkItems(items);
    }
@Then("Then the transaction \"([^\"]*)\" is")
    public void ThenTheTransaction(String outcome) {
        if (StringUtils.equals(item, "petrol")) {
                assertEquals(transaction.isAllowedItem(item), outcome); }
    }
```

When applying a test-first approach, the tests are completed as the priority before writing the actual code. At this stage, the acceptance tests, defining the user expectation of the final software product, give the inspiration for the design and development of the desired software. The code that implements the fuel-card logic is as follows:

```
public void checkItems (String items) {
    try {
            WebElement payButton = getElement(payButtonLocator);
            payButton.click();
        } catch(ElementNotVisibleException e) {
            log.warn("Pay button not visible", e); }
}
```

A feature tends to be a "higher-level" objective than a requirement and is usually more in line with business requirements than implementation. Examples clarify features and provide means for testing the requirements.

3.5.2   Defining Features

Features are essential distinguishing characteristics, qualities, or aspects of software products derived from requirements which are in turn derived from the goals. For instance, Fig. 3.30 presents the "user views records" feature. The feature specifies a portion of the desires of a telemedicine system. The feature concerns the patients' desire to view their medical records.

Behave [9], Cucumber [12], and Specflow [11] are three well-known software tools for behavior-driven development (BDD) in Python, Java, and C# programming languages. These three BDD tools look for features in .feature files in a directory named feature. Specflow and Cucumber automatically generate the feature directory, while the user should create the directory manually to use Behave. However, Cucumber and Specflow generate the feature directory automatically. Right-click on the feature directory, select the "new file," and then the "feature file" option to create a new feature in Behave. When creating a new BDD project using Cucumber and

Fig. 3.30 This feature describes the requirement to view patients' records

Feature: User views records
  As a user
  I want to be able to view the medical records related to me
  So that I am informed about the records

Scenario Outline: User asks for records
  Given the <username> is signed in
  When the user asks for the records of a <patient>
  Then the user sees the patient records
  Examples:
username	patient
@daviddoc	@jamesp
@jamesp	@jamesp

SpecFlow, two directories named "feature" and steps are created next to each other. The "step" directory in Behave is created manually inside the "feature" directory:

Features should be written in plain English or any other native language to facilitate communication between stakeholders and the project team members. Even step definitions are named after the "given," "when," and "then" clauses in the scenarios. Therefore, the BDD tools must support Unicode and accept features in any native language. Fortunately, Behave, Cucumber, and SpecFlow support Unicode.

The BDD tools automatically convert feature scenarios into executable specifications. Executable specifications illustrate and verify how a specific business requirement is delivered. The main idea behind behavior-driven development is to define the acceptance criteria for the customers' requirements in the form of the stakeholders' expectations of the application's behavior. It should then be possible to execute the acceptance criteria thus written and execute them directly against the application to verify that the application meets its specifications.

3.5.3 Executable Specifications

The acceptance criteria defined for a user story are the tests that an application must pass to demonstrate that it has met the requirement of the user story. BDD tools transform scenario-based acceptance criteria into executable specifications of the behavior of the software under test. It is worth mentioning that Dan North's suggestion is to use "behavior" instead of "test" to point out how acceptance tests express system behaviors. An executable specification can be automated by writing test code corresponding to each scenario step, i.e., given, when, and then. The test code comprises a method corresponding to each step. The BDD tools automatically generate a method for each step of a given feature.

To add step definitions, place the caret at a step in your .feature file and press Alt+Enter. The list of suggested intention actions opens. Select Create step definition to create a definition only for one step, or select Create all step definitions to add definitions for all steps in a scenario. Cucumber and Specflow create a class for each

Fig. 3.31 Step definitions
generated by behave for the
feature in Fig. 3.6

```
from behave import *
@given(u'the {username} is signed in')
def step_impl(context, username):
    pass
@when(u'the user asks for the records of a {patient}')
def step_impl(context, patient):
    pass
@then(u'the user sees the patient records')
def step_impl(context):
    pass
```

feature and wrap the step definitions inside the class. Behave generates a function for each method definition solely. Below are the step definitions generated by Behave, Cucumber, and SpecFlow for the "user views records" feature in Fig. 3.30.

1. **Behave for Python**

Behave generates step functions as executable specifications and saves them in the "steps" directory, created manually. All the Python files (files ending in ".py") will be imported to find step implementations in the "steps" directory. Behave loads all the generated step functions before it starts executing the feature tests. Figure 3.31 represents the functions generated for the feature given in Fig. 3.30.

The context object passed as a parameter to the step functions is an instance of the "behave.runner.Context" class. Context may be used to share values between the steps. Any table defined in a scenario step is available to the Python step code as the ".table" attribute in the Context variable passed into the step definition. For instance, consider the following step:

```
Scenario: some_scenario
Given a bunch of specific users
    | name       | department |
    | Big billy  | Barrel     |
    | Robbert    | Rabbit     |
    | Armestrang | Moon       |
```

The context object can be used to access the table as follows:

```
@given('a set of specific users')
    def step_impl(context):
    for row in context.table:
        model.add_user(name=row['name'], department=row['department'])
```

Decorators are used to identifying the scenario steps: given, when, then, and step. Behave prefaces each step definition with a @given, @when, or @then decorator. The steps themselves are the parameter of these decorators. Step decorators match the step to any step type, "given," "when," or "then." Steps containing an "and" or "but" are renamed internally to account for the preceding step's keyword.

Behave uses three types of matches ParseMatcher (parse), extended Parse-Matcher(cfparse), and RegexMatcher (re). One of the three parsers can be selected by invoking the behave.use_step_matcher(name) function with the name of the matcher to use. When running a feature, the parser replaces regular expressions for step parameters with a readable syntax like {param: Type}.

2. Cucumber for Java

Cucumber-JVM step definitions are annotated Java methods, similar to the ones used by SpecFlow and Behave. Cucumber generates the following code for the scenario in Fig. 3.30. The "when" step of the scenario is modified as follows:

When the user asks for events of " <box_views>" at "<data>"

When using Cucumber, the feature is rewritten as follows:

```
Scenario Outline: User asks for events
    Given the "<username>" is signed in
    When the user asks for events of "<box_view>" at "<date>"
    Then the user sees the calendar details
    Examples:
    | username | box_view | date       |
    | d2       | day      | 2022/02/25 |
    | d3       | day      | 2022/02/25 |
    | d1       | month    | 2022/03/12 |
```

Cucumber generates the class EventStepDefs, including a step definition for each step in the above scenario outline. The class is presented in Fig. 3.32.

Cucumber uses an @Given, @When, or @Then annotation to match a step definition method with a line in the feature file. The method name itself is not essential.

```
public class EventStepdefs {
    @Given("^the \"([^\"]*)\" is signed in$")
    public void theIsSignedIn(String username) throws Throwable {
    }
    @When("^the user asks for events of \"([^\"]*)\" at \"([^\"]*)\"$")
    public void theUserAsksForEventsOfAt(String box_view, String date)
                                                    throws Throwable {
    }
    @Then("^the user sees the calendar details$")
    public void theUserSeesTheCalendarDetails() {
    }}
```

Fig. 3.32 Step definitions generated by Behave

Behave, Cucumber, and SpecFlow scan each step's text for patterns they recognize. The patterns are defined using regular expressions. Regular expressions are used in compilers to define lexical rules as patterns based on which lexicons are recognized. The backslash (01XB in hexadecimal) character in regular expressions represents the literal escape characters in the strings. For instance, as Cucumber executes the feature in Fig. 3.30, it will come to the given step of the scenario:

Given the <username> is signed in

It generates a step definition annotated with the "given" attribute:

```
@Given("^the \"([^\"]*)\" is signed in$")
public void theIsSignedIn(String username) throws Throwable {
}
```

The Gherkin language used by Cucumber depends on regular expressions to identify variables in the steps. Cucumber will match the regular expressions in parentheses and assign them to the method parameters. For instance, the expression \"([^\"]*)\" detects the user name, "username," as a sequence of any character apart from the double quotation.

3. **SpecFlow for C#**

Behave, SpecFlow, and Cucumber match the text in each scenario step to the appropriate step definition. A Step Definition file is a piece of code with a pattern attached to it. In other words, a step definition is a C# or Java method or a Pythom function in a class with an annotation above it. The annotation is used to link the step definition to all the matching steps when observing a Gherkin Keyword. Below is the feature defined in a format acceptable for SpecFlow:

```
Feature: User views records
  @Record
  Scenario: User asks for records
    Given the <username> is signed in
    When the <username> asks for the records of a <patient>
    Then the user sees the patient records
    Examples:
      | username | patient |
      | Jack     | Joe     |
      | d1       | u1      |
```

Below is the step definitions generated by SpechFlow for the above scenario:

```
namespace Telemedicine_specflow.StepDefinitions
{
  [Binding]
  public class RecordSteps {
    String response;
    String expected;
    [Given(@"the (.*) is signed in")]
    public void User_Signed_In(string name) {
      throw new NotImplementedException("The step is not implemented");
    }
    [When(@"the (.*) asks for the records of a (.*)")]
    public void theUserAsksForTheRecords(string user,string patient_name) {
      throw new NotImplementedException("The step is not implemented");
    }
    [Then("the user sees the patient records")]
    public void theUserSeesThePatientRecords() {
      throw new NotImplementedException("The step is not implemented");
    }
  }
}
```

Like Behave and Cucumber, the SpecFlow software tool supports shared data between steps during a scenario execution, even if the step definitions are not in the same class. The two static classes, ScenarioContext and FeatureContext, in SpecFlow may hold a shared state/context between steps at runtime. This pair of classes is somewhat of a key-value dictionary, with the key representing the state variable and the value representing the actual object to be shared. The syntax for setting a value in ScenarioContest is as follows:

ScenarioContext.Current["key"] = value;

For instance, consider the following a value set in a step definition is used in another one:

```
[Given(@"the (.*) is signed in")]
    public void User_Signed_In(string name){
      ScenarioContext.Current["name"] = name.ToLower();
    }
    [When(@"the (.*) asks for the records of a (.*)")]
    public void theUserAsksForTheRecords(string user,string patient_name) {
            var registeredUserName =
                            (String)ScenarioContext.Current["name"];
    }
```

A table defined in a scenario step can be accessed directly in the step's function. Behave provides the context.table() method to access tables, defined in step via the method implementing the step. When using SpecFlow to pass a data table to a step definition method, the type "Table" parameter should be added to the method. The class "Table" encapsulates the contents of the tabular parameter with a convenient API that lets both compare and extract data from tables. For instance, consider a step definition for the following step:

Scenario: current_account
Given the following current_accounts

Account no.	balance
666.666.	666666666666
666.123	1230000
123;234	230000000

A table parameter:

```
[Given(@"the following current_accounts:")]
public void givenTheFollowingAccounts(Table current_accounts){
        foreach (var row in accounts.Rows){
            var owner = row["owner"]
            var points = row["points"]
            var statusPoints = row["statusPoints "]
    }
}
```

As shown above, SpecFlow precedes each method with an expression matching the method with the corresponding step in the feature. SpekFlow generates a distinct regular expression to identify each variable used in the steps of the scenario. It matches the regular expressions in parentheses and assigns them to the method parameters. For instance, the following step:

When the <user> asks for the records of a <patient>

Is translated into:

```
[When(@"the (.*) asks for the records of a (.*)")]
public void theUserAsksForTheRecords(string user,string patient_name) {
    throw new NotImplementedException("The step is not implemented"); }
```

As described above, BDD tools are automation frameworks that pull the inputs from the specification and validate the expected outputs without requiring them to change the specification document. As soon as the validation is automated, the specification becomes executable.

The test based on the executable specification will initially fail because it simply specifies an acceptance test of the behavior as a method/function. The developers should implement the relevant code before the acceptance test can be performed appropriately. This is where the low-level specification of the executable specification comes to play.

3.5.4   Low-Level Specification

Each executable functionality implemented as a step method/function represents a required functionality to be implemented by the developer. The low-level specification in the BDD process concerns using a TDD method to implement the code. When using a test-first approach, the acceptance tests should be completed before starting with the detailed implementation of the actual code. After the low-level specifications for all of the classes involved in the high-level acceptance criteria are written and implemented, the acceptance criteria should pass.

– Behave for Python

BDD suggests using a test-first practice to develop code like the TDD approach. The test first practice stresses that tests come before implementation. Therefore, before writing the low-level specifications, acceptance tests should be completed. The implication is that when writing tests systematically after the production code, they exercise, even if both production and testing code are built iteratively and incrementally, the resulting technique would not be compliant. In Fig. 3.33, the step definitions generated by Behave are completed.

In Python, the assert statement carries out assertions. When Python encounters an assert statement, it evaluates the accompanying expression. An AssertionError exception is raised if the expression is false. The syntax for the assert statement is as follows:

assert Expression[, Arguments]

```
from behave import *
@given(u'the {username} is signed in')
def step_impl(context, username):
    context.username = username
@when(u'the user asks for the records of a {patient}')
def step_impl(context, patient):
    url = f"http://localhost:5000/{context.username}/api/v1/record/"
    response = requests.get(url, json={"patient_username": patient})
    context.response = response
@then(u'the user sees the patient records')
def step_impl(context):
    status_code = context.response.status_code
    records = context.response.json()["records"]
    assert status_code == 200
    assert len(records) != 0
```

Fig. 3.33 Python implementation of the acceptance tests

For instance, consider the following assert statement:

assert (Temperature >= 0),"Colder than absolute zero!"

Programmers often place assertions after a function call to check for valid output and at the start of a function to check for valid input.

2. **Cucumber for Java**

BDD uses an outside-in approach to implement acceptance criteria. Therefore, when using a BDD framework such as Cucumber, the acceptance tests are designed based on business behavior (outside) rather than technical implementation (inside). Figure 3.34 presents the EvenStepdefs class implementing the acceptance criteria for the feature given in Fig. 3.30.

In the "Then" step of the code presented in Fig. 3.34, the assert statement checks whether the calendar is accessed appropriately. Then steps should make assertions

```java
public class EventStepdefs {
    Response response;
    String uname;
    Calender calender = new Calender();
    String expected;

    @Given("^the \"([^\"]*)\" is signed in$")
    public void theIsSignedIn(String username) throws Throwable {
        uname = username;
    }
    @When("^the user asks for events of \"([^\"]*)\" at \"([^\"]*)\"$")
    public void theUserAsksForEventsOfAt(String box_view, String date)
                                                        throws Throwable {
        calender.setBox_view(box_view); calender.setDate(date);
        response = given().contentType(ContentType.JSON).body(calender).
        post(String.format("http://localhost:8080/%s/event/", uname));
    }
    @Then("^the user sees the calendar details$")
    public void theUserSeesTheCalendarDetails() {
    System.out.println( String.format("Username %s",   uname));
    System.out.println( String.format("TestCase status
            code: %s", response.statusCode())));
    Assert.assertEquals(200, response.statusCode());

    }
}}
```

Fig. 3.34 Executable specifications implementing acceptance criteria comparing expected results to actual results from the application. The *assertEquals* method compares expected and actual values. The assertNotEquals method is just opposite the assertEquals. The other types of assertions are assertTrue and assert-False statements. Cucumber does not come with an assertion library. Instead, use the assertion methods from a unit testing tool.

Assert is a Java keyword used to declare an expected boolean condition to be checked at runtime. If the condition is not satisfied, the Java runtime system throws an AssertionError. The syntax of the assert statement is as follows:

$$\text{assert expression\_1 [: expression\_2];}$$

Where expression1 is a boolean that will throw the assertion if it is false. When it is thrown, the assertion error exception is created with the parameter expression2 (if applicable). For instance, consider the following assertion statement:

```
assert my_list != null && my_list.size() > 0 : "my_list variable is null or empty";
Object value = my_list.get(0);
```

Assertions are enabled with the Java-ea or -enableassertions runtime option. See a Java environment documentation for additional options to control assertions.

3. **SpecFlow for C#**

Figure 3.35 represents the completed step definitions. A test-first approach is used to write the acceptance tests. Therefore, at first, the acceptance tests fail. The developer should write the minimum amount of code to get the tests to pass. Once all the acceptance tests pass, one may refactor the code to get it clean before selecting the next feature.

The Assert.AreEqual(expected, response) statement checks whether the value of the variable "expected" is equal to the "response" value. Assertions are not a standard language feature in C#. Instead, several classes provide functions for assertion handling in C#.

3.5.5 Hooks

Hooks allow developers to specify actions to execute before or after a feature, a scenario, steps, or test cases.

– **Hooks in Behave**

Behave supports hooks to handle the environment for the steps. Environment.py is a file Environment.py is the name of the file in which the Python Behave hooks can be defined. Environment.py can be used to define code that should be executed before and after certain events. As shown below, environment.py is saved in the feature directory:

```
namespace Telemedicine_specflow.StepDefinitions
{
  [Binding]
  public class RecordSteps {
    String response;
    String expected;
    [Given(@"the (.*) is signed in")]
    public void User_Signed_In(string name)  {
      var uname = name;
    }
    [When(@"the (.*) asks for the records of a (.*)")]
    public void theUserAsksForTheRecords(string user,string patient_name)  {
      var client = new HttpClient();
      string url =
        $"https://localhost:7295/api/Patient/Records/{user}?
          name={patient_name}";
      WebRequest request = WebRequest.Create(url);
      WebResponse response = request.GetResponse();
      var responseStream = response.GetResponseStream();

      var reader = new StreamReader(responseStream);
      this.response = reader.ReadToEnd();
    }
    [Then("the user sees the patient records")]
    public void theUserSeesThePatientRecords()
    {
      expected = "[{\"title\":\"Cold\",\"symptoms\":
                  [\"Runny/Stuffy nose\",\"Sore throat\",\"Sneezing\"]},
                  {\"title\":\"Influenza\",\"symptoms\":[\"Fever\",\"Sore
                  throat\",\"Headache\"]}]";
      Assert.AreEqual(expected, response);
    }
  }
}
```

Fig. 3.35 Executable specifications written in C#

```
└── features/
    ├── steps/
    │   ├── common.py
    │   ├── events.py
    │   ├── records.py
    │   └── symptoms.py
    │
    ├── environment.py
    ├── events.feature
    ├── recordSubmission.feature
    ├── records.feature
    └── symptom.feature
```

For instance, below are the hook functions, defined in prperty.py, for the feature in Fig. 3.30.

```
from threading import Thread
from werkzeug.serving import make_server
from src.app import app
def before_all(context):
    server = make_server("0.0.0.0", 5000, app)
    server_thread = Thread(target=server.serve_forever)
    server_thread.start()
    context.server = server
def after_all(context):
    context.server.shutdown()
```

– Hooks in Cucumber

Hooks in Cucumber may control the flow of the program and optimize lines of code before and after a scenario. When using @before and @after in a step definition file, Cucumber creates a block to write the code. The Cucumber hooks facilitate handling the code workflow and help reduce code redundancy.

```
public class EventStepdefs {
    Response response;
    String uname;
    Calender calender = new Calender();
    String expected;
```

```
@Before
public void beforeScenario(){
    server = make_server("0.0.0.0", 8080, app);
    server_thread = Thread(target=server.serve_forever);
    server_thread.start();
    context.server = server;
}

protected String mapToJson(Object obj) throws JsonProcessingException {
    ObjectMapper objectMapper = new ObjectMapper();
    return objectMapper.writeValueAsString(obj);
}


# Copy Given, Then, and When steps from Fig. 9.10
@After
public void afterScenario(){
    context.server.shutdown();
}
}
```

Hooks in SpecFlow

As a Given/When/Then tool, Specflow contains hook features, which allow developers to specify actions to perform before or after certain features, scenarios, or even specific steps.

```
using System.Diagnostics;
using TechTalk.SpecFlow;
namespace Telemedicine_specflow {
    [Binding]
    public sealed class Hooks_records {
        [BeforeScenario]
        public void BeforeScenarioWithTag() {
            Process.Start(@"C:\Users\mohammad\source\repos\old\
                          Telemedicine\bin\Debug\net6.0\Telemedicine.exe");
            Console.WriteLine("Server  is Starting!!!");
        }

        [AfterScenario]
        public void AfterScenarioWithTag()
        {
            Console.WriteLine("The  Test Is Over!!");
        }
    }
}
```

Hook attributes supported by SpecFlow are generally defined as BeforeX and AfterX, where X can be "TestRun," "Feature," "Scenario," "Step," and "ScenarioBloc." For instance, in the class HookRecords, the [AfterScenario] attribute is used.

The question is how to restrict hooks to a specific feature, scenario, or step definition. An answer is to use tags. For instance, consider the tag "@register" in the following feature:

Feature: Registering
  As Customer, I want to regester in app why to login in app
@register
Scenario: Registering
  #Given open the app
  #And clicl on the MainPage Menu
  #And click on LogOn Menu
  And Enter *<UserName>*,*<PhoneNumber>*,*<City>*
  #When click on Register button
  Then print *<UserName>* ,your registering is *<Result>* !!! now login
Examples:

UserName	PhoneNumber	City	Result
samira	1369	Tehran	Succssesfully
sara	8521	Ahvaz	Succssesfully
shiva	1372	Tabriz	Succssesfully

The hook class is saved in a directory called "Hooks" next to the "Feature" directory as follows:

```
using System;
using System.Collections.Generic;
using System.Data.SqlClient;
using System.Linq;
using System.Text;
using TechTalk.SpecFlow;
namespace OnlineShopping_SpecFlow.Hooks {
  [Binding]
  public sealed class Prerequisite {
    [BeforeScenario("register")]
    public void BeforeScenario() {
      // write code for (Current user = null
      //check configuration
      Console.WriteLine("configuration your SQL Server");
    }
    [AfterScenario]
    public void AfterScenario() {
      //write
    }
  }
}
```

3.6   Using the Behave Software Tool for BDD in Python

Behave is a software tool for behavior-driven development (BDD) in Python [11]. BDD has based the software development process on the requirements through which acceptance tests are formalized and facilitated as the core of Agile methodologies. Behave allows Python programmers to develop code derived from the user requirements. It is similar to SpecFlow and Cucumber. Behave allows writing test cases in easily readable language.

The Behave plugin requires Python 2.7.14 or any above version, Python package manager or pip, and any IDE like Pycharm or other. Section 3.6.1 shows how to install the Behave plugin.

3.6.1   Installing Behave

The package installer for the Python language, pip, provides a command interface to install Python plugins. Issue the following command to install the Behave package:

```
$ pip install behave
#or:
$ conda install behave -c conda-forge
```

After installing the Behave, you can run it by entering the "behave" command in the command line. Notice that you must enter the command inside the project's root directory. Behave will find the proper files and directories to run the tests if there are any.

3.6.2   Tutorial

Behave searches for a 'features' directory containing all the '.feature' files of the user stories specified in the Gherkin format. All scenario steps should be implemented under the 'steps' directory as follow:

```
.
└── features/
    ├── steps/
    │   └── steps_of_feature.py
    │
    └── a_feature.feature
```

For instance, below is a general example of what a feature looks like in Behave.

Feature: Demonstrating Behave
Scenario: run a simple acceptance test
Given that the Behave library installed
When a test is implemented
Then Behave will test it for us!

Features are saved in a directory named "features." The above feature is saved in "features/example.feature". In the next step, an executable specification implementing the feature is developed as follows:

```python
from behave import *
@given('we have behave installed')
def step_impl(context):
    pass
@when('we implement a test')
def step_impl(context):
    assert True is not False
@then('behave will test it for us!')
def step_impl(context):
    assert context.failed is False
```

The Python file, e.g., tutorial.py, including the executable specification, should be saved in a directory, "features\steps." As shown above, a separate function should be created for each scenario step. To start testing all the features, run the following command in the terminal:

$behave

Behave will match the specified steps in feature files with their implementations and run them step by step:

```
Feature: showing off behave          # features/tutorial.feature:1
Scenario: run a simple test          #features/tutorial.feature:3
Given we have behave installed       #features/steps/tutorial.py:3
When we implement a test             #features/steps/tutorial.py:7
Then Behave will test it for us!     #features/steps/tutorial.py:11
1 feature passed, 0 failed, 0 skipped
1 scenario passed, 0 failed, 0 skipped
3 steps passed, 0 failed, 0 skipped, 0 undefined
```

3.6.3   A Telemedicine System

Video calls between doctors and patients are the basis for telemedicine platforms, which may integrate with patient information systems. The Tele-Medicine System implementation described in this section is a simple Restful web application back-end, expected to provide some features based on the user stories specified in the Gherkin format. The web application is implemented using flask, an easy-to-extend web application framework. This implementation aims to augment the patient experience of virtual doctor's appointments and make virtual visits as effective as in-person visits. Some specific features will be discussed later in this section.

3.6.4   Acceptance Tests

As a formal description of a software product's behavior, acceptance tests generally express an example of a usage scenario to guarantee the software product is performing what the customers require. Acceptance tests are described as scenarios in the Gherkin language.

Like any other BDD libraries, behave uses feature files created by the Business analyst or stakeholders. Below is a user story for scheduling an appointment with a doctor:

> User Story: As a patient,
>   I want to view any of the doctor's available time slots,
>   so that I can schedule an appointment suitable for both sides.

Acceptance criteria are added to the user story to facilitate acceptance tests. For instance, the acceptance criteria for the above user story can be defined as follows:

> Given that patients are logged in,
> when they navigate to the 'Schedule Appointment' page, they should be able to see the doctor's available time slots on a calendar to select a doctor
> And schedule an appointment in that time slot.
> Then the slot is marked as unavailable in the doctor's calendar,
> And both the patient and doctor can see the appointment set up at that time.

Specification by example bridges the communication gap between business stakeholders and software development teams. For further clarification, the acceptance criteria are rewritten as follows:

> Scenario: User asks for events
>   Given the "@jamesp" is signed in
>   When the user asks for events of "day" at "10/11/2022"
>   Then the user sees the calendar details

Behave uses a Gherkin syntax that is practically compliant with the official Cucumber Gherkin standard. Behave takes features from .feature files saved in the features directory in the project directory.

Fig. 3.36 Directory
hierarchy for a Behave
project

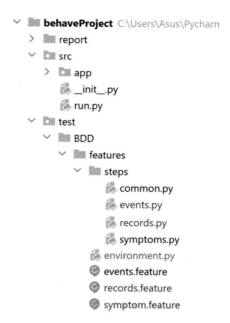

A .feature file has Feature sections, which in turn have Scenario sections with Given-When-Then steps. Below, the above user story is defined as a feature in the Gherkin language. The feature should be written in a text file with a .feature extension, and the file should be saved in a directory named "features" within the project root directory. Figure 3.36 illustrates the directory hierarchy for the Telemedicine project.

As shown in Fig. 3.36, there are three features: event .feature, records .feature, and symptom .feature in a directory called "features". Healthcare is difficult to access because of the need to commute, wait times, and scheduling constraints.

As a solution to such difficulties, telemedicine makes use of virtual appointments. The feature presented in Fig. 3.37 is concerned with scheduling an appointment with a doctor.

Once a patient clicks on the "symptoms" feature, a symptom page to checkbox what the patient is experimenting with should be displayed. The scenario presented in Fig. 3.38 shows how select the symptoms".

When writing behavioral tests, one should strive to write features fluently to be easily readable for anyone working on the codebase. When writing tests, it is desired to test the same behavior with many different parameters and check the results. *Behave* makes this easier by providing tools to create an example table instead of writing out each test separately. A scenario including an example table is called *Scenario Outline*. In a *scenario outline, each row of the example table* shows the input and the expected output of the test. As an example, consider the examples table in the above features.

Fig. 3.37 Play cards feature
in the Gherkin language

Feature: user asks for events
    As a user
        I want to be able to view a calendar of my appointments
        So that I am informed about my workday calendar
    Scenario Outline: User asks for events
        Given the <username> is signed in
        When the user asks for events of <box_view> at <date>
        Then the user sees the calendar details
        Examples:
        | username | box_view | date |
        | @jimi | day | 1990-01-13 |
        | @jacki | month | 1990-01-23 |
        | @Judi | year | 1990-05-23 |

Fig. 3.38 Allowing patients
to check their symptoms
before their appointments

Feature: Patient views default symptoms
    As a patient
    I want to be able to view the default symptoms
    So that I can checkbox my symptoms easily
    Scenario Outline: User asks for events
        Given the <username> is signed in
        When the user asks for the list of default symptoms
        Then a page to checkbox the symptoms is displayed
        Examples:
        | Symptoms |
        | @Garbled Speech |
        | @wheezing |
        | @Abdonimal pain |
        | @Headache |
        | @Ear/Sinus pain |
        | @Fever above 40° |
        | @Abdonimal pain |
        | @Other |

3.6.5   Step Definition

Typically, a scenario has three steps: given, when, and then. All the steps defined in the scenarios of a feature should be saved in a Python file in the "features/steps" directory. To start the test in Pycharm Professional, click on the green play button on the gutter.

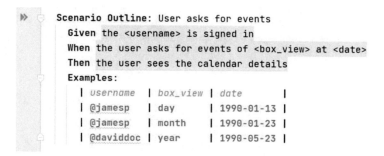

The first code, written in a behavior-driven development (BDD) approach, is always the test code. This phase fails, which is why it is called the red phase. The test fails because there is no code written to make it pass.

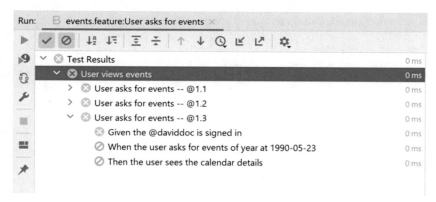

The following figure shows that one should right-click on the Scenario and select "Generate step Definitions" to convert a BDD feature file into a step definition. Also, it is possible to write the step template manually.

As shown below, there are two options to create a definition for one step or all the selected scenario steps. After selecting the "create all step definition" option, the following window is displayed to enter the address and file name of the target step definition:

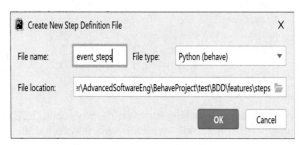

You may enter the desired code in the generated step definition file in the editor. The generated step definition templates are as follows:

```
use_step_matcher("re")
@given("the (?P<username>.+) is signed in")
def step_impl(context, username):
    """
    :type context: behave.runner.Context
    :type username: str
    """
    raise NotImplementedError(u'STEP: Given the <username> is signed in')

@when("the user asks for events of (?P<box_view>.+) at (?P<date>.+)")
def step_impl(context, box_view, date):
    """
    :type context: behave.runner.Context
    :type box_view: str
    :type date: str
    """
    raise NotImplementedError(u'STEP: When the user asks for
                                         events of <box_view> at <date>')

@then("the user sees the calendar details")
def step_impl(context):
    """
    :type context: behave.runner.Context
    """
    raise NotImplementedError(u'STEP: Then the user sees the calendar details')
```

To start the test in Pycharm Professional, click on the green play button on the gutter. Since the code is not ready yet, the test fails, and the following messages are shown:

Test results

⊗ User views events	⊛ Test failed 3, ignored 6 of 9 tests, 1ms
> ⊗ User asks for events - e1.1	File "G:\behave\model.py
> ⊗ User asks for events - e1.2	line 1329 in run match.run(numer,context)
> ⊗ User asks for events - e1.3	File ""G:\behave\model.py
⊗ Given the @daviddoc is signed in	line 98 in run self.func(context,args,**kwargs)
⊗ When the user asks for events of	File ""G:\behave\model.py
@year at @1990-05-23	line 30 in step.implraise NotImplemetedError(
⊗ Then the user sees the calendar details	u'STEP: Given the <username> is signed in')

3.6.6 Step Implementation

Step functions are implemented in the Python modules in the "steps" directory. Step functions are identified using step decorators defined by Behave. The following line imports decorators defined by Behave to relate given, when, and then steps in feature files,.feature, in the features directory with their corresponding function,.py, in the steps directory:

```
from behave import *
```

This line imports the decorators defined by Behave to identify step functions. The decorators take a single string argument to match against the feature file step text. So the following step implementation code:

```
@given(u'the {username} is signed in')
def step_impl(context, username):
    context.username = username
```
will match the "Given" step from the following feature:
```
Scenario: User asks for events
    Given the <username> is signed in
    When the user asks for a <patient's> records
    Then the user sees the patient's records.
```

Similarly, the following step implementation code for the "when" and "then" steps matches their corresponding definitions in the feature file:

```
@when(u'the user asks for a {patient's} records ')
def step_impl(context, patient):
    url = f"http://localhost:5000/{context.username}/api/v1/record/"
    response = requests.get(url, json={
        "patient_username": patient
    })
    context.response = response
```

```
@then(u'the user sees the patient's records')
def step_impl(context):
    status_code = context.response.status_code
    records = context.response.json()["records"]
    assert status_code == 200
    assert len(records) != 0
```

Decorators are automatically available to step implementation modules as global variables, and it is not necessary to import them. Steps beginning with "but" or "and" are renamed to take the name of their preceding keyword.

At least an argument that is always the context variable is passed to the function decorated by the step decorator. After the context variable, additional arguments come from step parameters, if any. Custom context fields may be added, too, to share data between steps. Using context to share data is always suggested—never use global variables! The context object is passed from step to step, where we can store information to be used in other steps.

3.6.7 Unit Testing in Python

Unit testing 'unittest' module is a built-in Python code testing tool. The 'unittest' module provides a rich set of tools for constructing and running tests. The 'unittest' unit testing framework was initially inspired by 'JUnit' and had a similar flavor as powerful unit testing frameworks in other languages.

As a convention, 'unittest' module forces us to prepend the 'test' word to the module names. It searches the whole directory, then finds and runs all the test cases by this convention.

```
├── __init__.py
├── dummy_model_objects.py
├── stubs.py
├── test_auth.py
├── test_event_controller.py
├── test_record_controller.py
├── test_symptom_controller.py
└── test_views.py
```

A test case is created by subclassing 'unittest.TestCase'. The individual tests are defined with methods whose names start with the letters test. This naming convention informs the test runner about which methods represent tests.

When a 'setUp()' method is defined, the test runner will run that method prior to each test. Likewise, if a 'tearDown()' method is defined, the test runner will invoke that method after each test.

```
class TestEventController(TestCase):
    def setUp(self):
        self.controller = EventController()
        self.user_name = dummy_model_objects.random_string()
        self.zone_date = date.today()
```

The test results are shown below. Considering the test results, this is the red phase of the test process:

Python tests for unit_test.test_e

Test Results
unit_test
  test_event_controller
    TestEventController
      test_day_view
      test_month_view
      test_year_view

```
Error
Traceback (most recent call last):
  File "C:\Users\babay\projects\Master\AdvancedSoftwareEng\BehaveProject\test\unit_test\test_event_controller.py",
    with patch("src.app.controllers.EventController._retrieve_corresponded_events",
  File "C:\Users\babay\miniconda3\envs\BehaveProject\lib\unittest\mock.py", line 1393, in __enter__
    original, local = self.get_original()
  File "C:\Users\babay\miniconda3\envs\BehaveProject\lib\unittest\mock.py", line 1366, in get_original
    raise AttributeError(
AttributeError: <class 'src.app.controllers.EventController'> does not have the attribute '_retrieve_corresponded_
```

```
Error
Traceback (most recent call last):
  File "C:\Users\babay\projects\Master\AdvancedSoftwareEng\BehaveProject\test\unit_test\test_event_controller.py",
    with patch("src.app.controllers.EventController._retrieve_corresponded_events",
  File "C:\Users\babay\miniconda3\envs\BehaveProject\lib\unittest\mock.py", line 1393, in __enter__
    original, local = self.get_original()
```

3.6.8   Project File Structure

The application makes use of the MVC architectural pattern. The end-points of the rest APIs are defined in the 'resources.py' file. As the names suggest, controllers and the architecture models are defined in the 'controllers.py' and the 'models.py' files, respectively. Below is the MVC architecture used to organize the directories:

```
src/
├── app/
│     ├── __init__.py
│     ├── authentication.py
│     ├── controllers.py
│     ├── views.py
│     ├── models.py
│     └── resources.py
├── __init__.py
└── run.py
```

The test first approach is followed to complete the code. All the classes used to implement the program are listed below:

```python
class PatientRecord(Resource):
    get_parser = reqparse.RequestParser()
    ...
    def post(self, user_name):
        ....
    def get(self, user_name):
        ...

class RecordController:
    view = RecordView()
    def create_record_from(self, user_name, record_title, symptoms):
        patient = Patient.query.filter_by(user_name=user_name).first()
    ....

    def records_of(self, patient_user_name):
        ...


class RecordView:
    @classmethod
    def jsonify_records(cls, records):
        return {"records": [
            {
                "title": record.title,
                "symptoms": [symptom.title for symptom in record.symptoms]
            }
            for record in records]}


class Event(db.Model):
    def __repr__(self):
        .....
```

After completing the code, the Features are executed. This time all the tests pass. The acceptance test results after completing the code are shown below:

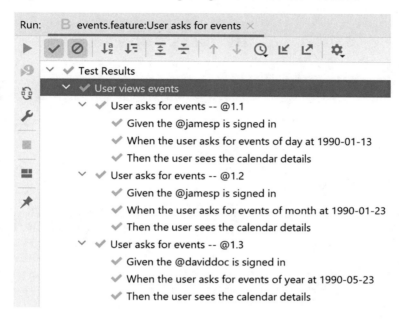

3.6.9 Report Generation Using Allure

Allure Framework is a powerful multi-language test report tool. It supports well-known languages such as Python, Java, and all the .Net. With Allure, the users may group tests by features, stories, or assertions. The testers can mark flaky tests that pass and fail periodically and display them on the reports. The following command installs' allure.

$pip install allure-behave
$scoop install allure

By executing the following command, a report collection will be generated:

$behave -f allure_behave.formatter:AllureFormatter -o report ./features

Allure provides a dashboard to analyze the report collection:

$allure serve report

Allure generates the following report. The report shows that the "user asks for events" step has been tested successfully three times, each time with a different row of the examples table.

✓	**#1 User asks for events -- @1.1**	day, 1990-01-13, @jamesp
✓	**#2 User asks for events -- @1.2**	month, 1990-01-23, @jamesp
✓	**#3 User asks for events -- @1.3**	year, 1990-05-23, @daviddoc

When clicking on each row of the test results in the above Figure, the exact detail of the test will be shown.

3.6.9.1 Hooks in Behave

Hooks support handling automation outside of the steps of Gherkin in Behave. The hook functions run before or after a step, scenario, feature, or whole test suite. Hooks should be placed in a file called environment.py under the features/directory.

Hook functions can also check the current scenario's tags so that logic can be selectively applied. The example below shows how to use hooks to set up and tear down a Selenium WebDriver instance for any scenario tagged as @web.

```
from threading import Thread
from werkzeug.serving import make_server
from src.app import app
# Before doing the acceptance tests
# Makes a server Flask App object and brings it up in a separate threat
def before_all(context):
    server = make_server("0.0.0.0", 5000, app)
    server_thread = Thread(target=server.serve_forever)
    server_thread.start()
    context.server = server

# Shutdown after the acceptance tests are completed
def after_all(context):
    context.server.shutdown()
```

3.7 Using SpecFlow for BDD

SpecFlow is a BDD open-source framework for DOT-NET that helps write feature files and automation code using C# and DOT-NET methods.

3.7.1   Installation

You may install SpecFlow by either directly downloading the Specflow extension for the Visual Studio from the marketplace:

https://marketplace.visualstudio.com/items?itemName=TechTalkSpecFlo wTeam.

or via the Manage Extensions option from within your IDE. You must close the Visual Studio if you download it from the marketplace before the installation.

As shown below, to install Specflow via the extensions manager, navigate to Extensions-> Manage Extensions, then select "Online" on the left panel, search for "SpecFlow for Visual Studio" in the Online Extensions, install the same, and finally restart your visual Studio.

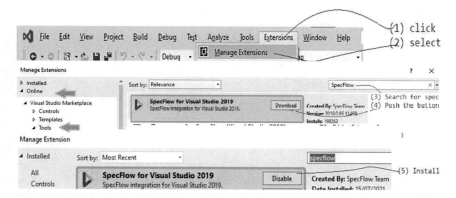

BDD uses TDD to develop code. Therefore, a unit testing tool should be installed for TDD. We preferably install the Nunit3 test adaptor and test generator extension for Visual Studio. To install NUnit, do the same as we did for SpecFlow but this time, instead of SpecFlow, search for NUnit and install it.

3.7.2   Create SpecFlow Project

To add a SpecFlow project to your project, say OnlineShopping, in the solution explorer, right-click on your project name and then select add==>New project and then search for SpecFlow:

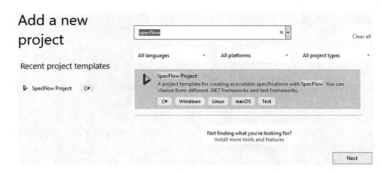

Enter the project name "OnlineShopping_SpecFlow.Specs". Keep the solution folder and click. You can configure the Test Framework you want to use on the next screen. We suggest using the free NUnit3 Runner.

Create a new SpecFlow project

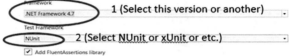

Visual Studio creates the new SpecFlow project and resolves the NuGet packages in the background. You should see the new SpecFlow project in Solution Explorer.

The question is how to link the SpecFlow project, OnlineShopping_SpecFlow, to the OnlineShopping project. To add references and link the projects, as shown below, in the solution explorer, expand the node OnlineShopping_SpecFlow, click on dependencies, and then select "Add Project Reference."

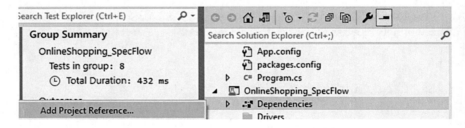

In the "Reference Manager" dialog, check the "OnlineShopping_SpecSFlow" class library and click OK. Now, the solution is set up with a class library containing the implementation of the OnlineShopping and OnlineShopping_SpecFlow projects. OnlineShopping_SpecFlow has the executable specification of requirements and tests of the OnlineShopping project.

3.7.3   Create Feature Files

Open your project in Visual Studio and navigate to solution explorer. As shown below, right-click on the "Feature" folder to add a new feature file and select Add==>new item.

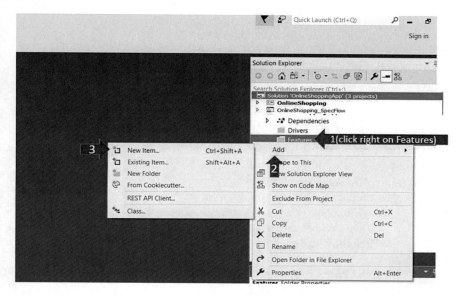

The following window opens. Select "new Specflow feature file," enter the feature file's name (e.g., Registeringfeayure), and then click on the Add button.

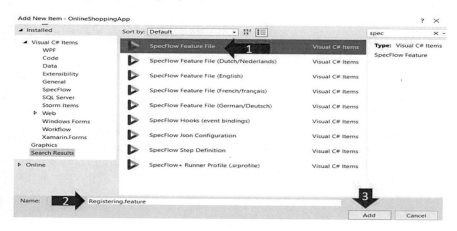

Specflow creates a feature file including a default feature description shown below:

```
Feature: SpecFlowFeature1
    Simple calculator for adding two numbers
@mytag
Scenario: Add two numbers
    Given the first number is 50
        And the second number is 70
    When the two numbers are added
    Then the result should be 120
```

Override the default feature with your feature description. The "AddAdvertisemet" feature defines the story of advertising an item for sale on the Internet. Each item's category, price, description, and title are entered.

```
Feature: AddAdvertisement
    As a vendor, I would like to create an advertisement

@add_Adv
Scenario: Add Advertisement
    #Given click on selling menu
        And EnterAdv <Category>,<Price>,<Description>,<Title>
    #When click on create advertisement button
        Then show the <result>
Examples:
    | Category | Price  | Description | Title             | result        |
    | Housing  |778686  | Good Quality| Ekbatan Apartment| Succssesfully|
    | Clothing | 65258  | Luxuray     | Shirt             | Succssesfully|
```

With this feature file, an intended behavior of AdvertiseAndSell is documented in a human-readable format that can be used for automated testing using Given/When/Then steps.

3.7.4   Create Executable Specifications

Next, create step definitions that bind the steps in the feature files to automation code. If you do not want to test some step definitions, you can comment on them out. As shown below, right-click the And step "And EnterAdv <Category>, <Price>, <Description>, <Title>" and select "Generate Step Definition" or the "Go To Step Definition" command.

Feature: AddAdvertisement
    As a vendor I want to create an advertisement

@add_Adv
Scenario: Add Advertisement
    #Given click on selling menu
    And EnterAdv <Category>,<Price>,<Description>,<Title>
    #When click on create advertisement button
    Then show the <result>
Examples:

Category	Price	Description	Title	result
Housing	778686	Good Quality	Ekbatan Apartment	Succssesfu
Clothing	65258	Luxuray	Shirt	Succssesf

Rename...	Ctrl+R, Ctrl+R
Go To Definition	F12
Breakpoint	▶
Run To Cursor	Ctrl+F10
Cut	Ctrl+X
Copy	Ctrl+C
Paste	Ctrl+V
Annotation	▶
Outlining	▶
Generate Step Definitions	
Go To Step Definition	Ctrl+Shift+Alt+S

Selecting the "Generate Step Definition", Specflow generates a class with the same name as the feature, AddAdvertisment. In fact, the step definitions implement the acceptance criteria defined by the feature.

```
using System;
using TechTalk.SpecFlow;
namespace OnlineShopping_SpecFlow.Steps
{
    [Binding]
    public class AddAdvertisementSteps
    {
    [Given(@"EnterAdv (.*),(.*),(.*),(.*)")]
    public void GivenEnterAdv(OnlineShopping.Category Category, int Price,
                                        string Description, string Title)
    {
        ScenarioContext.Current.pending();
    }

    [Then(@"show the (.*)")]
    public void ThenShowThe(string result)
    {
            ScenarioContext.Current.pending();
    }
    }
}
```

The method "GivenEnterAdv()" has four input parameters, Category, Price, Description, and Title, taken from the definition of the And step in the AddAdvertisement feature definition. Also, the method "ThenShowThe()" has one output parameter, "result", taken from the Then step in the feature definition. The test cases are taken from the Examples table, shown below, in the AddAdvertisement feature definition. Since there are two rows in the Examples table, the methods EnterAdv() and ThenShowThe() are invoked twice with the two rows as the test parameters.

And EnterAdv <Category>,<Price>,<Description>,<Title>
        #When click on create advertisement button
          Then show the <result>
Examples:
Category	Price	Description	Title	result
Housing	778686	Good Quality	Ekbatan Apartment	Succssesfully
Clothing	65258	Luxury	Shirt	Succssesfully

The two test cases taken from the above Examples table are:

Test-case-1:
   (Housing, 778,686, Good Quality, Ekbatan Apartment, Successfullyl)

Test-case-2:
   (Clothing, 65,258, Luxury, Shirt, Successfullyl)
   BDD suggests working on one example at a time. The examples provide the basis
for acceptance tests. Therefore, the business owner should confirm the examples.

3.7.5 Low-Level Specifications

Automating the scenarios becomes executable specifications, but they will be
reported as pending until you write the code that exercises the application and verifies
its results. The methods are initially empty and should be completed while writing
the code to satisfy the requirement of the feature. Below, the completed class is listed:

```
using System;
using TechTalk.SpecFlow;
namespace OnlineShopping_SpecFlow.Steps
{
  [Binding]
  public class AddAdvertisementSteps
    {
      OnlineShopping.Connection connection =
                    new OnlineShopping.Connection();

   [Given(@"EnterAdv (.*),(.*),(.*),(.*)")]
    public void GivenEnterAdv(OnlineShopping.Category  Category, int Price,
                                   string Description, string Title)
    {
    OnlineShopping.Advertisement adv =
       new OnlineShopping.Advertisement(Category,Price,Description,Title);
```

```
    Console.WriteLine("Watting");
    connection.addAdvertisement(Category, Price, Description, Title);

    }
[Then(@"show the (.*)")]
public void ThenShowThe(string result)
    {
      if (connection.IsCreatedAdvertisement)
        Console.WriteLine(" Advertisement created " + result);
      else
        {

          Console.WriteLine("Advertisement  created " + result);
          throw new Exception("test is fail");

        }
      }

    }

}
```

The class Connection implements the OnlineShopping System. BDD suggests using a test-first approach to implement the class. According to the single responsibility principle Considering Single Responsibility Principle (SRP)[], each class should be developed in response to a single requirement and not multiple requirements. Therefore, when developing code, care should be taken so that each class does not implement more than one requirement.

3.7.6 Code Implementation

To be cohesive and follow the single responsibility principle, a class should be responsible to a single actor, and only that actor should force the class to change its behavior. An actor is a user role. The user role in the user story definition is the one that is actually getting the value described in the story. It is worth mentioning that a user story is defined with the following template:

> As a <role>
> I want <something>
> So that <I can achieve some business goal

Therefore, when developing a class to satisfy a requirement, care should be taken to restrict the class methods to fulfill solely that requirement scenario and not any other requirements. The class should be responsible for satisfying a single and not many requirements. In this way, only changes to a single requirement, defined by a particular role, trigger the changes in the class. A single reason to change is what the single responsibility principle instructs the developers to take care of when developing a class. Before carrying on with the implementation of the system, let us have a look at the system specification provided by the business owner.

Design a system so that people can advertise new and second-hand stuff for sale. After advertising for the retail sale, people may contact the vendors. People may visit the site and register with the system if they want to buy items. The registration requires the name of the city, phone number, and the customer's name and address. A brief description, price, and a link to chat with the advertiser should be provided for each advertised item. In addition, a link to similar items advertised on the website should be available.

As shown below, the MVC architecture is used to implement the system. All the classes initially include empty methods. The method's signature and the method bodies are completed during the TDD process.

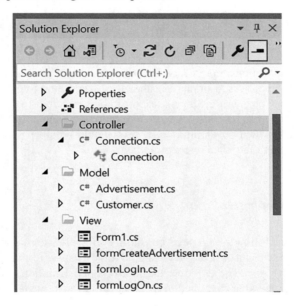

Before completing the method bodies, a test class, including test methods for each class, is generated. Here, Nunit3 is used to generate test classes. Below is a test class, ConnectionTests, generated for the class "Connection."

```csharp
using NUnit.Framework;

using OnlineShopping;
…
namespace OnlineShopping.Tests
{
  [TestFixture()]
  public class ConnectionTests
  {
    [Test()]
    [TestCase("samad", "2468", City.London, Result = true)]
    [TestCase("simin", "1357", City.Paris, Result = true)]
    public bool CreateUserTest(string name, string phoneNumber,
                                           City city)
    { bool result = false;
      Connection cnn = new Connection();
      result = cnn.CreatedUser(name, phoneNumber, city);
      return result;
    }


    [Test()]
    [TestCase("Jhon ", "1236", Result = true)]
    [TestCase("Sami", "1234",  Result = true)]
    [TestCase("Sara", "4563",  Result = true)]
    public bool LoginTest(string name, string phonNumber)
    { bool result = false;
      Connection cnn = new Connection();
      result = cnn.Login(name, phonNumber);
      return result;
    }


    [Test()]
    [TestCase(Category.Vehicle, 20000, "gooood","405 SLX",
     Result = true)]
    public bool addAdvertisementTest(Category advcategory, int advprice,
                                          string advdescription, string advtitle)
    {
       …
```

Typically, BDD practitioners use an outside-in approach. They begin by building whatever is necessary to pass the acceptance criteria as they write a feature. The design and development of code should be up to passing acceptance tests and implementing the features. A TDD approach is followed to implement the code for the online shopping system. As described in Chap. 2, the test-driven development process generates test classes before actually completing the methods. In fact, in TDD, the design is conducted by the tests. Initially, all the tests fail because all the methods are empty. Once all the tests pass, the code is refactored, and the following user story is selected and implemented. Below is the code.

```
using System;
using System.Collections.Generic;
using System.Collections.Specialized;
using System.Data;
using System.Data.SqlClient;
using System.Linq;
using System.Text;
using System.Threading.Tasks;

namespace OnlineShopping
{
  public class Connection
  {
     bool isCreatedUser;
     bool isLogin;
     bool isExists;
     bool isCreatedAdvertisement;
     public static int LoggedInCustomer;
     public static Customer customer= new Customer();

     public bool CreateUser(string name, string phoneNumber, City city)
     {  isCreatedUser = false;
        isExists = false;
        SqlConnection cnn = SQLconnectoin();
        cnn.Open();
        string selectQuery = $"SELECT * FROM [dbo].[User] " +
                    $"WHERE (PhoneNumber = '{phoneNumber}')";
           ...
```

```
public bool Login(string name, string phoneNumber)
{
    isLogin = false;
    SqlConnection cnn = SQLconnectoin();
    cnn.Open();
    string selectQuery = $"SELECT * FROM [dbo].[User] " +
        $"WHERE (Name = '{name}' AND
                    PhoneNumber = '{phoneNumber}')";
    ...


public bool addAdvertisement(Category advcategory, int advprice,
                            string advdescription, string advtitle)
{
```

While developing the code, the methods are unit tested several times. In fact, tests conduct the design and development of code.

3.7.7 Acceptance and Unit Tests

All the tests, including unit and acceptance tests, can be executed through the test explorer. Click on the test from the menu bar in the C# integrated development environment (IDE) and then select the "run all tests" option. To view the results, select the "Test Explorer" option:

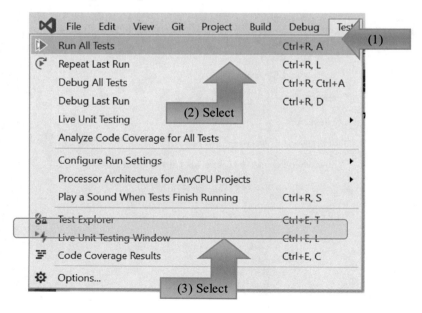

The final test results are shown below. It includes both acceptance and unit test results:

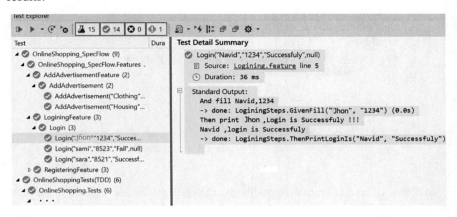

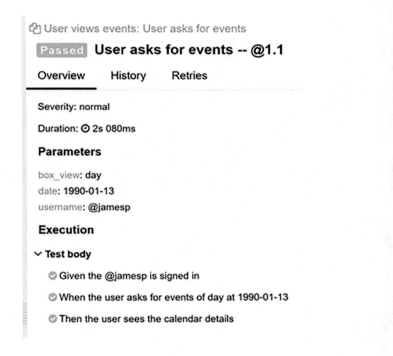

3.8 Summary

Feature definitions provide a straightforward method of expressing requirements. The Gherkin language provides a simple format for defining features in plain natural language. A feature definition comprises three main components: (1) a short text describing the user story, (2) the exchange of perspectives between the business owner, tester, and the developer teams on the user story, and (3) acceptance criteria.

The first component identifies the objective, what the system is expected to accomplish, and (optionally) why it is essential. The acceptance criteria are defined as one or more scenarios where a scenario itself is expressed in a given-when-then format.

Behavior-driven development (BDD) tools map the given, when and then steps of a feature scenario into executable methods. A test-first approach to software development is used to complete the step methods as acceptance tests. The acceptance tests initially fail. Test-driven development (TDD) methods are employed to implement the feature and pass the tests.

3.9 Exercises

[1] Apply a BDD tool to implement the following feature and then use the outcomes to answer the following exercises.

> User Story: Flight Tickets Search
> #1. Story (actor – requirement – rationale):
>> As a frequent traveler
>> I want to be able to search for tickets, providing locations and dates
>> So that I can obtain information about rates and times of the flights.
> #2. Acceptance criteria (Given-when-then)
> Scenario: One-Way Tickets Search
>> # Given -- the beginning state of the scenario.
>> Given I go to "Find flights."
>> # When -- a specific action that the user takes.
>> When I choose "One way,"
>> And I type <source> in the field "From,"
>> And I type <destination> in the field "To,"
>> And I choose the option of value <np> in the field "Number of passengers,"
>> And I set <date> in the field "Depart," And I click on "Search,"
>> #Then — a testable outcome, usually caused by the action in When
>> Then will be displayed "Choose Flights."

Scenario: Return Tickets Search
    Given I go to "Find flights"
    When I choose "Round trip"
    And I type \<source\> in the field "From"
    And I type \<destination\> in the field "To"
    And I choose the option of value \<np\> in the field "Number of passengers"
    And I set \<departure_date\> in the field "Depart"
    And I set \<return_date\> in the field "Return"
    And I click on "Search"
    Then will be displayed "Choose Flights"
Examples: booking a one-way flight

Source	destination	np	departure_date	retunr_date
Paris	Nakazaki	11	05/10/2022	8/10/2022
Newyork	Washington	04	06/12/2022	8/12/2022
London	Adsisababa	11	05/10/2022	8/05/2023
Dalas	Manchester	03	06/12/2022	8/12/2023

1.1   Use "background" in Gherkin to share steps between the two scenarios,
1.2   Describe how to share data between steps,
1.3   Describe how to use stubs and mocks for unit testing.

[2]   Behave uses decorators in Python to match each step with its corresponding step definition. Considering the actions of the decorator in Python, implement the behave approach to match steps.

[3]   If a test fails at an assertion, it will not continue with the following assertion. Devise a way to get around this problem. For instance, when using NUnit or Xunit for a unit test containing a sequence of test assertions as follow:

```
{
  Assert.IsTrue({condition 1})
  Assert.IsTrue({condition 2})
    ..
  Assert.IsTrue(condition N})
}
```

An error is thrown when one of the assertions fails. Give an example, using fluent assertions to to combine the assertions and have any errors be silently collected at the end of an assertion scope block:

```
using (new AssertionScope())
{
  Subject1.Should();
  Subject2.Should()
  ..
  SubjectN.Should().
  Assert.Pass();
}
```

When the unit test is run, any errors that show up in the test explorer will be combined in a summary message.

[4] Use the Background keyword in SpecFlow to share an Example table between different scenarios of a given feature. The background keyword can be used as follows:

```
Background:
  Given my-table looks like
  | .... | .... |
```

[5] Implement a feature with a table shared between the scenarios. It is possible to access the current table row without explicitly mentioning all parameters in the given/when/then step via context.active_outline, returning a behave.model.row object, which can be accessed in the following ways:

- **context.active_outline.headings**: Regardless of whether the current row is iterated, it returns a list of the table headers.
- **context.active_outline.cells**: Returns a list of cell values for the currently iterated row.
- **index-based access:** similar to context.active_outline[0] returns the cell value from the first column.
- **named-based access**: returns the cell value for the column with the first_thing header.

[6] Give an example of using hooks to disable a feature in spec flow (Gherkin) without deleting the feature.

[7] Specify particular setup and tear-down steps for each specific feature file. Some hooks allow code to execute before every scenario and hooks to execute code before each feature. Specify code to run once before and after all the scenarios

run for one specific feature. Implement some feature files of your choice. Try the following tag in one of the feature files:

@tagToIdentifyThatBeginAfterShouldRunForThisFeatureOnly
Feature : A new feature ....

In Stepdefinitions_file.java :

```
@Before("@tagToIdentifyThatBeginAfterShouldRunForThisFeatureOnly")
public void testStart() throws Throwable {
    }
```

```
@After("@tagToIdentifyThatBeginAfterShouldRunForThisFeatureOnly")
Public void testStart() throws Throwable {
    }
```

A not ideal but reasonable solution is to provide a specific setUp and tear-Down for each feature by defining a JUnit test class for each feature. The standard @Before stuff goes in the steps class, but the @BeforeClass annotation can be used in the main unit test class.

References

1. Palmer, S.R., Felsing, J.M.: A Practical Guide to Feature-Driven Development. Prentice Hall (2002)
2. Nagy, G., Rose, S.: The BDD Books—Discovery: explore Behavior Using Examples. Packt Publishing—ebooks Account (2019)
3. Engel, J., Rice, B., Jones, R.: Behave Documentation Release 1.2.7.dev2 (2022)
4. Smart, J.F.: BDD in Action: behavior-Driven Development for the Whole Software Lifecycle. Manning Publications Co., (2015)
5. The SpecFlow Team.: Welcome to SpecFlow's Documentation! Accessed 03 May 2022
6. Dees, I., Wynne, M., Hellesøy, A.: Cucumber Recipes Automate Anything with BDD Tools and Techniques. Potomac Indexing (2013)
7. Wynne, M., Hellesøy, A.: The Cucumber Book Behaviour-Driven Development for Testers and Developers. The Pragmatic Bookshelf (2012)
8. Sale, D.: Testing Python; Applying Unit Testing, TDD, BDD, and Acceptance Testing. Wiley (2014)
9. Fowler, M.: Refactoring: improving the Design of Existing Code, 2nd edn. Addison-Wesley (2019)
10. Suryanarayana, G., Samarthyam, G., Sharma, T.: Refactoring for Software Design Smells: Managing Technical Debt. Elsevier (2015)
11. Bridging the Communication Gap: Specification by Example and Agile Acceptance Testing (2023) https://www.softwaretestinghelp.com/specflow-tutorial/
12. Smart, JF.: BDD in Action: Behavior-driven development for the whole software lifecycle, 1st edn. Manning Publishing Co., (2015)

Chapter 4
Testability Driven Development (TsDD)

4.1 Introduction

The cost of the testing rises exponentially as the size of the code under test grows. TDD and BDD aggravate the cost by using failing tests to conduct the design and implementation code. Testability-Driven Development (TsDD) suggests repeated measurement of testability and refactoring for testability while applying a test-first approach to design and develop code. Testability is the ease with which a software artifact can be tested efficiently and effectively. The more testable a code fragment, the less effort is required to test it effectively.

This chapter shows the readers how to build a machine learning model to predict class testability. The tool supports software development using a TsDD-based approach. To avoid high testing costs, one should ensure that the code under test is testable. Section 4.2 introduces testability-driven development or, in short, TsDD as a new approach to developing testable code. TsDD uses unit tests, which means it tests tiny bits of functionality in isolation.

4.2 Testability-Driven Development

Software testing is essential to delivering high-quality systems, yet it could be unnecessarily challenging, expensive, and time-consuming if care is not taken when developing code. These factors have motivated researchers to seek ways to improve software testability. As shown in Fig. 4.1, the main essence of testability-driven development or, in short, TsDD is to get the code ready for the test. Before describing the TsDD methodology in Sect. 4.2, an elaboration of TDD is given in Sect. 4.2.1.

© The Author(s), under exclusive license to Springer Nature Switzerland AG 2023
S. Parsa, *Software Testing Automation*,
https://doi.org/10.1007/978-3-031-22057-9_4

Fig. 4.1 TsDD supports test
first approaches by getting
the code ready for the test

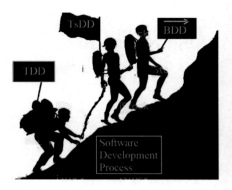

4.2.1 TDD Elaboration

Most programming faults do not do anything with requirements. For instance, repair
patterns such as IFChecker, Mutate Data Type, VarReplacer, and binary checker
focus on programmers' mistakes concerning the use of operators and variables [1,
2]. Therefore, even if all the tests derived from the requirements pass, it is a bit
clumsy to claim that the code is correct and the green step is reached.

Despite the high cost of software testing, TDD encourages incremental development of software based on the test results. In a test-first approach such as TDD, test
cases derived from the system requirements should be prepared for unit testing of
the class or function to be implemented before starting to write code. In effect, test
failures evolve the software design. Failing tests gives the developer the inspiration
of how to get around failures by appropriately designing and writing the code.

With increasing software size, the number of paths and, consequently, the cost
of software testing increases exponentially [3]. The test data generation could be
time-consuming for projects with hundreds of complex classes [4]. In addition, a
developing code that is not executable yet or depends on the code that is not complete
cannot be tested. A study by Khanam and Ahsan [5] on the advantages and pitfalls
of TDD reveals the incapability of TDD in the rapid and cost-effective design of
influential test cases for the projects under test.

As shown in Fig. 4.2, the TDD cycle starts with the red step, in which the next
feature is selected, and test data is generated manually to run tests while completing
the code. The target of the red and green steps is to write the minimum code to pass
the tests. The blue step is concerned with refactoring the code. The refactoring could
be either manual or automated.

The catch behind blind refactoring, i.e., refactoring without evaluating source code
quality, is that refactoring does not necessarily improve the quality attributes such as
testability. Refactoring operations might improve or deteriorate the software quality
[6], and testability is no exception. Especially when applying batch refactoring [7],
the order of refactoring operations also matters. Hence, testability measurement,
even after applying each refactoring, is required to achieve the appropriate batch of

Fig. 4.2 The TDD
red-green-blue cycle

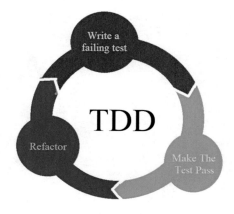

refactoring operations, maximizing the testability. When testability measurement is used, in the cases that refactoring reduces testability, developers may roll back and discard them rapidly.

4.2.2   Testability-Driven Paradigm

It is not uncommon for software testing to consume more than half of the total development time. Especially when it comes to cyber-physical and scientific software, testing could be incredibly labor-intensive and expensive. TsDD advocates reducing the test burden and cost by ensuring code is testable before a test runs.

The TsDD process begins by forward-engineering class and component diagrams onto source code. Afterward, generate test classes for every class in the project using a unit testing tool such as JUnit, NUnit, or XUnit. The code is ready for the test, even before it is executable, by frequently predicting its testability and refactoring to improve testability when required. In fact, along with the testing results, testability improvements conduct the incremental development process. Test suites could be created manually and automatically when the code is ready for an efficient and effective test. After a successful test, additional code is written to fulfil the next system requirement, and the process is repeated until the code is complete. The rhythm of TsDD is as follows:

1.   Add a test quickly,
2.   Run all tests and see the new one failing,
3.   Make a little change.
4.   A sure the code is testable.

    a.   Measure testability.
    b.   Refactor for testability.

Fig. 4.3 Testability-driven
development life cycle

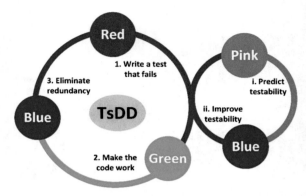

The mantra of TsDD is "red, pink, blue, green, blue."

5. Run all tests and see that they all pass.
6. Refactor to remove duplication.

As shown in Fig. 4.3, the mantra of TsDD is: write a test, make it effective, make it run, and make it right.

The rationale behind testability-driven development is effectively using the testability measurement information in the software development lifecycle (SDLC). It encourages the developers to minimize the test cost and effort by emphasizing refactoring for testability. The most promising benefit of TsDD is that the whole testability measurement and improvement process can be automated.

TsDD allows automated test data generation by postponing the test data generation to when the code is ready to test. Thus, TsDD enables quality software development by conducting programmers towards automated refactoring to improve testability and other quality factors such as modularity, maintainability, and scalability.

Compared to TDD, TsDD has better conformance in leveraging automated software engineering techniques, including automated test data generation and automated refactoring tools. TsDD can be applied to a vast range of developed and developing codebases with minimum human effort. Refactoring for testability improves testability and other software quality factors such as modularity and maintainability.

TsDD proposed model has made it possible to develop testable code by measuring testability before and after refactoring [][]. Frequent refactoring followed by testability measurement leads to efficient and effective tests, reducing testing costs. Moreover, as testability improves, some quality attributes, including reusability, functionality, extensibility, and modularity.

4.3 TsDD in Practice

Before describing details of the machine learning model to predict software testability in Sect. 4.3.2 to motivate the reader, experimental results demonstrate the testability prediction's applicability. A motivating example is given in Sect. 4.3.1.

4.3.1 Motivating Example-1

As shown in Fig. 4.3, TsDD improves TDD by inserting a pink-blue cycle into the red-green-blue cycle of TDD. The pink-blue cycle recommends predicting and refactoring for testability until the code gets ready for the unit test. For instance, consider the GridGenerator class of a Java project, Water Simulator, in the SF110 corpus [8]. First, as depicted in Fig. 4.4, the class diagram is drawn in the Enterprize Architect (EA) environment.

The class diagram is forward-engineered into Java code in the second step. The code is presented in Fig. 4.5.

In the third step, JUnit is used to generate a test class for the GridGenerator class. JUnit is a tool for unit testing in Java. The fourth step starts with writing code and completing the class. Below, in Fig. 4.6 is the code for the GetNeighbor() method.

In the fifth step, after completing all the methods of the GridGenerator class and before unit testing, I use our testability predictor tool [9], described in Sect. 4.6, to measure the testability of the class components. The GridGenerator class testability is 6.57% which is relatively low. Therefore, a refactoring tool is employed to detect

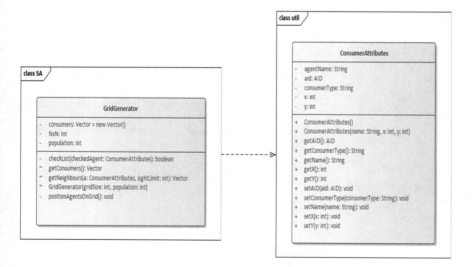

Fig. 4.4 A class diagram including two classes GridGenerator and ConsumerAttributes

Fig. 4.5 Class
`GridGenerator` with
empty methods, generated
by EA

```java
package water1.simulator.SA;
import water1.simulator.util.ConsumerAttributes;
/**
* @since 2003-2006
* @author Vartalas Panagiotis
* @version 1.9
* @created 18-Aug-2021 17:55:10
*/
public class GridGenerator {
 /** * all agents' attributes */
   private Vector consumers = new Vector();
 /** * grid size (one dimension) */
   private int NxN;
 /** * agents' population */
   private int population;
   public GridGenerator(){   }
   public void finalize() throws Throwable { }
/**
   * @param gridSize
   * @param population
   */
   GridGenerator(int gridSize, int population){ }
   /**
   * CheckList: check if there is an agent in this position on grid
   * @return TRUE if there is an agent in this position FALSE if
     the agent can be put in this position
   *
   * @param checkedAgent: the agent whose position is checked
   */
   private boolean checkList(ConsumerAttributes checkedAgent){
       return false; }
   /**
   * getConsumers
   * @return : a Vector with all created Consumers
   */
   Vector getConsumers(){ return null; }
   /** * @return * @param a * @param sightLimit */
   @SuppressWarnings("unchecked")
   Vector getNeighbours(ConsumerAttributes a, int sightLimit){
           return null; }
   /** * Get random positions on a grid and name agents after
   these eg consumer(2,3) */
   @SuppressWarnings("unchecked")
   private void positionAgentsOnGrid(){ }
}//end GridGenerator
```

```
Vector getNeighbours(ConsumerAttributes a, int sightLimit){
int x0,y0;
int x,y;
Vector myNeighbours = new Vector();
x0 = a.getX();
y0 = a.getY();
for (int j=0; j <consumers.size(); j++){
  if(!(a.getName().equals(((ConsumerAttributes)
        consumers.elementAt(j)).getName())))){
    x = ((ConsumerAttrutes)
              consumers.elementAt(j)).getX();
    y = ((ConsumerAttributes)
              consumers.elementAt(j)).getY();
    if ((Math.abs(x0-x) <= sightLimit)
        && (Math.abs(y0-y) <= sightLimit)){
      myNeighbours.addElement(
          (ConsumerAttributes)consumers.elementAt(j));
    }
  }
}
return myNeighbours;
}
```

Fig. 4.6 The getNeighbors method with the extract method refactoring opportunity

the code smells in the getNeighbours method and remove the smell by refactoring. A code smell is not a fault in code. It is a design and quality issue stemming from poor programming practices. Detecting a code smell, you may remove the smell by applying appropriate refactorings. Code refactoring is a technique to clean up code to improve its quality and remove code smells.

As shown in Fig. 4.6, the getNeighbours method suffers from the long method smell [10] detected by the JDeodorant [11] refactoring tool. The long method design smell concerns the method length. Once a method surpasses a certain size, it performs too many operations and should be split or shortened. The smell can be resolved by applying the extract method refactoring. Extract Method refactoring is a powerful technique that can be used to simplify your code by turning a section of code into a new method.

The long method smell in the getNeighbours method is removed by applying the extract method refactoring. The method is refactored by moving the lines that check whether two given grid cells are adjacent to a new method, isNeighbour(). As a result, the GridGenerator class testability increases from 11.18 to 18.90%. Even the new testability value, 18.90, is meager. It means that the class is almost untestable.

During a study to find the factors that affect this low level of testability, we found that the newly created method, isNeighbour(), suffers from feature envy code smell [10]. The feature envy smell is present when a method from a class use accesses the attributes or methods from another class excessively. A solution to the Feature Envy smell is to move the method to the class to which it is more closely related.

The isNeighbour() method has six accesses to attributes of the ConsumerAttributes class, while it does not access any member of the GridGenerator class. By applying the move method refactoring [11], we move the isNeighbour method from the GridGenerator class to the ConsumerAttributes class. In this way, the move method refactoring removes both the feature envy and data class smells simultaneously. It also eliminates the first parameter of the isNeighbour method, an instance of the class ConsumerAttributes to which the method is moved. This way, the attained coverage for the ConsumerAttributes class increases to 100%.

The GridGenerator class testability is improved from 18.90 to 20.26%. However, still, the testability is very low. Nevertheless, what is the reason? The reason is the while loop with two if statements in its body in the positionAgentsOnGrid() method, shown in Fig. 4.7.

The value of the boolean variable "done" is determined dynamically and depends on the program's runtime behavior. Therefore, the number of execution paths in the program is not predictable because the number of iterations of the while loop is not predictable, and each iteration may go through one of the five different new execution paths caused by the three nested "if" statements within the body of the "while".

Fig. 4.7 An example of the loop-assigned flag problem

```
private void positionAgentsOnGrid(){
  // help variables
  boolean done = false;
  int x,y;
  String agentName = "consumer_";
  Random gen = new Random();  // number Generator
  while (!done){
    if ((( x = gen.nextInt(NxN)) != 0) &&
      (( y = gen.nextInt(NxN)) != 0)){
      agentName += x + "," + y;
      ConsumerAttributes a = new
              ConsumerAttributes(agentName,x,y);
    if (!checkList(a)){
      consumers.addElement(a);
      if (consumers.size() == population){done = true;}
    }
    agentName = "consumer_";
    }
  }
}
```

Table 4.1 Test results and testability measures computed for the class GridGenerator

Refactoring	Coverage				Test suite		Testability
	Branch (%)	Statement (%)	Mutation (%)	Average (%)	Size	Time (min)	predicted (%)
None	12.50	29.73	6.10	13.14	2	6	11.18
Extract method	11.54	33.33	8.93	15.09	2	6	18.90
Move method	15.00	34.38	12.35	18.54	2	6	20.26
Simplifying conditional logic	72.20	93.33	65.33	76.07	7	6	42.39

This issue is referred to as the flag problem [12]. With the flag problem, search-based test data generators can not provide effective test data sets. Testability transformation techniques have been used to resolve the problem. Testability transformation techniques substitute the unpredictable flag with a predictable expression such that test data can be generated effectively. In testability transformation, the aim is not to preserve the semantics of a program but to preserve test sets' effectiveness according to some chosen criterion for adequacy. That is why, after generating test data, the transformation is discarded.

Instead of testability transformation, I modify the method poitionAgentsOnGrid without altering its semantics by simplifying conditional logic refactoring. As a result, as shown in Table 4.1, the testability increased from 20.26 to 42.39%, and the number of test data generated in 6 min increased from two to seven. In 0 4.8, you can see that the code is cleaned up by flattening the nested structure caused by an "if" statement in the "while" construct. As a result, it is observed the mean code coverage is improved from 18.54 to 76.07 (Fig. 4.8).

The size of the test suite generated by running the EvoSuite [12] automat test data generator for six minutes and the coverage obtained when running the class GridGenerator with the test suite is shown in Table 4.1. The last column shows the testability measure predicted by the testability regression model, described in Sect. 4.4.

4.3.2 Experimental Results

As described before, there is no guarantee that the code quality, especially testability, will improve after refactoring. The experimental results in Table 4.1 show that the same refactoring may impact different classes differently. The impact depends on how the refactored code is composed. For instance, as shown in the first row of Table 4.1, applying the extract method refactoring on a method in the first row of the Weka project [13] reduces its testability by about 0.023. The reason is that the extracted method contains eight parameters, leading to a long parameter list smell

```
private void positionAgentsOnGrid(){
  int x,y;
  String agentName = "consumer_";
  Random gen = new Random();  // number Generator
  while (consumers.size() < population){
    if ((( x = gen.nextInt(NxN) ) != 0) && (( y = gen.nextInt(NxN) ) != 0)){
      agentName += x + "," + y;
      ConsumerAttributes a = new ConsumerAttributes(agentName,x,y);
      if (!checkList(a)){
        consumers.addElement(a);
      }
      agentName = "consumer_";
    }
  }
}
```

Fig. 4.8   After refactoring the positionAgentsOnGrid(){

[10], which makes it difficult to test. In addition, in this specific case, the extract method refactoring makes some duplicate codes that enhance the complexity of the DecisionStump class. On the other hand, as shown in the second row of Table 4.2 method of weka.classifiers.evaluation.ThresholdCurve class in the Weka project [13] improves testability by 0.1229 (a relative improvement of 166.53%). In this case, extract method refactoring reduces the size and complexity of the refactored method. Similar behaviors are observed when applying other types of refactoring operations. For instance, the third row of Table 4.2 shows that move method refactoring reduces the testability of weka.classifiers.evaluation.ThresholdCurve class. However, the move method improves the testability of another method of the same class. The remaining rows of Table 4.2 show examples of the refactoring operations that decrease or do not change the testability. In Table 4.2, the ↓ sign indicates testability improvements while the ↑ sign indicates the opposite.

(a)   weka.classifiers.trees.DecisionStump
(b)   weka.classifiers.evaluation.ThresholdCurve
(c)   weka.classifiers.evaluation.ThresholdCurve
(d)   weka.classifiers.evaluation.ThresholdCurve
(e)   weka.classifiers.pmml.consumer.TreeModel
(f)   weka.classifiers.pmml.consumer.Regression
(g)   weka.classifiers.trees.DecisionStump
(h)   weka.core.pmml.Apply
(i)   weka.classifiers.pmml.consumer.RuleSetModel.

Table 4.2 Refactoring impact on testability value in different situations

Class	Refactoring	Refactored method	Testability Before	Testability After
(a)	Extract method	findSplitNumericNumeric	0.2168	0.1941 ↓
(b)	Extract method	sourceClass	0.0738	0.1967 ↑
(c)	Move method	makeInstance	0.0738	0.0688 ↓
(d)	Move method	getProbabilities	0.0738	0.2114 ↑
(e)	Make field final	m_classLabel	0.1164	0.1075 ↓
(f)	Simplifying conditional logic	determineNormalization	0.0134	0.0123
(g)	Move method	printDist	0.2168	0.1924 ↓
(h)	Extract method	toString	0.0221	0.0125 ↓
(i)	Extract method	score	0.0608	0.0552 ↓

4.4 Testability Mathematical Model

This section offers a mathematical model to compute testability based on the runtime information. The mathematical model is used to label samples used for building a machine learning model to predict testability statically without any need to run the class under test.

The mathematical model presented in this section stems from the definition of testability provided by the system and software quality models standard ISO/IEC25010: 2011 [1]. The standard defines testability as the" degree of effectiveness and efficiency with which test criteria can be established for a system, product or component and tests can be performed to determine whether those criteria have been met."

The assumption is that the standard is correct. Therefore, testability is formulated by analyzing and interpreting this standard definition. Based on this definition, there are two factors affecting testability:

1. The efficiency of test criteria establishment
2. The effectiveness of test criteria establishment.

The extent of code coverage achieved by a test suite provides a sound indication of test effectiveness and quality [14]. A test suite that covers more code has a greater chance of detecting faulty executions if any.

4.4.1 Test Effectiveness

Test effectiveness depends on the average coverage provided by a test case belonging to the test suite collected for the test. The more coverage a test suite provides, the more effective the test will be. The test effectiveness, $T_{\text{effectiveness}}(X)$, of the class,

X, is computed as the average of different coverage criteria, TestCriteria, considered for a given test suite:

$$T_{\text{effectiveness}}\ (X) = \frac{1}{|\text{TestCriteria}|} \sum_{i=1}^{|\text{TestCriteria}|} \text{TestCriteria}_i^{\text{level}} \qquad (4.1)$$

where the level of a test criterion indicates the percentage of the coverage provided by the test.

Test effectiveness depends on the average coverage provided by each influential test case belonging to a minimized test suite collected for the test. A test case is influential if it enhances the test coverage. A minimized test suite includes only non-overlapping influential test cases.

Research is underway on the impact of various combinations of coverage metrics on the effectiveness and effort of test and fault localization [6]. A significant finding is that combining coverage criteria is more effective than a single criterion [7]. For instance, the EvoSuite test data generator tool [8] incorporates eight code coverage metrics when generating tests. Our experiments with EvoSuite revealed that combining statement, branch, and mutation coverage boosted test efficiency with a fixed budget.

It should be noted that the branch coverage does not necessarily subsume statement coverage unless it is 100%. However, most of the time, test suites do not provide complete branch coverage. As a result, due to the imbalanced distribution of the source code statements in branches, there are many cases where the percentage of branches covered by a test suite is very different from the percentage of statements it covers. For instance, out of 16 branches, a test suite may cover only 12 branches, and a second test suite covers the remaining four branches. However, it can not be guaranteed that the 12 branches covered by the first test suite are more than the statements covered by the second test suite. Our experiments with EvoSuite revealed that combining statement, branch, and mutation coverage boosted test efficiency with a fixed budget.

4.4.2 Test Efficiency

The more coverage a test suite provides, the more effective the test will be. On the other hand, as the number of test cases increases, more effort in terms of time and cost will be required to test. Hence, the test effort is inversely proportional to the test coverage percentage:

$$T_{\text{effort}}(x) = \frac{1}{T_{\text{efficieny}}(x)} \qquad (4.2)$$

Relation (4.3) stems from the evidence that the more coverable code, the less testing effort would be required. Suppose a program has 200 statements. The goal is to test these 200 statements. For sure, three test runs of this program take less effort than four. For instance, running the program with the fourth test case is unnecessary if these 200 statements are covered with three test cases. Therefore, it can be said that the higher the percentage of coverage, the less test effort is needed.

The minimized test suite size seems to be the main factor used to measure test effort in software testing literature because the more test data, the more effort will be needed to run and evaluate the test results. In addition, as the number of tests increases, the growth rate of the code coverage decreases.

Our empirical observations [14, 15] show that as the number of influential test data increases, the growth rate of the code coverage decreases because as the number of test data increases, it takes more effort and time a new influential test data that does not overlap with the existing ones. Therefore, considering the average time, ω, needed to generate a minimized test suite, $\tau(X, C)$, to obtain the coverage C for the methods of class X, the test effort $T_{\text{effort}}(x)$ is computed as follows:

$$T_{\text{effort}}(x) = (1 + \omega)^{\frac{|\tau(X,C)|}{\text{no.methods}(X)} - 1} \tag{4.3}$$

The average time, ω, required to generate test data is computed as follows:

$$\omega = \frac{t - 1}{|\tau(X, C)|} \tag{4.4}$$

The parameter ω in Eq. (4.3) regulates the effort, time, and cost required to generate, execute, and evaluate a single test case. Generating, executing, and evaluating a test case for all software types does not have equal time, cost, and effort. For example, generating and running a test case for an autonomous car controller most presumably requires more time and effort than generating and running a test case for a simple accounting system.

This hyperparameter must be determined by either the test engineer or project manager, considering the tester's experience to provide the most informative results. However, for our experiments, we considered ω the minimum cost of generating an influential test case. To this aim, we ran EvoSuite on the whole corpus (6 min for each class), and the number of influential test cases generated for a class under test was at most 410. Dependent on the class under test, EvoSuite might spend most of its time in the search phase but may also spend a non-negligible time in these other phases. Therefore, to ensure that the experiments finished within a predictable amount of time, we gave a general timeout of six minutes per class []. We computed ω as follows:

$$\omega = \frac{\text{time budget in minutes}}{\text{Maximum no. influential test cases}} = \frac{6}{410} \approx 0.015$$

In fact, 0.015 is the minimum cost of generating an influential test case, affecting the dataset coverage. According to EvoSuite creators, generated test suites are minimized at the end of the search process such that only statements contributing to coverage remain [15]. Therefore, the remaining test suite contains only the influential test cases affecting the testability while all generated test case does not affect it.

4.4.3 Testability

Based on the ISO/IEC25010: 2011 [1] standard, test efficiency, and effectiveness are the two main factors affecting testability. Formally, for a given class, X, testability, $T(X)$, is defined as the product of its test effectiveness, $T_{effectiveness}(X)$, and test efficiency, $T_{efficiency}(X)$ [15]:

$$T(X) = T_{effectiveness}(X) \times T_{efficiencx}(X) \tag{4.5}$$

Substituting test effectiveness and efficiency Eqs. (4.1) and (4.3) in (4.5), the mathematical model for computing testability will be as follows:

$$T(X) = \frac{1}{|\text{TestCriteria}|} \sum_{i=1}^{|\text{TestCriteria}|} \text{TestCriteria}_i^{\text{level}} + T_{\text{effort}}(x) = (1+\omega)^{\frac{|\tau(X,C)|}{\text{no. methods}(X)} - 1} \tag{4.6}$$

The testability, $T^-(M)$, of a component, M, including n classes, can be computed as the average testability of its classes:

$$T(M) = \frac{1}{n} \sum_{i-1}^{n} T_i(X) \tag{4.7}$$

When generating test data, the observation was that the growth rate of code coverage decreased as more test cases were generated. As a matter of fact, the more test cases are generated, the longer it takes to generate the next influential test case. This is because as the number of test cases grows, a newly generated influential test case is more likely to cover the same parts as previously generated test cases.

Therefore, as the number of test cases increases, the effort to create new influential test cases increases. In contrast, if the number of test cases generated for a class, X1, is higher than the class, X2, for a given time budget T, then it can be inferred that class X1 is relatively more testable than class X2.

4.4.4 Coverageability

Detecting all bugs in a program is not a decidable problem, so it is vital to predicting test effectiveness and quality. As described in Sect. 4.4.1, the test effectiveness depends on the coverage provided by the collected test data set. The point is that measuring test coverage after each change to the code is costly and time-consuming because the coverage reveals only after running the code with the selected test data set. In addition, the code has to be executable.

Coverageability is a new concept I introduced in [16]. Coverageability is the extent to which any test data set may cover a source code with a limited test budget. Measuring the test coverage after every change can be time-consuming and costly because it needs to rerun the program after each change to its source code.

Formally, Coverageability, $C_\mu(X, C)$, is the mean coverage, $C(X)$, gained for a class, X, by an influential test case in a minimized test suite, $\tau(X, C)$, multiplied by the number of influential test cases, b_B, that can be generated considering the test budget, B:

$$C_\mu(X, C) = \frac{C(X)}{|\tau(X, C)|} \times b_B \tag{4.8}$$

The test coverage, $C(X)$, is computed in terms of the coverage metrics $C_1(X)$ to $C_n(X)$ for $n \geq 1$.

$$C(X) = \sqrt[n]{(C_1(X) \times C_2(X) \times \ldots \times C_n(X))} \tag{4.9}$$

Relation (4.4) computes the geometric mean of the coverage metrics. When the variables are dependent and widely skewed, the geometric mean is more suitable for calculating the mean and provides more accurate results.

The size of an automatically generated test suite is a meaningless metric because different test data may cover the same execution path without increasing the test coverage. However, most test data generators, such as EvoSuite, perform a minimization step to reduce the number of test cases while achieving the same coverage.

The experimental results with EvoSuite, show that Coverageability increases logarithmically as the time budget increases. The graph in Fig. 4.9 illustrates the Coverageability values attained while repeatedly configuring and running EvoSuite with successive timeouts to generate test cases for some complex classes in the SF110 dataset.

While EvoSuite generates test cases, the growth rate of the test cases affecting the coverage decreases with the order of t^{-2}, where t is the elapsed time.

$$\text{Rate/speed} = \frac{dC_\mu(X, t)}{dt} = \frac{d(\ln(t))}{dt}$$

Fig. 4.9 Coverageability versus test data generation time

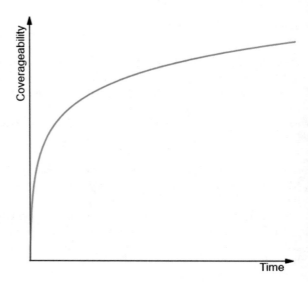

$$\text{Acceleration} = \frac{d^2\big(C_\mu(X,t)\big)}{dt^2} = -\frac{1}{t^2}$$

As time passes, the cost of generating a new influential test case and improving the code coverage increases exponentially, and Coverageability increases logarithmically. Therefore, in terms of time:

$$T(X,t) = \frac{\ln(t)}{e^t}$$

The growth rate of $T(X,t)$ is computed as follows:

$$\text{Rate} = \frac{dT(X,t)}{dt} \approx \frac{d\Big(\frac{\ln(t)}{e^t}\Big)}{dt} = \frac{\frac{e^t}{t} - e^t \times \ln(t)}{e^{2t}}$$

Figure 4.10 illustrates the plot of testability versus the elapsed time. The testability is observed to be steady after about six minutes.

4.5   Testability Prediction Model[1]

The desired testability prediction model estimates testability statically and without running the programs. This way, there will be no need to run the class under test. This Section describes how to develop an ensembled regressor model to predict

---

[1] Download the source code form: https://doi.org/10.6084/m9.figshare.21258609

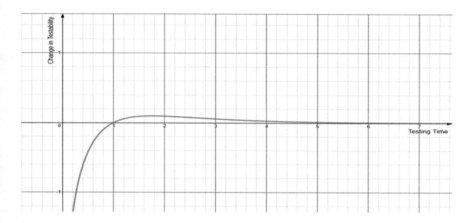

Fig. 4.10 Testability variations in the order of the elapsed time

testability at a class level statistically. One way to categorize Machine Learning systems is by how they generalize. Most Machine Learning tasks are about making predictions. This means that given some training examples, the System needs to be able to generalize to examples it has never seen before.

Equation (4.7) defines testability in terms of test effectiveness (Covergeability) and test effort (cost), both of which are computed using runtime information collected while actually testing the class under test. The computed testability value for each class is used to label a vector representing the class structural attributes, measured by source code metrics, described in Sect. 4.5.1. The labeled vectors are used as samples for building the testability prediction model.

The aim is to learn a real-valued function f that maps the vector of metrics, v, of a given class, χ, to its testability value, $T(\chi)$, with minimum possible error, i.e., $f: v \in R^n \longrightarrow T \in R$. The learning process consists of five steps, as follows:

1. **Data collection**: SF110 corpus [17] containing more than 23,000 classes from 110 different Java projects can be used as the initial dataset to learn the model.
2. **Target value computation**: The target value for the regression model is computed using the testability Eq. (4.8).
3. **Feature vector construction**: Each class in the data set, SF110, is converted to a feature vector, in which each feature indicates a source code metric computed by the Understand static tool [18].
4. **Regression model training**: 12,000 samples [19] were used to train an ensemble of the Random-forest, Multilayer perceptron, and Histogram-based gradient boosting regressors. Each of these regressors is trained using a five-fold cross-validation method. Each training sample is a vector of source code metrics labeled by the testability of a class in the dataset.

Regression model evaluation: We used 4000 samples to test the accuracy and effectiveness of the trained model. **Infer the models**: The learned model is used to predict the testability of a given class based on static metrics.

A detailed description of the abovementioned steps is given in [15]. The learning process begins with computing source code and runtime metrics used as independent and dependent variables.

4.5.1 Feature Selection

For each class in the collected dataset, a test data set is automatically generated for a given time budget, e.g., 2 min. The test data generator also reports the branch and statement coverages obtained by the test data set. Afterward, for each class, the testability value is computed by Eq. (4.7). The computed value is used to label the feature vector for the class.

Source code metrics listed in Table 4.3 can be used as features to measure testability. A source code metric is a quantitive measure of a software attribute, and a feature is a numeric representation of raw data. Therefore, source code metrics may provide a suitable repository of features to vectorize any source code fragment. The source code metrics in Table 4.3 are used to convert a Java class into a feature vector. In Table 4.3, "CS" and "PK" refer to class and package level metrics, respectively.

As the number of relevant independent features increases, the accuracy of the machine learning models may improve. The feature vectors provide a relatively more accurate representation of a class source code as the number of independent metrics used as features of the class increases. Lexical metrics, such as the number of operators and identifiers capturing lexical properties, are given in Table 4.4. The metrics were first introduced in [15].

Feature construction is building new features from existing ones. New sub-metrics can be created by applying the statistical Minimum, Maximum, Mean, First Quartile, Third Quartile, Standard deviation, and Range of the existing metrics. Sub-metric is a new concept introduced in [15]. It is observed that the system accuracy improves as the number of statistical features increases [20]. For instance, the seven metrics in blue in Table 4.3 are increased by applying statistical operations, including minimum, maximum, mean, sum, and standard deviation.

In addition, some metrics have different definitions and hence different computations. For instance, cyclomatic complexity has four definitions and, thereby, four different computations, each considered a sub-metrics. In addition, a class may include several setters and getter methods. These methods do not affect class functionality. Therefore, sub-metrics can be derived for each class, once with the setters and getter methods and once without them. The number of sub-metrics for each metric at both the class level, 'Cs,' and package level, 'Pk,' is given in Table 4.3. For instance, as shown in Fig. 4.5, the number of sub-metrics for cyclomatic complexity considering the four different computations, five different statistical sub-metrics for each

Table 4.3 Source code metrics

Subject	Metric	Full name	CS	PK	Sum
size	LOC	Line of code	36	15	
	NOST	Number of (NO.) statements	36	15	
	NOSM	NO. static methods	1	1	
	NOSA	NO. static attributes	1	1	128
	NOIM	NO. instance methods	1	1	
	NOIA	NO. instance attributes	1	1	
	NOM	NO. methods	1	1	
	NOMNAMM	NO. not accessor or mutator methods	1	1	
	NOCON	NO. constructors	1	1	
	NOP	NO. parameters	10	0	
	NOCS	NO. classes	0	1	
	NOFL	NO. files	0	1	
Complexity	CC	Cyclomatic complexity	40	20	
	NESTING	Nesting block level	4	4	88
	PATH	NO. unique paths	10	0	
	KNOTS	NO. overlapping jumps	10	0	
Cohesion and Coupling	NOMCALL	NO. method calls	1	0	
	DAC	Data abstraction coupling	1	0	
	ATFD	Access to foreign data	1	0	
	LOCM	Lack of cohesion in methods	1	0	
	CBO	Coupling between objects	1	0	
	RFC	Response set for a class	1	0	11
	FANIN	NO. incoming invocations	1	0	
	FANOUT	NO. outgoing invocations	1	0	
	DEPENDS	All dependencies of class	1	0	
	DEPENDSBY	Entities depended on class	1	0	
	CFNAMM	Called foreign not accessors or mutators	1	0	
Visibility	NODM	NO. default meth-	1	1	
	NOPM	ods NO. private	1	1	
	NOPRM	methods	1	1	10
	NOPLM	NO. protected methods	1	1	
	NOAMM	NO. public methods	1	1	
		NO. accessor methods			
Inheritance	DIT	Depth of inheritance tree	1	0	8
	NOC	NO. children	1	0	
	NOP	NO. parents	1	0	
	NIM	NO. inherited methods	1	0	
	NMO	NO. methods overridden	1	0	
	NOII	NO. implemented interfaces	1	0	
	NOI	NO. interfaces	0	1	
	NOAC	NO. abstract classes	0	1	
Sum		40	175	70	245

Table 4.4 Lexical metrics

Metric	Full name
NOTK	Number of (NO.) tokens
NOTKU	NO. unique tokens
NOID	NO. identifiers
NOIDU	NO. unique identifiers
NOKW	NO. keywords
NOKWU	NO. unique keywords
NOASS	NO. assignments
NOOP	NO. operators without assignments
NOOPU	NO. unique operators
NOSC	NO. semicolons
NODOT	NO. dots
NOREPR	NO. return and print statements
NOCJST	NO. conditional jumps
NOCUJST	NO. unconditional jumps
NOEXST	NO. exceptions
NONEW	NO. new objects instantiation
NOSUPER	NO. super calls

computation, and the enclosing classes with and without setter and getter methods is 4 * 5 * 2 = 40 (Fig. 4.11).

Around 23,886 Java classes in the SF110 dataset have been used to build feature vectors [18]. Each class in the SF110 data set is represented as a feature vector labeled by the testability value computed. Each feature is a source code metric computed by the Understand API for Python [17].

4.5.2   Dataset Preparation

In the first step, the metrics for each class are computed. To this end, a software tool called SourceMeter to extract class-level metrics can be used. The following command is used to analyze a project, e.g., apache-log4j-1.2.17, and compute source code metrics for each class:

SourceMeterJava.exe-projectName = log4j-1.2.17-projectBaseDir = apache-log4j-1.2.17-resultsDir = Results

The above command is repeated for each class in each project in the SF110 dataset. As a result, a file including source code metrics for all the classes in the project is created. As sown in Table 4.5, there is a row for each class in the dataset whose features are source code metrics, and the label is the testability value computed by

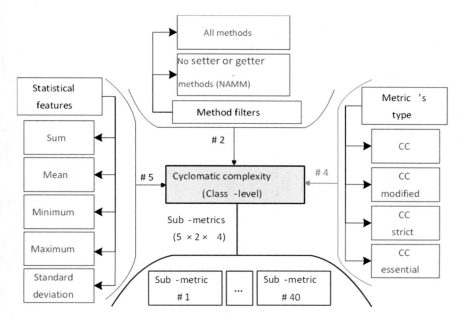

Fig. 4.11 40 sub-metrics for the cyclomatic complexity software metric

the testability Eq. (4.7). The size of the test suite and some test adequacy criteria, such as statement coverage, branch coverage, and mutation score for each class, are needed to compute the testability value.

The EvoSuite [21] test data generator has been used to generate test data for each of the 23,886 Java classes in the SF110 dataset. EvoSuite uses a genetic algorithm to generate a minimized test data suite for each class in a Java project. It also computes branch, statement, and mutation coverages obtained by the test suite for each class. The following command is used to generate tests for a given class X in package p of a given project:

java-jar evosuite.jar-projectCP.\build\classes-class p.X

where the-projectCP point to the location of the .class file of the class under test. In the second step, regression models to learn testability prediction are created. This is accomplished using the Python machine-learning library Scikit-learn [19]. Scikit learn has implemented various regression algorithms.

Figure 4.12 presents a Python class called 'RegressionModel.' The class's constructor takes learning samples from the dataset given by 'learning_samples'. The testability metrics are defined as a python dictionary containing pairs of metrics and full names, listed in Table 4.3.

A Scikit-learn library function called train_test_split is invoked to split the learning samples dataset into the train and test datasets. In the code snippet shown in Fig. 4.12, 20% of samples are taken for the test. The random state parameter of train_test_split allows reproducing the split in the subsequent executions. The last

Table 4.5 Classes metrics for Apache Log4j project

Class long name	CC	CCL	CCO	CI	CLC	CLLC	LDC	LLDC	LCOM5	NL
AppenderTable	0.16	1	1	1	0.09	0.09	11	7	1	1
AppenderTableModel	0	0	0	0	0	0	0	0	1	0
JTableAddAction	0	0	0	0	0	0	0	0	1	0
JMSQueueAppender	0.16	1	5	1	0.093	0.13	19	16	2	3
LoggingOutputStream	0	0	0	0	0	0	0	0	1	2
Log4JTest	0.84	1	2	1	0.70	0.83	22	15	1	1
Log4JTest	0	0	0	0	0	0	0	0	1	1
Class long name	NLE	WMC	CBO	CBOI	NII	NOI	RFC	AD	CD	
AppenderTable	1	5	6	1	1	6	9	0.5	0.35	
AppenderTableModel	0	5	2	1	2	4	9	0.8	0.46	
JTableAddAction	0	2	4	1	1	3	5	0.33	0.2	
JMSQueueAppender	2	27	5	0	0	6	21	0.85	0.31	
LoggingOutputStream	2	17	2	0	0	1	7	1	0.56	
Log4JTest	1	2	3	0	0	7	8	0	0.24	
Log4JTest	1	2	3	0	0	7	8	0	0.39	

```
class RegressionModel:
    def __init__(self, learning_samples = None, testability_metrics = {} ):
        self.df = pd.read_csv(learning_samples, delimiter=',', index_col=False)
        cols = []
        for i, metric_name_ in enumerate(testability_metrics):
            cols.append(testability_metrics[metric_name_][0])
        self.X_train1, self.X_test1, self.y_train, self.y_test =
                train_test_split(self.df[cols], self.df['Testability'],
        test_size=0.20,
        random_state=13)
        # Standardization
        self.scaler = preprocessing.StandardScaler()
        self.scaler.fit(self.X_train1) #computes mean and std of the train set
        #Standardize the train set
        self.X_train = self.scaler.transform(self.X_train1)
        #Standardize the test set
        self.X_test = self.scaler.transform(self.X_test1)
```

Fig. 4.12 Constructor of the class "Regression"

operation performed in the constructor is standardization. Source code metrics have different values with different ranges. It makes the training of machine learning algorithms difficult. A standard preprocessing step in the machine learning pipeline is to transfer the numerical values within a fixed range. The StandardScaler class of the Scikit-learn library [5] is used for the standardization. The standard score, z, of a sample, x, is calculated as follows:

$$z = (x - u)/s$$

where u is the mean of the training samples or zero if with_mean = False and s is the standard deviation of the training samples that is equal to one if with_std = False. Centering and scaling are applied independently to each feature by computing the relevant statistics, i.e., standard deviation and mean, on the samples in the training set. Mean and standard deviations are then stored to be used later to transform the learning samples.

It is worth mentioning that it is assumed that when training, the test samples are not available. Hence, the standard scaler object, self.scaler, parameters, i.e., train and test samples, are fitted on the training dataset to avoid any methodological mistakes, i.e., information leakage of the train data to the test. The complete implementation of the testability prediction model is publically available in: https://codeocean.com/capsule/7939508/tree/v1.

4.5.3   Learning Models

After the data preprocessing, a machine learning model should be configured and used to learn testability. There are many different machine learning regression algorithms for predicting testability. One may test different algorithms to find the best model. Here, a multi-layer perceptron is used. To this aim, we a method, regress, to the class Regression. This method implements a machine learning model based on a multi-layer perceptron. The only required argument for this method is the model_patch which is used to save the learned model for the static prediction of testability. The source code of the "regress" method is given in Fig. 4.13.

The 'regress' method creates an instance of MLPRegressor class, regressor. The sklearn.neural_network.MLPRegressor is a multi-layer perceptron regression system within sklearn.neural_network module. MLPs belong to the class of artificial neural networks (ANN). MLP, in its simplest form, has at least three layers of nodes: an input layer, a hidden layer, and an output layer.

Multi-layer perceptrons have different hyper-parameters to be configured before training. The most important ones are hidden layer sizes, activation function, learning rate, optimization algorithm, and the number of iterations. The parameters should be tuned. A grid search with cross-validation can be used to find the best set of hyperparameters using a list of predefined values for each hyperparameter. The parameters

```python
def regress(self, model_path: str = None):
    """
    input: 1. param model_path: the path to save the learned model
    Output: A testability prediction model
            Creates an instance of the Multi-layer perceptron regressors in Scikit-learn
    """
    # random_state determines weight and bias initialization for random number
    regressor = MLPRegressor(random_state=13)
    # Some of the parameters used in grid search to find the best configuration
    parameters = {
    # NN no.1 has two hidden layers of 256 and 100 neurons, and NN no. 2 has 3 ...
        'hidden_layer_sizes': [(256, 100), (512, 256, 100)],
        # activation functions are tangent hyperbolic and rectified linear unit
        'activation': ['tanh', 'relu' ],
        # The solvers for weight optimization are: adam and SGD
        'solver': ['adam', 'SGD'],
            # The no. iterations of NN training
        'max_iter': range(50, 250, 50)}
    # Set the objectives which must be optimized during parameter tuning
    scoring = ['neg_root_mean_squared_error', ]
    # CrossValidation iterator object
    cv = ShuffleSplit(n_splits=5, test_size=0.20, random_state=42)
    # Find the best model using grid-search with cross-validation
    clf = GridSearchCV(regressor, param_grid=parameters,
                        scoring=scoring, cv=cv, n_jobs=18,
                        refit='neg_root_mean_squared_error')
    print('fitting model number', model_number)
    clf.fit(X=self.X_train, y=self.y_train)
    print('Writing grid search result ...')
    df = pd.DataFrame(clf.cv_results_, )
    df.to_csv(model_path[:-7] + '_grid_search_cv_results.csv', index=False)
    df = pd.DataFrame()
    print('Best parameters set found on development set:', clf.best_params_)
    df['best_parameters_development_set'] = [clf.best_params_]
    print('Best classifier score on development set:', clf.best_score_)
    df['best_score_development_set'] = [clf.best_score_]
    print('best classifier score on test set:', clf.score(self.X_test, self.y_test))
    df['best_score_test_set:'] = [clf.score(self.X_test, self.y_test)]
    df.to_csv(model_path[:-7] + '_grid_search_cv_results_best.csv', index=False)
    # Save and evaluate the best obtained model
    print('Writing evaluation result ...')
    clf = clf.best_estimator_
    joblib.dump(clf, model_path)
    self.evaluate_model(model=clf, model_path=model_path)
```

Fig. 4.13 Xtestability regression model with grid-search and cross-validation

```
def evaluate_model(self, model=None):
    y_true, y_pred = self.y_test, model.predict(self.X_test)
    df = pd.DataFrame()
    df['r2_score_uniform_average'] = [r2_score(y_true, y_pred,
                                      multioutput='uniform_average')]
    df['mean_absolute_error'] = [mean_absolute_error(y_true, y_pred)]
    df['mean_squared_error_MSE'] = [mean_squared_error(y_true, y_pred)]
    df['mean_squared_error_RMSE'] = [mean_squared_error(y_true, y_pred,
                                     squared=False)]
    df['median_absolute_error'] = [median_absolute_error(y_true, y_pred)]
    if min(y_pred) >= 0:
        df['mean_squared_log_error'] = [mean_squared_log_error(y_true, y_pred)]
    df.to_csv(model_path[:-7] + '_evaluation_metrics_R1.csv',
              index=True, index_label='Row')
```

Fig. 4.14 Evaluation of the learned model on test data

used in the grid search are defined in the 'parameters' dictionary defined in the *regress* method in Fig. 4.13.

The rest of the code uses grid search with cross-validation to find and fit the best testability prediction model. The model is then saved with *joblib.dump* command. The regress method calls the evaluate_model function to evaluate the performance, i.e., R^2-score, mean absolute error, mean squared error, and median absolute error, of the learned model on the test set. The resultant model is executed on the test set by calling the model's prediction method inside the body of the evaluate_model, shown in Fig. 4.14. The prediction result is stored in the y_pred list. By comparing the prediction results with the actual result, y_true, the machine learning model's performance is determined based on standard error metrics. The Scikit-learn library provides functions to compute error metrics.

4.5.4 Important Testability Metrics

One benefit of using machine learning models to predict testability is determining the most crucial source code metrics affecting testability. The permutation importance technique [6] can be applied to indicate the importance of each feature in the learned model. The permutation feature importance is defined as the decrease in a model score when a single feature value is randomly shuffled. This procedure breaks the relationship between the feature and the target. Thus the drop in the model score indicates how much the model depends on the feature. This technique benefits from being model agnostic and can be calculated many times with different feature permutations.

```
def compute_permutation_importance(self, model_path=None, model=None,
                                   n_repeats=10, scoring='r2', top_n=10):
    if model is None:
        model = load(model_path)
    result = permutation_importance(model, self.X_test, self.y_test,scoring=scoring,
                                    n_repeats=n_repeats,random_state=13,)
    perm_sorted_idx = result.importances_mean.argsort()
    result_top_features = result.importances[perm_sorted_idx].T
    print('Top ten metrics:\n', result_top_features[:, - top_n:])
    df1 = pd.DataFrame(data=result_top_features)
    df1.to_csv(r'results.csv')
```

Fig. 4.15 Computing the permutation importance

To compute permutation importance, we add a method compute_permutation_importance, shown in Fig. 4.15, to the Regression class. It takes as input an instance of the learned model or the path of the saved model, the number of repeats of the permutation process, n_repeats, the score used to compute the model's performance after permutations, scoring, and the number of top n important feature to be saved, top_n. The result is the sorted list of important top_n features, which is saved in a CSV file. We provide default values for n_repeats, scoring, and top_n. However, they can be set to other valid values. Specifically, n_repeats, can increase to improve the reliability of results.

The topmost source code metrics affecting testability are shown in Fig. 4.16. These metrics are found when executing the function compute_permutation_importance in Fig. 4.15.

Figure 4.17 shows the correlation between the testability values and the influential source code metrics. The value of source code metrics is normalized between 0 and 1 to enhance visualization. The correlation coefficient and p-value corresponding to each metric has been shown on each plot.

It is observed that most source code metrics (10 out of 15) negatively correlate with testability. In contrast, five metrics increase class testability, including the number of inherited, public, static, constructors, and print and return statements. As a general result, classes with few lines of code, few jump statements, and many visible methods are more testable. These results provide informative clues about the automatic modification of source code to enhance testability. It also confirms the results of previous measures proposed for testability, particularly the visibility measure.

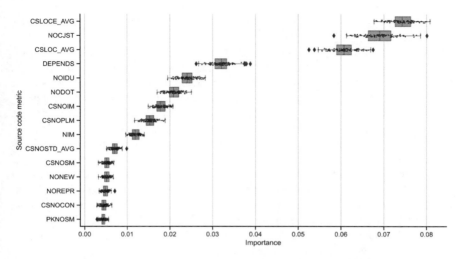

Fig. 4.16 Impacts of the elected source code metrics on testability prediction

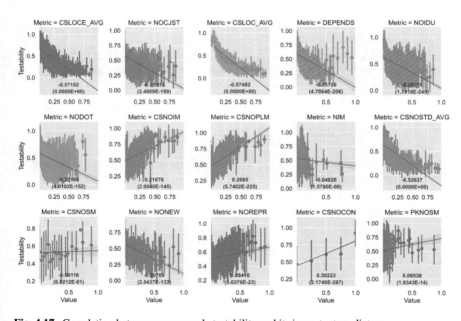

Fig. 4.17 Correlation between source code testability and its important predictors

4.5.5 Why Predicts Testability?

Predicting testability is significantly faster than actual testing, which generates and executes test cases. It is handy in search-based evolutionary refactorings, where numerous sequences of refactoring operations must be examined to find the best ones.

To clarify the impact of testability prediction on the efficiency of search-based refactoring, consider an ideal case in which it takes at least two minutes for EvoSuite to generate and evaluate a suitable test suite for a unit under test (UUT). A moderate program with 100 classes will take about 200 min (more than 3 h) to be tested. Each individual containing a sequence of refactoring operations needs to compute the testability in search-based refactoring. For 100 individuals in one generation, finishing one iteration takes about 300 h, making the search process practically impossible.

However, our testability prediction model could predict the testability (with an acceptable error) of all classes within a project in a few minutes. Testability prediction drastically improves the efficiency of search-based refactoring applied to improve testability. Previous studies on improving testability with automated refactoring have yielded inconclusive results [20, 21]. I believe it is because of the lack of an appropriate testability measurement model.

In summary, the testability model presented in this chapter provides the following benefits and applications:

1. To be used as an objective function in search-based refactoring and guide the process of refactoring towards improving testability.
2. To identify software components requiring more test effort, developers must focus on ensuring software quality and reliability early in the software development lifecycle (SDLC) stage.
3. To estimate the technical code debt relevant to testing activities in a meaningful and interpretable interval of [0, 1].
4. To plan testing activities and make optimal decisions for resource allocation in large software projects.
5. To measure the testability of a piece of code before it is completed or even compiled while developing software.
6. To indicate the most crucial source code metrics affecting testability.
7. To designate whether a class is ready to test and establish an improvement goal.

4.5.6   Summary

This chapter presents a new software development process called TsDD. The corn stone of TsDD is to conduct the agile software developers towards developing testable code. As the testability of code improves, the test effectiveness and efficiency increase. In a test-first method such as BDD or TDD, failing test results give the developer the inspiration of how to design and write the code to get the right results and pass the test. The emphasis is on passing the tests; once the code is correct in the next stage, the developer should optimize the code using different refactoring techniques. The point I am making is that the developers are better off writing clean and testable code rather than just code that passes tests created manually based on the requirements.

A new approach for testability measurement based on source code metrics using machine learning regression techniques is introduced in this chapter. The testability value computed by our approach is based on the actual test criteria, including statement coverage, branch coverage, mutation score, and test suite size. These criteria are available only after testing performed on the class under test, which presumably required considerable time and effort. Once we compute the testability values by actual testing for a large corpus of classes, we can use machine learning to learn to predict the value of testability based on source code metrics. As a result, testability can be measured before testing efficiently. Machine learning helps identify the importance of each source code metric in predicting testability. The complete source code of testability prediction can be found on GitHub [9].

4.6 Assignment

1. We trained and tested a multi-layer perceptron. However, there are various machine learning regressions algorithms. Find and use other machine learning regression techniques to predict testability. What is the best model? What are the best hyperparameters for this model?
2. The discussed testability prediction model predicts the testability of the class. Design and implement a method-level testability prediction.
3. The discussed testability prediction model predicts the testability of Java classes. Design and implement a testability prediction for Python and C# programs.
4. Use transfer learning to predict the testability of C# and Python programs based on a model trained in Java. Report the performance of the best model and discuss the benefits and pitfalls of transfer learning in testability prediction.
5. The power of machine learning models can be increased by improving the feature space. In our case, the feature space is software metrics. We can generate new features using different feature generation techniques [15, 16]. For instance, we may generate new metrics by applying statistical operations like mean, max, min, and standard deviation on method-level metrics, e.g., cyclomatic complexity. Create at least 30 new metrics and retain the testability prediction model with enhanced feature space. Compare the result with the previous models.
6. The performance of machine learning models is improved by increasing the trainset size. Use EvoSuite to generate new tests for new Java projects. Increase the size of the dataset to at least 50,000 classes. A good source for open-source Java projects is Qualitas Corpus [11]. Also, you can find a larger corpus of Java projects in [9].
7. We used permutation importance to rank source code metrics according to their importance and determine the most influential source code metrics in testability prediction. Numerous feature importance analysis techniques, such as the

impurity-based importance, are used in tree-based learning algorithms like the Decision Tree. Find and experiment with at least two other feature importance analyses to determine the most important source code metrics. Compare the result with metrics obtained by permutation importance analysis.

8. We used software metrics to vectorize source code. Other source code vectorization techniques, such as Code2Vec and Code2Seq, have been popularized recently. Use Code2Vec and Cod2Seq approaches to predict testability. Compare the results with the metric-based approach.

9. We used EvoSuite to compute the actual testability value for many projects to construct our dataset. Use other Java automatic test data generation tools, such as Randoop and JDart, to compute the actual value of testability. Build prediction models on the new datasets. Compare the performance of models for each approach.

10. We selected and applied refactoring operations manually in TsDD. Use the search-based refactoring techniques to find sequences of refactoring operations, maximizing the program's testability. You may implement your approach using a genetic algorithm with a testability prediction model as an objective. Execute your tool on different projects. Compare the best sequences of refactoring which each other. Is there any commonality between the best refactoring sequences of different projects?

11. Develop a simple software project with three development approaches described in Chaps. 2, 3, and this chapter: TDD, BDD, and TsDD. Measure some parameters, including development time, testing time, number of applied refactoring, and testability. Compare the results of different development approaches. What are the advantages and disadvantages of each approach?

12. The general idea of using code coverage as a concrete proxy for testability can be employed to measure other software quality attributes. For instance, the reuse rate, i.e., the number of reuses of a software component such as classes and packages, can be used as a concrete proxy to measure reusability. Propose a reusability prediction approach like testability prediction to measure software reusability.

References

1. ISO and IEC.: ISO/IEC 25010:2011 systems and software engineering—Systems and software quality requirements and evaluation (SQuaRE)—System and software quality models, p. 34. ISO (2011). https://www.iso.org/standard/35733.html
2. White, M., Tufano, M., Martinez, M., Monperrus, M., Poshyvanyk, D.: Sorting and transforming program repair ingredients via deep learning code similarities. In: 2019 IEEE 26th International Conference on Software Analysis, Evolution, and Reengineering (SANER), pp. 479–490. IEEE (2019)
3. Ammann, P., Offutt, J.: Introduction to Software Testing. Cambridge University Press, Cambridge (2016). https://doi.org/10.1017/9781316771273

4. Fraser, G., Arcuri, A.: A large-scale evaluation of automated unit test generation using EvoSuite. ACM Trans. Softw. Eng. Methodol. 24(2), 1–42 (2014). https://doi.org/10.1145/2685612; [20] SciTools.: Understand. https://scitools.com/. Accessed 11 Sep. 2020

5. Khanam, Z., Ahsan, M.N.: Evaluating the effectiveness of test-driven development: advantages and pitfalls. Int. J. Appl. Eng. Res. 12(18), 7705–7716 (2017)

6. Mkaouer, M.W., Kessentini, M., Bechikh, S., Cinnéide, M.Ó., Deb, K.: On the use of many quality attributes for software refactoring: a many-objective search-based software engineering approach. Empir. Softw. Eng. 21(6), 2503–2545 (2016). https://doi.org/10.1007/s10664-015-9414-4

7. Bibiano, A.C., et al.: A quantitative study on characteristics and effect of batch refactoring on code smells. In: 2019 ACM/IEEE International Symposium on Empirical Software Engineering and Measurement (ESEM), pp. 1–11. https://doi.org/10.1109/ESEM.2019.8870183

8. Zakeri-Nasrabadi, M., Parsa, S.: ADAFEST (2021). https://github.com/m-zakeri/ADAFEST. Accessed 26 Sep. 2021

9. Fowler, M., Beck, K.: Refactoring: improving the Design of Existing Code, 2nd edn. Addison-Wesley (2018). https://refactoring.com/

10. Tsantalis, N., Chaikalis, T., Chatzigeorgiou, A.: Ten years of JDeodorant: lessons learned from the hunt for smells. In: 2018 IEEE 25th International Conference on Software Analysis, Evolution, and Reengineering (SANER), pp. 4–14 (2018). https://doi.org/10.1109/SANER.2018.8330192

11. Baresel, A., Steamer, H.: Evolutionary testing of flag conditions. In: Lecture Notes in Computer Science (2003)

12. Long, F., et al.: Automatic patch generation via learning from successful human patches. Ph.D. dissertation, Massachusetts Institute of Technology (2018)

13. König, H., Heiner, M., Wolisz, A., (eds.): Numerical coverage estimation for the symbolic simulation of real-time systems. In: FORTE 2003. IFIP International Federation for Information Processing 2003, vol. 2767, pp. 160–176. LNCS (2003)

14. Zakeri Nasrabadi, M., Parsa, S.: An ensemble meta-estimator to predict source code testability. Appl. Soft Comput. J. (2022)

15. Zakeri-Nasrabadi, M., Parsa, S.: Learning to predict test effectiveness. Int. J. Intell. Syst. 37(8), 4363–4392 (2022)

16. Zakeri Nasrabadi, M., Parsa, S.: Testability prediction dataset (2021). https://zenodo.org/record/4650228

17. Badri, M., Badri, L., Hachemane, O., Ouellet, A.: Measuring the effect of clone refactoring on the size of unit test cases in object-oriented software: an empirical study. Innov. Syst. Softw. Eng. 15(2), 117–137 (2019)

18. Fraser, G., Arcuri, A.: EvoSuite: automatic test suite generation for object-oriented software. In: Proceedings of the 19th ACM SIGSOFT Symposium and the 13th European Conference on Foundations of Software Engineering—SIGSOFT/FSE '11, p. 416 (2011). https://doi.org/10.1145/2025113.2025179

19. Pedregosa, F., et al.: Scikit-learn: machine learning in python. J. Mach. Learn. Res. 12, 2825–2830 (2011)

20. Alshayeb, M.: Empirical investigation of refactoring effect on software quality. Inf. Softw. Technol. 51(9), 1319–1326 (2009)

21. Cinnéide, M.Ó., Boyle, D., Moghadam, I.H.: Automated refactoring for testability. In: 2011 IEEE Fourth International Conference on Software Testing, Verification and Validation Workshops, pp. 437–443 (2011)

Chapter 5
Automatic Refactoring

5.1 Introduction

Test cases are developed before writing the code in test-first techniques, such as TDD and BDD. Tests drive the design and development of software in such techniques. Once the code is correct, refactoring techniques are applied to optimize the code.

Refactoring is a well-known technique to improve the quality of software, in particular, the quality of source code. As a part of code refactoring, the code is rearranged in a series of more minor semantically-preserving transformations (i.e., it keeps working) to enhance its quality.

However, manual refactoring could be error-prone and time-consuming. That is why automatic refactoring tools are considered the most suitable way to manage the complexity that arises as a software project evolves.

Refactoring originates from William F. Opdyke's Ph.D. thesis [1], but it has already gained popularity with Martin Fowler's book "Refactoring: Improving the Design of Existing Code" [2]. Apparently, in the first position, the reader unfamiliar with refactoring should get insight into what refactoring is. Section 5.2 describes a very simple refactoring, encapsulated field. In this section, the intention is to describe the importance of using automated refactoring tools.

According to Fowler: "To do refactoring properly, the tool has to operate on the syntax tree of the code, not on the text. All in all, implementing decent refactoring is a challenging programming exercise—one that I'm mostly unaware of as I gaily use the tools". A parse tree, representing the syntactical structure of a source code, can be derived from the code using a compiler generator. ANTLR provides a tree walker to traverse the parse trees. The tree walker invokes functions to operate on the parse tree while traversing the tree created for a given source code. The use of ANTLR for source code analysis is described in Sect. 5.3.

In summary, this chapter aims to show the reader how to develop software tools to refactor a given source code. Throughout this chapter, the reader will gain insights into developing software tools to automate refactoring techniques.

© The Author(s), under exclusive license to Springer Nature Switzerland AG 2023   191
S. Parsa, *Software Testing Automation*,
https://doi.org/10.1007/978-3-031-22057-9_5

5.2   Encapsulate Field Refactoring

Good programming practice suggests that class attributes should be private; otherwise, the encapsulation principle will be violated. Encapsulation ensures that "sensitive" data hides from users. To achieve this, declare class variables/attributes as private (cannot be accessed outside the class). Encapsulate field refactoring makes a private field accessible only through a getter or setter method. For instance, consider the following class:

<u>Before refactoring</u>
class Person {
    public String name;
}

The following steps are typically taken to refactor a class to Encapsulate field:

1. Create a getter and setter for the public field.
2. Find all invocations of the public field.
3. Replace receipt of the field value with the getter, and replace the setting of the new field values with the setter.
4. After all field invocations have been replaced, make the field private.

After applying the above steps, the Person class is refactored as follows:

```
After refactoring
class Person {
        private String name;
        public String getName() {
                return name;
        }
    public void setName(String arg) {
                name = arg;
        }
    }
```

It is not a big deal to refactor the class Person manually. However, in the case of a big project including many classes with public attributes, it takes relatively long to refactor all the classes and change any direct access to the public attributes with a call to the getter and setter methods. Section 5.4 shows how to develop a refactoring tool to refactor to Encapsulate Field automatically [3].

5.3 Automatic Refactoring

Any changes, including refactoring, to particularly large-scale codes, can be problematic. Care must be taken to maintain the structural and semantics constraints of the code while refactoring it. Test-first methods attempt to get the code right and then focus on optimizing the correct code by refactoring it. For instance, TDD and BDD emphasize creating code during three stages in the term red–green–blue. The question is how to refactor the spaghetti code that is not clean, and the only goal of the developer has been to get it right. In this chapter, solutions based on the structural analysis of the code using compiler-assisted techniques are described.

The ANTLR 4.8 parser generator enables the analysis and rewriting of Java source code while walking through the parse tree [4]. Before getting started with using ANTLR to develop different refactoring tools, it is necessary to go through the following steps:

1. Get a brief insight into ANTLR,
2. Install Java compiler and ANTLR,
3. Prepare the Java language grammar,
4. Use Listener to implement encapsulate field refactoring,
5. Override the resulting listener class for refactoring.

5.3.1   The ANTLR Parser Generator

ANTLR, Another Tool for Language Recognition, is a language tool that provides a framework for constructing recognizers, compilers, and translators from grammatical descriptions [5]. As shown in Fig. 5.1, ANTLR takes a grammar (.g4) file as input and generates a lexer and parser for the grammar as output.

The ANTLR tool generates a top-down parser from the grammar rules defined with the ANTLR meta-grammar [3]. ANTLR is an open-source parser generator written in Java. The initial version of ANTLR generated the target parser source code in Java. In the current version (version 4), the parser source code can be in various programming languages listed on the ANTLR official website [6]. It provides a "tree walker" to traverse parse trees, shown in (Fig. 5.2).

As shown in Fig. 5.2, ANTLR Provides two ways of traversing the parse tree in its runtime library called listener and visitor. The *listener* walker traverses the parse tree generated for a given source code in depth-first order. The generated class, GrListener, is unique for each grammar, Gr.g4, with 'enter' and 'exit' methods for each rule. The walker invokes the enter method, enterR(), for a rule R from the grammar when it encounters the node for the rule. As the walker visits all children of the rule, R, it triggers the exit method, exitR, for that rule.

Like the listener, the visitor walker triggers the 'enter' and 'exit' methods as it visits the parse tree nodes. However, in the case of the visitor, the triggered methods determine the tree traversal order.

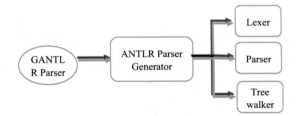

Fig. 5.1   ANTLR parser generator main activities

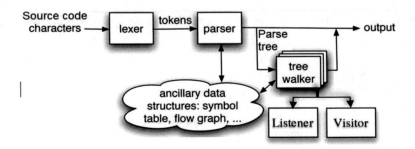

Fig. 5.2   The parser invokes the lexer to parse a given source code

Fig. 5.3 Assignment
statements grammar in
ANTLR format

```
grammar AssignmentStatement1;
start: prog EOF ;
prog: prog assign | assign;
assign: Id ':=' expr (NEWLINE | EOF);
expr:  expr '+' term
   | expr '-' term
   | term
   ;
term:  term '*' primary
   | term '/' primary
   | primary
   ;
primary: factor '^' primary
     |  factor
     ;
factor:
   '(' expr ')'
   | Id
   | number
   | String
   ;
number  : INT | FLOAT;

/* Lexical Rules */
Id    : LETTER(LETTER|DIGIT)*; //identifiers
INT   : DIGIT+;                // integer numbers
FLOAT:                         // Floating points
   DIGIT+ '.' DIGIT*           // e.g. 07.666
   |'.' DIGIT+ ;               // e.g. 00.666
String :
      '"' (ESC|.)*?'"';        //e.g. "abab"
fragment
      DIGIT: [0-9];            //e.g. 123
fragment
      LETTER: [a-zA-Z];        //
fragment
      ESC : '\\"' | '\\\\' ;
WS: [ \t\r]+ -> skip ;         //white spaces
NEWLINE: '\n' ;
RELOP: '<=' | '<' ;
```

Example Figure 5.3 represents a grammar for assignment statements, defined in
ANTLR syntax. The grammar rules allow writing one or more assignment statements
separated with the newline character and ending with the end of the file marker.

As shown in Fig. 5.3, the grammar is named AssignmentStatement1. Hence, by
the ANTLR conventions, the grammar should be saved in a file named Assign-
mentStatement1.g4. The lexer rule names start with a lower-case letter, while the
parser rule names start with an upper-case letter.

Antlr allows testing the grammar rules by generating a parse tree representing
the syntactic structure of a string (program) according to the grammar. The program

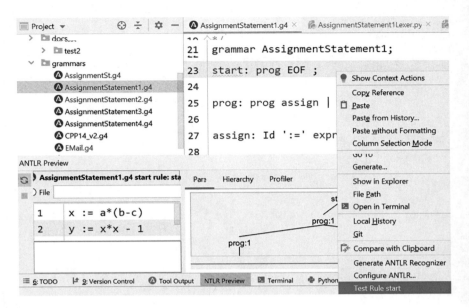

Fig. 5.4 Testing the grammar rules in PyCharm

can be written using the text editor located on the left side of the ANTLR preview window. The corresponding Parse Tree will be generated on the right-hand side of the ANTLR Preview window. The parse tree is generated by selecting the rule in the PyCharm edit pane, right-clicking on the rule of the grammar, and selecting the "Test Rule" option. For instance, in Fig. 5.4, the start rule is selected to depict the parse tree for the two assignment statements shown in the preview window.

In Fig. 5.4, the "start" grammar rule is selected. The parse tree generated by ANTLR is shown in Fig. 5.5.

The ANTLR recognizer tool generates a lexer and a parser for a given grammar. Figure 5.6 shows how to use the ANTLR recognizer in PyCharm [7] to generate a lexer and parser for the assignment statements grammar in Fig. 5.3.

The ANTLR recognizer takes as input the AssignStatement1 grammar and generates the following Python files:

AssignStatement1Lexer.py
AssignStatement1Parser.py
AssignStatement1Listener.py
AssignStatement1Visitor.py

In this chapter, the listener class generated by ANTLR is overridden to look for refactoring opportunities and apply the refactorings accordingly. As shown in Fig. 5.7, the listener class generated for AssignmentStatemen1 includes an enter and an exit method for each rule in the grammar.

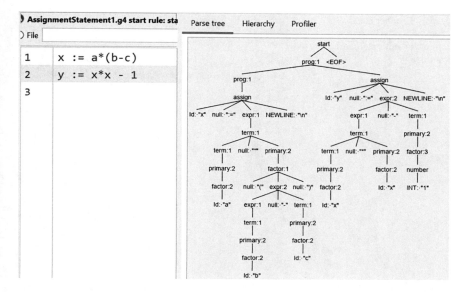

Fig. 5.5 The parse tree generated by ANTLR

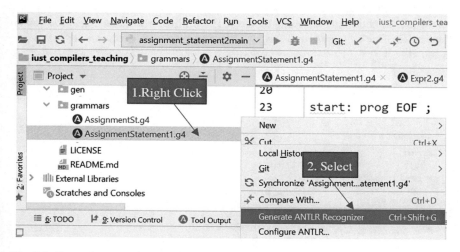

Fig. 5.6 Generating recognizer in Pycharm environment

5.3.2 Installing the ANTLR Parser Generator

ANTLR is written in Java, so Java must be installed even if it is going to be used with, say, Python. ANTLR requires a Java version of 1.6 or higher. The ANTLR plugin antlr-4.8-complete.jar (or whatever version) is available from:

Fig. 5.7 Listener class generated by PyCharm for the AssignmentStatement1 Grammar

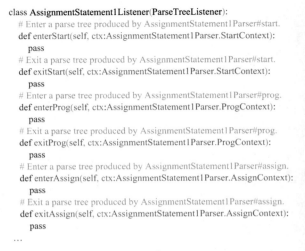

```
class AssignmentStatement1Listener(ParseTreeListener):
    # Enter a parse tree produced by AssignmentStatement1Parser#start.
    def enterStart(self, ctx:AssignmentStatement1Parser.StartContext):
        pass
    # Exit a parse tree produced by AssignmentStatement1Parser#start.
    def exitStart(self, ctx:AssignmentStatement1Parser.StartContext):
        pass
    # Enter a parse tree produced by AssignmentStatement1Parser#prog.
    def enterProg(self, ctx:AssignmentStatement1Parser.ProgContext):
        pass
    # Exit a parse tree produced by AssignmentStatement1Parser#prog.
    def exitProg(self, ctx:AssignmentStatement1Parser.ProgContext):
        pass
    # Enter a parse tree produced by AssignmentStatement1Parser#assign.
    def enterAssign(self, ctx:AssignmentStatement1Parser.AssignContext):
        pass
    # Exit a parse tree produced by AssignmentStatement1Parser#assign.
    def exitAssign(self, ctx:AssignmentStatement1Parser.AssignContext):
        pass
    ...
```

https://www.antlr.org/download/

Open a new project in the PyCharm environment and press CNTL + ALT + S to open the PyCharm setting window. As a result, the setting window opens. Click on the plugin option in the setting window. The following window appears; click on the Marketplace tab and search for ANTLR. Press the install button after the plugin is found.

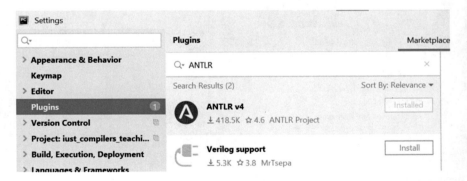

After installing the plugin, install the ANTLR4 library for Python 3 via the PyCharm IDE as follows:

(a) Open Settings
(b) On the left, find the current project, and under the project name, choose the Python Interpreter tab and click on the + icon on the left side of the window. Select the antlr4-python3-runtime package and install it.

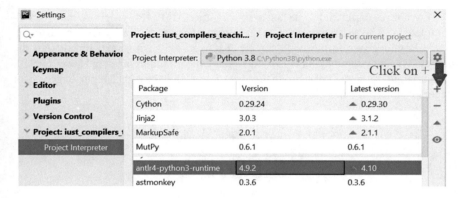

In summary, to install ANTLR in PyCharm, do as follows:

- File ⇒ Setting ⇒ Plugin ⇒ ↻Search for(ANTLR) ⇒ ⊚ Install
- File ⇒ Setting ⇒ Project ⇒ Project Interpreter ⇒ + (add) ⇒ ↻Search for(ANTLR) ⇒ ⊚ Install

As mentioned earlier, the grammar specification written in the ANTLR meta-grammar is required to generate a parser for a programming language. ANTLR grammar files are named with the ".g4" suffix. In this chapter, the Java 8 grammar is used. The grammar can be downloaded from ANTLR 4 grammar repository on:

GitHub: https://github.com/antlr/grammars-v4

Once the ANTLR tool is installed and the grammar files are prepared, lexer, parser, and listener class can be generated as follows:

```
> java -Xmx500M -cp antlr-4.9.3-complete.jar org.antlr.v4.Tool
                -Dlanguage=Python3 -o . JavaLabeledLexer.g4
> java -Xmx500M -cp antlr-4.9.3-complete.jar org.antlr.v4.Tool
                -Dlanguage=Python3 -visitor -listener -o . JavaLabeled.g4
```

The first command generates the lexer from the JavaLabeledLexer.g4 description file and the second command generates the parser from the JavaLabeledParser.g4 description file. It is worth noting that if the lexical rules are copied after the grammatical rules in one file, then a single command will generate all required codes in one step.

200 5 Automatic Refactoring

The above commands use Java to run the ANTLR4 program, antlr-4.9.3-complete.jar. The parameter -Dlanguage denotes the programming language in which ANTLR will generate the code for the lexer, parser, visitor, and listener. In the commands, the target language is Python3, and the grammar is named JavaLabeledGrammar.g4. As a result of running the above command, the ANTLR parser generator generates the following files:

JavaLabeledLexer.interp
JavaLabeledLexer.py
JavaLabeledLexer.tokens
JavaLabeledParser.interp
JavaLabeledParser.py
JavaLabeledParser.tokens
JavaLabeledParserListener.py
JavaLabeledParserVisitor.py

5.3.3 Preparing the Grammar

ANTLR provides the lexical rules and grammar for all the well-known languages, such as C#, C++, Java, and Python. Grammars can be downloaded from the following address:

https://github.com/antlr/grammars-v4

For each programming language, there are two files, one including the lexical rules and the other one including the language grammar. The following minor modifications are applied to the grammar to prepare it for efficient use:

1. Combine the lexical and grammatical rules
2. Label rule alternatives
3. Keep white spaces

1. **Combine rules**

The lexical and grammar rules on this site are in two different files. For instance, there are two files, Java8Lexer.g4, and Java8Parser.g4, for Java version 8. The file's contents, including lexical rules, may be either copied after the grammatical rules in the combined file or a link to the Java8Lexer.g4 file can be inserted into the Java8Parser.g4 file.

parser grammar JavaParserLabeled;
options { tokenVocab=JavaLexer; }

2. Label rule alternatives

Rule alternatives in grammar are better off being labeled; otherwise, ANTLR generates only one listener and one visitor method per rule. Labels appear on the right edge of alternatives, starting with the # symbol in the grammar. For instance, consider the following rule for expressions:

```
expr:   expr '+' term
    |  expr '-' term
    |  term
    ;
```

The following listener methods are generated for the above rule:

```
# Enter a parse tree produced by AssignmentStatement1Parser#expr.
def enterExpr(self, ctx:AssignmentStatement1Parser.ExprContext):
    pass
# Exit a parse tree produced by AssignmentStatement1Parser#expr.
def exitExpr(self, ctx:AssignmentStatement1Parser.ExprContext):
    pass
```

The grammar for expressions is labeled as follows;

```
expr :                          // Labels
    expr '+' term                   #expr_term_plus
    | expr '-' term                 #expr_term_minus
    | expr RELOP term               #expr_term_relop
    | term                          #term4
    ;
```

A label is, in a sense, a name given to an alternative rule. As a result, it is observed that distinct functions are generated for each labeled rule as follows:

```
# Enter a parse tree produced by AssignmentStatement2Parser#expr_term_minus.
def enterExpr_term_minus(self, ctx:AssignmentStatement2Parser.Expr_term_mi-
nusContext):
    pass
# Exit a parse tree produced by AssignmentStatement2Parser#expr_term_minus.
def exitExpr_term_minus(self, ctx:AssignmentStatement2Parser.Expr_term_mi-
nusContext):
    pass
# Enter a parse tree produced by AssignmentStatement2Parser#expr_term_plus.
def enterExpr_term_plus(self, ctx:AssignmentStatement2Par-
ser.Expr_term_plusContext):
    pass
# Exit a parse tree produced by AssignmentStatement2Parser#expr_term_plus.
def exitExpr_term_plus(self, ctx:AssignmentStatement2Par-
ser.Expr_term_plusContext):
    pass
# Enter a parse tree produced by AssignmentStatement2Parser#term4.
def enterTerm4(self, ctx:AssignmentStatement2Parser.Term4Context):
    pass
# Exit a parse tree produced by AssignmentStatement2Parser#term4.
def exitTerm4(self, ctx:AssignmentStatement2Parser.Term4Context):
    pass
```

3. **Keep white spaces**

Parsers do not use white spaces (as token separators) and comments. As a result, the white spaces and comment definitions are followed by "->skip" in the lexical rules, which means the ANTLR lexer ignores them. For instance, consider the definition of white spaces and comments in Java 8 grammar provided by the ANTLR:

```
// Whitespace and comments in Java8Lexer.g4
WS  : [ \t\r\n\u000C]+ -> skip
    ;
COMMENT
    : '/*' .*? '*/' -> skip
    ;
LINE_COMMENT
    : '//' ~[\r\n]* -> skip
    ;
```

However, when refactoring a source code, the white spaces are required to rewrite the parts of the code that are not altered. The secret to preserving but ignoring comments and whitespaces is to send them to the parser on a "hidden channel." Therefore, the lexical rules should be modified as follows:

```
// Whitespace and comments in Java8Lexer.g4
WS  : [ \t\r\n\u000C]+ -> channel (hidden)
    ;
COMMENT
    : '/*' .*? '*/' -> channel (hidden)
    ;
LINE_COMMENT
    : '//' ~[\r\n]* -> channel (hidden)
    ;
```

ANTLR has two channels called default token channel and hidden. The default token channel is the channel usually used by the lexer, and the hidden channel keeps comments and white space characters.

5.4 Implementing Encapsulate Field Refactoring[1]

The encapsulate field refactoring is described in Sect. 5.2. The refactoring tool is supposed to refactor Java classes. Therefore, Java 9 grammar should be used. As described in Sect. 5.3.3, first of all, the grammar should be labeled and get ready for use. Then the ANTLR recognizer in PyCharm is executed. Figure 5.8 shows how to generate the ANTLR recognizer for Java9_v2.g4. As a result, the following Python files are generated:

1. Java9_v2Lexer.py
2. Java9_v2Parser.py
3. Java9_v2Listener.py
4. Java9_v2Visitor.py

The three Python files include Python's lexer, parser, listener, and visitor classes. It is possible to configure the ANTLR recognizer to generate these classes in any other languages, such as C#, C++, or Java, via the "Configure ANTLR" option in Fig. 5.8.

5.4.1 The Main Function

Figure 5.9 represents the software tool's main function for encapsulating field refactoring.

---

[1] Download the source code form: https://doi.org/10.6084/m9.figshare.21270120

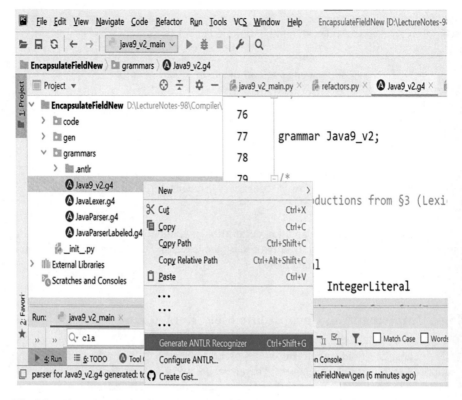

Fig. 5.8 Using ANTLR recognizer to generate lexer.py, parser.py and listener.py

As shown in Fig. 5.9, the main function accepts a Java file, sample.java, as input and refactors it based on the encapsulate field strategy. Step 1 of the main() function moves the input file sample.java into a buffer stream. In Step 2, the ANTLR-generated lexer, Java9_v2Lexer, performs lexical analysis on the character input stream (the source code file) and creates a stream of tokens to be passed onto the parser for syntactic and semantic analysis. Essentially, the lexer, Java9_v2Lexer, reads from the file and matches text patterns according to the rules of the lexer in Java9_v2.g4. As a result, a CommonTokenStream is created and passed to the ANTLR-generated parser, Java9_v2Parser. In Step 4 of the main function in Fig. 5.9, the ANTLR-generated parser moves along the token stream, both syntactic and semantic analysis.

CompilationUnit is the starting symbol in ANTLR grammar. The root of the parse tree always addresses the starting symbol. In step 6, the parser.CompilationUnit() method call returns the root address of the parse tree built for the input class. In Step 6, as the tree_walker traverses the execution tree parse tree, the listener methods are invoked to apply the encapsulate field refactoring to the input class. Here, the input to the refactoring tool is the source file name, e.g., sample.java, at line 33 of the main function in Fig. 5.9.

```
1.  from antlr4 import *
2.  from code.java9_v2.gen.Java9_v2Lexer import Java9_v2Lexer
3.  from code.java9_v2.gen.Java9_v2Parser import Java9_v2Parser
4.  from code.java9_v2.refactors import EncapsulateFiledRefactoringListener
5.  import argparse
6.  def main(args):
7.      # Step 1: Load input source into a stream
8.      stream = FileStream(args.file, encoding='utf8')
9.      # Step 2: Create an instance of AssignmentStLexer
10.     lexer = Java9_v2Lexer(stream)
11.     # Step 3: Convert the input source into a list of tokens
12.     token_stream = CommonTokenStream(lexer)
13.     # Step 4: Create an instance of the AssignmentStParser
14.     parser = Java9_v2Parser(token_stream)
15.     parser.getTokenStream()
16.     # Step 5: Create a parse tree
17.     parse_tree = parser.compilationUnit()
18.     # Step 6: Create an instance of AssignmentStListener
19.     my_listener = EncapsulateFiledRefactoringListener(
20.         common_token_stream=token_stream)
21.     walker = ParseTreeWalker()
22.     walker.walk(t=parse_tree, listener=my_listener)
23.     print('Compiler result:')
24.     print(my_listener.token_stream_rewriter.getDefaultText())
25.     refactored_filename = args.file.split(".")        #e.g. refactored = smple.java
26.     refactored_filename.insert(1, "refactored")    #e.g. filename = B.refactored
27.     refactored_filename = ".".join(refactored_filename)        # B.refactored.Java
28.     with open(refactored_filename, mode='w', newline='') as f:
29.         f.write(my_listener.token_stream_rewriter.getDefaultText())
30.  if __name__ == '__main__':
31.     argparser = argparse.ArgumentParser()
32.     argparser.add_argument(
33.         '-n', '--file', help='Input source', default=r'sample.java')
34.     args = argparser.parse_args()
35.     main(args)
```

Fig. 5.9 The main function of the encapsulate field refactoring

5.4.2   Using the Listener Class

Refactoring is restructuring (rearranging) code in a series of small, semantics-preserving transformations to improve the code quality. In order to perform the transformations, use the ANTLR visitor/listener to walk through the tree and make the necessary changes with a TokenStreamRewriter.

TokenStreamRewriter is an ANTLR class used to add text after specific tokens. Before an instance of the TokenStreamRewiter class can be created, the token stream buffer should be filled with the tokens:

```
CharStream input = ANTLRFileStream("inputSourceFile.java");
lexer = Java9_v2Lexer(input);
token_stream = new CommonTokenStream(lexer);
token_stream_rewriter = TokenStreamRewriter(token_stream);
```

As described, ANTLR generates a listener class, e.g., Java9_v2Listener, for a given grammar. The main loop of the program starts at line 22 of the main function in Fig. 5.9:

```
21.  walker = ParseTreeWalker()
22.  walker.walk(t=parse_tree, listener=my_listener)
```

The tree walker method, walker.walk, takes as input the root address of the parse tree, pars_tree, and an instance, my_listener, of the listener class built by the user. While traversing the tree, the walker invokes methods of the my_listener object derived from the ANTLR-generated listener class. As the walker visits, a node labeled R invokes the method enterR() of the class my_listener. After visiting all the children of R, the walker invokes the methods exitR() of my_listener. The listener class for developing the refactoring tool is described in the next section.

5.4.3 Encapsulate Field Refactoring Listener

As described in Sect. 5.2, encapsulate field refactoring replaces all references to a public field with accesses through setter and getter methods. This refactoring takes as input a class name, finds a public field's name to be encapsulated, and performs the following transformations:

- It creates a public-getter method that returns the field's value,
- Creates a public setter method that updates the field's value to a given parameter's value,
- Replaces the field reads and writes with calls to the getter and setter methods, respectively.
- Changes the field's access modifier from 'public' to 'private.'

A listener class, EncapsulateFiledRefactoringListener, implements the above transformations to perform the encapsulate field refactoring automatically. The constructor of this class is shown in the following. The class takes an instance of CommonTokenStream class, the source class name. The parameter is used to initialize an instance of TokenStreamRewriter class which provides a set of methods to manipulate the ANTLR-generated parse tree. The second and third parameters specify the entity to be refactored.

Below is a simple listener class's initializer (constructor) to apply the encapsulate field refactoring to a given Java class. The initializer creates an object token_stream_rewriter to modify the Java class.

```
class EncapsulateFiledRefactoringListener(Java9_v2Listener):
    """
    To implement the encapsulate filed refactoring.
    Encapsulate field: Make a public field private and provide accessors.
    """

    def __init__(self, common_token_stream: CommonTokenStream = None,
                 field_identifier: str = None):
        """
        :param common_token_stream:
                CommonTokenStream is a stream of tokens generated by parsing the
                main file using the ANTLR generator.
        """

        if common_token_stream is None:
            raise TypeError('common_token_stream is None')
        # Move all the tokens in the source code to a buffer, token_stream_rewriter.
        self.token_stream = common_token_stream
        self.token_stream_rewriter = TokenStreamRewriter(common_token_stream)
        self.properties = []
```

The class Java9_v2Listener includes empty method, enterR() and exitR(), for each rule, R, in the Java 9 grammar. The question is which class methods to override in the EncapsulateFiledRefactoringListener, derived Java9_v2Listener class, to apply to encapsulate field refactoring to a given class automatically.

5.4.4   Overriding the Listener Methods

Encapsulated field refactoring deals with a class and its attributes. Hence, the first thing to do is to look for private fields in a given class. The parse tree helps to find out how to track the rules to reach the target node. Therefore, as shown in Fig. 5.10, the normalClassDeclaration rule of the Java grammar, Java9_v2, is selected. In the PyCharm edit pane, Right-click on the normalClassDeclaration rule of the grammar and select the "Test Rule normalClassDeclaration" option. The previewer panel opens at the bottom.

On the left is a big text box where some sample code can be typed. On the right is a visualization of a parse tree. Beneath the tree, a slider can zoom in and out. Below is a sample function and its corresponding parse tree generated in the PyCharm environment.

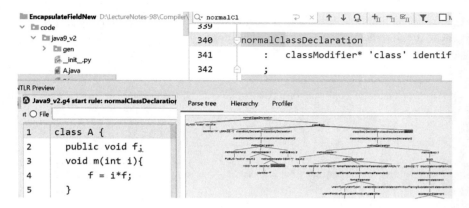

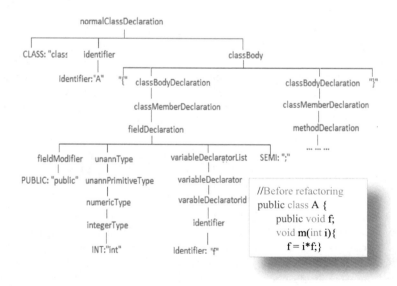

Fig. 5.10 Using the ANTLR previewer panel to generate a parse tree for a given class

Unrelevant parts of the tree are removed in Fig. 5.11 to provide a clear image of the parse tree. The question is how to find the methods to be overridden and how to override the methods. The immediate answer is that look for the rules defining the syntactic structure of a class.

Fig. 5.11 Parse tree to be traced to find the sequence of method calls to reach the desired node

```
normalClassDeclaration
  : classModifier* 'class' identifier typeParameters? superclass?
    superinterfaces? classBody
  ;
classModifier
  : annotation  | 'public'  | 'protected'  | 'private'  | 'abstract'
  | 'static' | 'final' | 'strictfp'
  ;
classBodyDeclaration
  : classMemberDeclaration #classBodyDeclaration1
  | instanceInitializer #classBodyDeclaration2
  | staticInitializer #classBodyDeclaration3
  | constructorDeclaration #classBodyDeclaration4
  ;
classMemberDeclaration
  : fieldDeclaration #classMemberDeclaration1
  | methodDeclaration #classMemberDeclaration2
  | classDeclaration #classMemberDeclaration3
  | interfaceDeclaration #classMemberDeclaration4
  | ';' #classMemberDeclaration5
  ;
fieldDeclaration
  : fieldModifier* unannType variableDeclaratorList ';'
  ;
fieldModifier
  : annotation | 'public' | 'protected' | 'private' | 'static' | 'final' | 'transient'
  | 'volatile'
  ;
```

Therefore, considering the grammar rules, the enterNormalClassDeclaration()
method should be modified to detect a new class in the given program. After finding
the class name in the next step, the fiedDeclaration methods should be overridden
to find public fields. The problem is that there are too many details in the grammar
rules. It is better to use the parse tree, as shown in Fig. 5.11.

Below is the exitFieldDeclaration method, copied from Java9_v2Listener, generated by the ANTLR recognizer:

```python
# Exit a parse tree produced by Java9_v2Parser#fieldDeclaration.
def exitFieldDeclaration(self, ctx:Java9_v2Parser.FieldDeclarationContext):
    pass
```

The "ctx" parameter in the listener methods references the corresponding parse tree node. The ctx parameter can be used to navigate the tree nodes from the current node by using ctx.getChild() or the children's names. The class field names are accessed by walking down the subtree rooted at fieldDeclaration in Fig. 5.11. The following sequence of listener functions are invoked while walking down the subtree to reach the field names:

```
Field_name = ctx.variableDeclrationList()
                .variableDeclator()
                .variableDeclaratorid()
                .identifier.getText()
```

Since there are no alternatives in the path from the variableDeclarationList to the Idebtifuer node, the call sequence can be shortened as follows:

```
ctx.variableDeclrationList().getText()
```

Listed below is the exitFieldDeclaration method of the EncapsulateFiledRefactoringListener class derived from the ANTLR generated class Java9_v2Listener. The exitFieldDeclaration method detects and converts public fields to private. The conversion is performed in four steps, separated by comments in the code.

```python
def exitFieldDeclaration(self, ctx: Java9_v2Parser.FieldDeclarationContext):
    # 1. flip public modifiers to private
    public_modifier = extract_public_modifier(ctx)
    if public_modifier is not None:
        replace_token(self.token_stream_rewriter, public_modifier, "private")
    # 2. generate accessor and mutator methods
    identifier = ctx.variableDeclaratorList().getText()
    type = ctx.unannType().getText()
    getter = build_getter_string(type, identifier)
    setter = build_setter_string(type, identifier)
    getter_setter_body = f"\n\n{getter}\n\n{setter}\n\n"
    # 3. Insert getter and setter methods body after the last token in fieldDeclaration
    self.token_stream_rewriter.insertAfter(ctx.stop.tokenIndex, getter_setter_body)
    #4. keep account of class properties to use in other trees
    self.properties.append(identifier)
```

The following method, extract_public_modifier, is invoked by exitFieldDeclaration() method. The method simply detects public fields:

ration() method. The method simply detects public fields:

```
def extract_public_modifier(ctx: Java9_v2Parser.FieldDeclarationContext) ->
Java9_v2Parser.FieldModifierContext:
    for modifier in ctx.fieldModifier():
        if modifier.getText() == "public":
            return modifier
```

The following methods build getter and setter methods for a given field, 'identifier' of the type, 'type.'

```
def build_getter_string(type: str, identifier: str) -> str:
    return dedent(f"""\
    |    public {type} get{identifier.capitalize()}() {{
    |        return this.{identifier};
    |    }}\
    """).replace("|", "")
def build_setter_string(type: str, identifier: str) -> str:
    return dedent(f"""\
    |    public void set{identifier.capitalize()}({type} {identifier}) {{
    |        this.{identifier} = {identifier};
    |    }}\
    """).replace("|", "")
```

The exitFieldDeclaration method walks through the subtree of the parse tree shown in Fig. 5.11. The method converts all the public fields of the class to private and creates setterF() and getterF() methods for each field 'f.' The methods are inserted after the last token in the field declaration. Figure 5.12 shows a class before and after refactoring.

As shown in Fig. 5.12, any access to a public field, f, should be replaced with the call to the setter, setF(), and getter, getF(), methods. Any statement, 'f = exp;,' should be replaced with 'setF(exp);.' The following method overrides the exitAssignment method.

```
// Before refctoring              //After refactoring
public class A {                  public class A {
    public int f;                     private int f;
    void m(int i){                    pubic void setF(int f){
        f = i*f;}                         this.f = f;
                                      }
                                      pubic int getF(){
                                          return this.f;
                                      }
                                      void m(int i){
                                          setF(i*getF(f));
                                      }
```

Fig. 5.12 The input and output of the proposed encapsulate field refactoring tool

```
#assignment : leftHandSide assignmentOperator expression :
#assignmentOperator :  '=' | '*=' | '/=' | '%=' | '+=' | '-=' | '<<=' | '>>='
                       | '>>>='  | '&=' | '^=' | '|=' ;

def exitAssignment(self, ctx: Java9_v2Parser.AssignmentContext):
    left_identifier: str = ctx.leftHandSide().getText().split(".")[-1]
    # 1. Is the identifier at the left-hand side not converted to private?
    if left_identifier not in self.properties:
        return
    #2. get the expression on the right-hand side of the assignment statement
    expression = extract_exact_text(self.token_stream_rewriter, ctx.expression())
    #3. Replace 'f = exp' with setF(exp)
    code = "this.set{}({})".format(left_identifier.capitalize(), expression)
    replace_token(self.token_stream_rewriter, ctx, code)
```

It is a common practice for programming languages' grammar to restrit any read access to the value of a variable to regular expressions. Any access to the value of a public field, f, as a variable, should be replaced with getF(f) after replacing the field modifier with 'private.' The following method is overridden to modify any access to public fields listed in an array called 'property':

```
def exitPrimary(self, ctx: Java9_v2Parser.PrimaryContext):
    if ctx.getChildCount() == 2:
        if ctx.getText() == 'this.' + self.field_identifier or \
        ctx.getText() == self.field_identifier:
            new_code = 'this.get' + str.capitalize(self.field_identifier) + '()'
            self.token_stream_rewriter.replaceRange(ctx.start.tokenIndex, \
                                        ctx.stop.tokenIndex, new_code)
```

In addition to modifying the code, it is possible to modify the comments using the tokenStreamRewriter class. For instance, in the following code, all the hidden characters, such as white spaces, blanks, and comments appearing before the starting symbol, called CompilationUnit, are replaced with a hard-coded string defined in the code. The starting symbol of the ANTLR grammar is always assumed to be CompilationUnit. Therefore, the first method invoked by the ANTLR tree walker will be enterCompilationInit.

```
def enterCompilationUnit1(self, ctx: Java9_v2Parser.CompilationUnit1Con-
text):
    hidden = self.token_stream.getHiddenTokensToLeft(ctx.start.tokenIndex)
    self.token_stream_rewriter.replaceRange(from_idx=hidden[0].tokenIndex,
                             to_idx=hidden[-1].tokenIndex,
                             text='/*After refactoring (Refactored version)*/\n')
```

In the above code, the getHiddenTokensToLeft method collects all tokens on the hidden channel to the left of the current token until it encounters a token that is not hidden. Then it replaces all the hidden tokens, such as white spaces and comments, with the comment, '/**After refactoring (Refactored version)*/\n'. Below is a class B before and after applying the proposed refactoring tool.

```
/* Before refactoring (Original version) */
class B {
    static int id;
    public static String name;
    public int age;
    int f;
    void growUp(int years) {
        int a, c;
        a = 2022;
        a = a + c;
        this.age = this.age + years;
    }
    void m(int i) {
        this.f = i * this.f;
    }
}
```

After refactoring the class B with the proposed refactoring tool, the class is modified as follows:

```java
/*After refactoring (Refactored version)*/
class B {
  static int id;
  public int getId() {
    return this.id;
  }
  public void setId(int id) {
    this.id = id;
  }
  private static String name;
  public String getName() {
    return this.name;
  }
  public void setName(String name) {
    this.name = name;
  }
  private int age;
  public int getAge() {
    return this.age;
  }
  public void setAge(int age) {
    this.age = age;
  }
  int f;
  public int getF() {
    return this.f;
  }
  public void setF(int f) {
    this.f = f;
  }
  void growUp(int years) {
    int a, c;
    a = 2022;
    a = a + c;
    this.setAge(this.getAge() + years);
  }
  void m(int i) {
    this.setF(i * this.getF());
  }
}
```

5.5   Rename Method Refactoring

Rename refactoring is all about selecting appropriate, meaningful names for program elements such as variables, methods, and functions [8]. Meaningful names affect source codes' cleanness because good names reveal intentions.

Clean code reveals bugs [9]. A good name tells the reader what is happening in the source code itself. The name of a variable, function, or class should indicate why it exists, what it does, and how it is used. However, it takes time to choose appropriate names, especially for those whose mother language is not English. This section presents a technique for automatic refactoring of method names.

The third king of Norway gave his name to the son of
Sigurðr jarl by sprinkling the child with water.

5.5.1   Meaningful Method Names

Method naming is one of the most critical factors in program comprehension. Misleading names for methods in a regular program or the APIs in a software library threaten understandability, readability, reusability, and testability and lead to issues and defects [10, 11]. Therefore, assisting programmers in choosing proper method names reduces the development cost and time while improving software quality. A function name should be a meaningful verb communicating its intent to the reader. For instance, consider the following function name:

ComputeEuclideanDiseance

The function and argument should form a complementary verb-noun pair. For instance, consider the following name:

CopyDocuments(String FromSource, String ToDestination)

5.5.2   *Method Vectorization*

A good selection of names to rename a method m is the names of the methods m'_i Very similar to m. The question is, which features of the methods should be considered as a basis for measuring their similarity? Source code metrics are essential to any software measurement [12]. Parser generators, such as ANTLR, are used to extract the metrics from the software's source code. However, the *Understand* software tool provides a library to extract valuable data from source code to compute the metrics. For instance, the following code fragment shows how to use the *Understand* [13] library to compute a software metric, Cyclomatic complexity, for a given Java project:

```
import understand
db=understand.open(project) // the address of the .udb file
for methodname in db.ents("Method"):
    cyclo_complexity =  CYCLO(methodname)
```

The de.ents("method") is a method call provided by Understand to access the next method name in one or more Java projects. All the projects, including methods for which source code metrics are to be computed, are saved in a Udbs database provided by Understand, as follows:

```
"""Save Udbs"""
    def create_understand_database_from_project(cls):
        #print("Start Process at " + str(datetime.now()))
        # Comand to create an Understand database.
        cmd = 'und create -db {0}{1}.udb -languages java add {2} analyze -all'
        root_path = os.getcwd()
        #print(os.getcwd())
        root_path = root_path +'\\'
        root_path = root_path.replace('\\','/')
        projects = [name for name in os.listdir(root_path) \
                if os.path.isdir(os.path.join(root_path, name)) \
                if not os.path.isfile(os.path.join(root_path, name+".udb"))]
        for project in projects:
            #print("\n_____ project : ",project ,"_____")
            #print(cmd.format(root_path, project, root_path + project))
        command_ = cmd.format(root_path, project, root_path + project)
    os.system('cmd /c "{0}"'.format(command_))
    #print("\n End Process at " + str(datetime.now()))
```

The following function, CYCLO, computes the cyclomatic complexity for a given method, function_entity, provided that the function is not abstract or interface. CYCLO invokes the Understand function metric with the parameter ["Cyclomatic"] to compute the Cyclomatic complexity software metric described in Table 5.1. The functiom_entity is an instance of the Ent class in Understand. The Ent class represents all kinds of entities in a program, including variable, method, class, and package.

```
def CYCLO(self, function_entity):
    try:
        if (self.is_abstract(function_entity)
            or self.is_interface(function_entity.parent())):
            return None
        else:
            return funcname.metric(["Cyclomatic"])["Cyclomatic"]
    except:
        return None
```

As another example, consider the following code fragment to compute another software metric called "Maximum nesting."

Table 5.1 The list of selected software metrics

#	Metric abbreviated name	Metric full name and description	Quality attribute
1	LOC	Lines of codes	Size
2	CountLineCodeDec	Number of lines containing declarative source code	Size
3	CountLineCodeExe	Number of lines containing executable source code	Size
4	CountLineComment	Number of lines containing the comment	Size
5	RatioCommentToCode	The ratio of the number of comment lines to the number of code lines	Size
6	CountLineBlank	Number of blank lines	Size
7	NL	Number of physical lines	Size
8	NOLV	Number of local variables	Size
9	NOP	Number of parameters	Size
10	CountStmt	Number of declarative plus executable statements	Size
11	CountStmtDecl	Number of declarative statements	Size
12	CountStmtExe	Number of executable statements	Size
13	CYCLO	Cyclomatic complexity	Complexity
14	CyclomaticStrict	Strict cyclomatic complexity	Complexity
15	CyclomaticModified	Modified McCabe cyclomatic complexity	Complexity
16	Essential	Essential cyclomatic complexity	Complexity
17	MAXNESTING	Maximum nesting level	Complexity
18	NPATH	Count path	Complexity
19	CountPathLog	The logarithm of the number of unique paths through a body of code	Complexity
20	Knots	The measure of overlapping jumps	Complexity
21	MaxEssentialKnots	Maximum Knots after removing structured programming constructs	Complexity
22	CLNAMM	Called local, not accessor or mutator methods	Cohesion
23	ATLD	Access to local data	Cohesion
24	NOAV	Number of accessed variables	Cohesion
25	FANIN	Number of calling classes	Coupling
26	FANOUT	Number of called classes	Coupling
27	ATFD	Access to foreign data	Coupling
28	FDP	Foreign data providers	Coupling
29	CFNAMM	Called foreign not accessor or mutator methods	Coupling

(continued)

Table 5.1   (continued)

#	Metric abbreviated name	Metric full name and description	Quality attribute
30	CC	Changing classes	Coupling
31	CINT	Coupling intensity	Coupling
32	CDISP	Coupling dispersion	Coupling
33	CM	Changing methods	Coupling
34	MaMCL	Maximum message chain length	Coupling
35	MeMCL	Mean message chain length	Coupling
36	NMCS	Number of message chain statements	Coupling

```
# Calculate maxnesting
maxnesting=obj_get_metrics.MAXNESTING(methodname)
maxnesting = float(0 if maxnesting is None else maxnesting)
#if maxnesting > 0:
MAXNESTING_List.append(maxnesting)
```

The following function invokes the Understand function metric with the parameter ["**MaxNesting**"] to compute the MaxNesting software metric described in Table 5.1. scitools.com

```
def MAXNESTING(self, function_entity):
    try:
        if (self.is_abstract(function_entity)
            or self.is_interface(function_entity.parent())):
            return None
        else:
            return function_entity.metric(["MaxNesting"])["MaxNesting"]
    except:
        return None
```

The software metrics listed in Table 5.1 have been found helpful in measuring similarity in clones [14–16], plagiarism [17], and malware detection [18].

For instance, consider the Java method in Fig. 5.13.

Computing 37 software metrics listed in Table 5.1 for the "releasePlayer" method and considering each computed metric as a feature of the method, the vectorized method is as follows.

Fig. 5.13 A method with an appropriate name

```
private void releasePlayer() {
    if (player != null) {
        isPaused = player.getPlayWhenReady();
        updateResumePosition();
        player.release();
        player.setMetadataOutput(null);
        player = null;
        trackSelector = null;
    }
    progressHandler.removeMessages(SHOW_PROGRESS)
    themedReactContext.removeLifecycleEventListener(this);
    audioBecomingNoisyReceiver.removeListener();
}
```

5.5.3   Cross-Project Naming

Cross-project naming techniques propose the appropriate names for a method depending on its similarities with the methods in the other projects. By representing each method in a large corpus of Java projects [19] as a feature vector, a reasonable benchmark for determining names based on similarity is formed.

The feature vector computed for a given method is compared with a relatively large corpus of feature vectors computed for 3512331 methods in JavaMed corpus [19]. The comparison between a given method m, with the methods, m'_i, in the corpus, is made by computing the Euclidean distance between the feature vectors:

$$\text{Euclidean} - \text{distance}\left(m^{vec}.m'^{\,vec}_i\right) = \sqrt{|m^{vec}|^2 - \left|m'^{\,vec}_i\right|^2} \qquad (5.1)$$

For instance, when comparing the feature vector of the releasePlayer method, in Fig. 5.13 with the feature vectors of 3512331 methods in JavaMed corpus, the distances of the methods' feature vectors, shown in Table 5.2, were relatively less than the others:

Table 5.3 shows the Euclidean distances of the feature vectors of the methods in Table 5.2 with the feature vector of the releasePlay method, shown in Fig. 5.13. Table 5.4 represents the Euclidean-distance of the functions closest to the releasePlayer.

Table 5.2 Representing the function "releasePlayer" as a feature vector

F1	F2	F3	F4	F5	F6	F7	F8	F9	F10	F11	F12	F13
0.00	0.00	0.00	0.00	0.00	0.17	0.04	0.00	0.08	0.04	0.29	0.41	0.00
F14	F15	F16	F17	F18	F19	F20	F21	F22	F23	F24	F25	F26
0.00	0.00	0.17	0.41	0.00	0.04	0.04	0.00	0.17	0.00	0.41	0.04	0.29
F28	F29	F30	F31	F32	F33	F34	F35	F36	F37			
0.33	0.04	0.29	0.08	0.08	0.04	0.00	0.00	0.00	0.00			

Table 5.3 Feature vectors for the closest vectors to the releasePlay

Sample	F1	F2	F3	F4	F5	F6	F7	F8	F9	F10	F11	F12	F13	F14	F15	F16	F17
releasePlay	0.00	0.00	0.00	0.00	0.00	0.17	0.04	0.00	0.08	0.04	0.29	0.41	0.00	0.00	0.00	0.17	0.41
releasePlay	0.00	0.00	0.00	0.00	0.00	0.15	0.03	0.00	0.06	0.03	0.30	0.39	0.00	0.00	0.00	0.21	0.39
Stop	0.00	0.00	0.00	0.00	0.04	0.17	0.04	0.04	0.04	0.04	0.30	0.39	0.00	0.00	0.00	0.22	0.39
initPlayer	0.00	0.00	0.00	0.00	0.00	0.12	0.04	0.00	0.04	0.04	0.27	0.39	0.00	0.00	0.00	0.19	0.39
setTextsize	0.00	0.00	0.00	0.00	0.08	0.30	0.03	0.08	0.07	0.04	0.33	0.39	0.00	0.00	0.00	0.23	0.39
initView	0.00	0.05	0.00	0.00	0.00	0.16	0.03	0.00	0.05	0.03	0.35	0.37	0.00	0.00	0.00	0.19	0.37

Sample	F18	F19	F20	F21	F22	F23	F24	F25	F26	F27	F28	F29	F30	F31	F32	F33	F34	F35	F36	F37
releasePlay	0.00	0.04	0.04	0.00	0.17	0.00	0.41	0.04	0.29	0.08	0.33	0.04	0.29	0.08	0.08	0.04	0.00	0.00	0.00	0.00
releasePlay	0.00	0.03	0.03	0.00	0.24	0.00	0.39	0.03	0.30	0.06	0.33	0.03	0.30	0.06	0.06	0.03	0.00	0.00	0.00	0.00
Stop	0.00	0.00	0.04	0.00	0.22	0.00	0.39	0.04	0.30	0.04	0.35	0.04	0.30	0.04	0.04	0.04	0.00	0.00	0.00	0.00
initPlayer	0.00	0.00	0.04	0.00	0.31	0.00	0.39	0.04	0.31	0.04	0.35	0.04	0.31	0.04	0.04	0.04	0.00	0.00	0.00	0.00
setTextsize	0.00	0.03	0.03	0.08	0.16	0.00	0.39	0.03	0.30	0.07	0.33	0.03	0.30	0.07	0.07	0.03	0.00	0.00	0.00	0.00
initView	0.00	0.03	0.03	0.00	0.32	0.00	0.37	0.03	0.29	0.05	0.32	0.03	0.29	0.05	0.05	0.03	0.00	0.00	0.00	0.00

Table 5.4 The distance of the recommended methods from the method under question

Method name	Euclidean-distance from releasePlayer()
releasePlayer	0.13
Stop	0.16
initPlayer	0.18
setTextSize	0.18
initView	0.19

Fig. 5.14 The most similar method to the method in Fig. 5.13

```
private void releasePlayer() {
    if (player != null) {
        mAutoplay = player.getPlayWhenReady();
        updateResumePosition();
        player.release();
        player = null;
        trackSelector = null;
        Log_OC.v(TAG, "playerReleased"); }
```

The nearest feature vector to the releasePlayer is the feature vector of a method with the same name in the JavaMed corpus. The body of the recommended method, relasePlayer, is given in Fig. 5.14.

The difficulty is that the name of the closest method to the method under question is not necessarily appropriate. That is why instead of one, k nearest neighbors are selected. Also, the recommendation of more than one name allows the developer to select the most appropriate name.

5.5.4   K-Nearest Neighbor Similarity

Suppose $m \in C$ is a method in a class, C. The aim is to find the k most similar methods, $m'_i \in C'_i$ in a reference dataset, D, to the method m. It is assumed that all the methods in D are named appropriately. The similarity is measured in terms of the software metrics. The set of software metrics, $r_i \in R$, listed in Table 5.3, are used as features to vectorize the methods. Each method, $m \in C$, is represented as a vector of the metrics, m^{vec}. The Understand API [13] is invoked to compute the selected metrics $r_i \in R$ for a given method, m.

$$\text{Sim}(m \in C, m'_i \in C'_I) = \text{Eucleadian} - \text{distance}(m^{vec}.m'^{vec}_i)$$
$$= \sqrt{|m^{vec}|^2 - |m'^{vec}_i|^2} \tag{5.2}$$

Equation 5.2 is used to compute the distance between the feature vector of each method m'_i in the JavaMed corpus with the vector of the method under question, m. By experience, $k = 10$ seems to provide enough options for selecting an appropriate

name. Now, assume M is the set of the methods in class C, and $M\backslash\{m\}$ denotes all
the methods in M, apart from the method m:

$$M\backslash\{m\} = M - \{m\} \tag{5.3}$$

After finding the k-most similar methods to a given method, We select from the
methods in k-sim the method $m' \in C'$ whose neighboring methods, $M'\backslash\{m'\}$, in the
class C' are, on average, the most similar to the neighboring methods of $m \in C$,
denoted by $M\backslash\{m\}$. The following relation is used to score the similarity of each
method $m' \in$ k-sim as follows:

Let M' = the set of methods including m' in class C',
    averageSim = MaxSim = 0
$\forall$ methods $q' \in M' \backslash\{m'\}$
        $\forall$ methods $q \in M\backslash\{m\}$ => MaxSim = Max(Sim(q', q), MaxSim)
            averageSIm += MaxSim
    averageSim = averageSim / $|M\backslash\{m\}|$
    Similarity_score(m', m) = β0 + β1*Sim(m',m) + β2*averageSim

A linear regression modeling method is used to compute the values of the coefficients β0 to β2. The estimated values for β0, β1 and β2 are 10, 2.5, and 3,
respectively.

It takes a long time to compare the feature vector of a method with the feature
vectors of 3512331 methods to find an appropriate name for the method. Therefore,
the KNN classifier is used to find the k nearest neighbors for a given method feature
vector in the benchmark data set. The benchmark test set has been built by vectorizing
3512331 Java methods in JavaMed corpus. The training set for building the KNN
model is 3091803 vectors, and the test set is 11.98% of the benchmark test set. The
Scikit-learn [20] machine-learning library is used to create KNN classifiers. Below
is a code snippet implementing the training and testing of the KNN classifier:

```
from sklearn.neighbors import KneighborsClassifier
# 1. Create an instance of the Kbeighbors classifier class
classifier = KNeighborsClassifier(n_neighbors=Neighbors, \
                algorithm=Algorithm, metric=Metric,n_jobs=N_jobs)
# 2. Train the Kbeighbors classifier
classifier.fit(learned_dataX, learned_dataY)
# 3. Test the Kbeighbors classifier
classifier_pred = classifier.predict(test_dataX)
```

Source code metrics are suitable for vectorizing each method in a large corpus of quality-named methods. The vectors are further applied as a benchmark to propose appropriate names for a given method. To this end, the K-Nearest Neighbor algorithm is applied to determine the k-most similar methods to the method under question in an unsupervised manner. The similarity between methods is also affected by the similarity between methods of the classes they belong to. The selected methods are further scored based on the similarity of their class methods.

5.6 Refactorings

Each refactoring operation has a definition and is clearly specified by the entities in which it is involved and the role of each. Table 5.5 shows the most well-known refactoring operations used by researchers and developers. Knowing when and where to start and stop refactoring is just as important as it is to know how to activate it [2]. The refactoring process consists of six different activities [21]:

1. Identify which parts of the software should be refactored.
2. Determine which refactoring(s) should be applied to the identified parts.
3. Guarantee that the applied refactoring preserves behavior.
4. Apply the refactoring.
5. Assess the effect of the refactoring on the quality characteristics of the software (e.g., complexity, understandability, maintainability) or the process (e.g., productivity, cost, effort).
6. Maintain the consistency between the refactored program code and other software artifacts (such as documentation, design documents, requirements specifications, tests, etc.).

The first decision that needs to be made is to determine which part of the software should be refactored. The most general approach to detecting program parts that require refactoring is the identification of code smells. According to Beck [2], bad smells are "structures in the code that suggest (sometimes scream for) the possibility of refactoring."

Code smells are code snippets with design problems. Their presence in the code makes software maintenance difficult and affects software quality. When a code smell is detected, it is suggested to refactor the code to remove the smell.

Software engineering researchers and practitioners propose various code smells with different names and definitions. Table 5.6 lists the 20 most well-known code smells associated with one or two refactoring operations in Table 5.5. A complete list of existing code smells, along with more information about their smells, features, and relation with refactorings, is discussed in Ref. [2].

Table 5.5 Well-known refactorings

Refactoring	Definition	Entities	Roles
Move class	Move a class from one package to another	Package Class	Source package, target package Moved class
Move method	Move a method from one class to another	Class Method	Source class, target class Moved method
Merge packages	Merge the elements of a set of packages in one of them	Package	Source package, target package
Extract/split package	Add a package to compose the elements of another package	Package	Source package, target package
Extract class	Create a new class and move fields and methods from the old class to the new one	Class Method	Source class, new class Moved methods
Extract method	Extract a code fragment into a method	Method Statement	Source method, a new method Moved statements
Inline class	Move all features of a class to another one and remove it	Class	Source class, target class
Move field	Move a field from one class to another	Class Field	Source class, target class Field
Push-down field	Move a field of a superclass to a subclass	Class Field	Superclass, subclasses Move field
Push-down method	Move a method of a superclass to a subclass	Class Method	Superclass, subclasses Moved method
Pull-up field	Move a field from subclasses to the superclass	Class Field	Subclasses, superclass Moved field
Pull-up method	Move a method from subclasses to the superclass	Class Method	Subclasses, superclass Moved method
Increase field visibility	Increase the visibility of a field from public to protected, protected to package or package to private	Class Field	Source class Source filed
Decrease field visibility	Decrease the visibility of a field from private to package, package to protected, or protected to public	Class Field	Source class Source filed
Make field final	Make a non-final field final	Class Field	Source class Source filed
Make field non-final	Make a final field non-final	Class Field	Source class Source filed

(continued)

Table 5.5 (continued)

Refactoring	Definition	Entities	Roles
Make field static	Make a non-static field static	Class Field	Source class Source filed
Make field non-static	Make a static field non-static	Class Field	Source class Source filed
Remove field	Remove a field from a class	Class Field	Source class Source filed
Increase method visibility	Increase the visibility of a method from public to protected, protected to package, or package to private	Class Method	Source class Source method
Decrease method visibility	Decrease the visibility of a method from private to package, package to protected, or protected to public	Class Method	Source class Source method
Make method final	Make a non-final method final	Class Method	Source class Source method
Make method non-final	Make a final method non-final	Class Method	Source class Source method
Make method static	Make a non-static method static	Class Method	Source class Source method
Make method non-static	Make a static method non-static	Class Method	Source class Source method
Remove method	Remove a method from a class	Class Method	Source class Source method
Make class-final	Make a non-final class final	Class	Source class
Make class non-final	Make a final class non-final	Class	Source class
Make class abstract	Change a concrete class to abstract	Class	Source class
Make class concrete	Change an abstract class to concrete	Class	Source class
Extract subclass	Create a subclass for a set of features	Class Method	Source class, a new subclass Moved methods
Extract interface	Extract methods of a class into an interface	Class Method	Source class, new interface Interface methods
Inline method	Move the body of a method into its callers and remove the method	Method	Source method, callers method
Collapse hierarchy	Merge a superclass and a subclass	Class	Superclass, subclass

(continued)

Table 5.5 (continued)

Refactoring	Definition	Entities	Roles
Remove control flag	Replace the control flag with a break	Class Method	Source class Source method
Replace nested conditional with guard clauses	Replace nested conditional with guard clauses	Class Method	Source class Source method
Replace the constructor with a factory function	Replace the constructor with a factory function	Class	Source class
Replace exception with test	Replace exception with precheck	Class Method	Source class Source method
Rename field	Rename a field	Class Field	Source class Source filed
Rename method	Rename a method	Class Method	Source class Source method
Rename class	Rename a class	Class	Source class
Rename package	Rename a package	Package	Source package
Encapsulate field	Create setter/mutator and getter/accessor methods for a private field	Class Field	Source class Source filed
Replace parameter with query	Replace parameter with query	Class Method	Source class Source method
Pull up the constructor body	Move the constructor	Class Method	Subclass class, superclass Constructor
Replace the control flag with a break	Replace the control flag with the break statement	Class Method	Source class Source method
Remove flag argument	Remove flag argument	Class Method	Source class Source method
Total	47		

5.7 Extract Class[2]

Extract class refactoring is often performed in response to the Large class, God class, or Blob smells. These smells appear when a class consists of many unrelated fields and methods that are not cohesive enough to act as a single responsibility component. Large classes often violate the single responsibility principle [9] and negatively affect many quality attributes, including changeability, understandability, reusability, and testability [27].

The extract class refactoring improves class cohesion by splitting a large class into two or more cohesive classes. A class is cohesive if each method in the class operates on all the class attributes. Figure 5.15 shows an example of extract class

---

[2] Download the source code form: https://doi.org/10.6084/m9.figshare.21270213

Table 5.6 Code smells associated with refactoring in Table 5.5

Code smell	Descriptions and other names
God class	The class defines many data members (fields) and methods and exhibits low cohesion. The god class smell occurs when a huge class surrounded by many data classes acts as a controller (i.e., takes most of the decisions and monopolizes the software's functionality) Other names: Blob, large class, brain class
Long method	This smell occurs when a method is too long to understand and most presumably performs more than one responsibility Other names: God method, brain method, large method
Feature envy	This smell occurs when a method seems more interested in a class other than the one it actually is in
Data class	This smell occurs when a class contains only fields and possibly getters/setters without any behavior (methods)
Shotgun surgery	This smell characterizes the situation when one kind of change leads to many changes in multiple classes. When the changes are all over the place, they are hard to find, and it is easy to miss a necessary change
Refused bequest	This smell occurs when a subclass rejects some of the methods or properties offered by its superclass
Functional decomposition	This smell occurs when experienced developers from procedural language backgrounds write highly procedural and non-object-oriented code in an object-oriented language
Long parameter list	This smell occurs when a method accepts a long list of parameters. Such lists are hard to understand and difficult to use
Promiscuous package	A package can be considered promiscuous if it contains classes implementing too many features, making it too hard to understand and maintain. As for god class and long method, this smell arises when the package has low cohesion since it manages different responsibilities
Misplaced class	A misplaced class smell suggests a class in a package containing other classes not related to it
Switch statement	This smell occurs when switch statements that switch on type codes are spread across the software system instead of exploiting *polymorphism*
Spaghetti code	This smell refers to an unmaintainable, incomprehensible code without any structure. The smell does not exploit and prevents the use of object-orientation mechanisms and concepts
Divergent change	Divergent change occurs when one class is commonly changed differently for different reasons Other names: Multifaceted abstraction

(continued)

Table 5.6 (continued)

Code smell	Descriptions and other names
Deficient encapsulation	This smell occurs when the declared accessibility of one or more members of abstraction is more permissive than actually required
Swiss army knife	This smell arises when the designer attempts to provide all possible class uses and ends up in an excessively complex class interface
Lazy class	Unnecessary abstraction
Cyclically-dependent modularization	This smell arises when two or more abstractions depend on each other directly or indirectly
Primitive obsession	This smell occurs when primitive data types are used where an abstraction encapsulating the primitives could serve better
Speculative generality	This smell occurs when abstraction is created based on speculated requirements. It is often unnecessary, which makes things difficult to understand and maintain
Message chains	A message chain occurs when a client requests another object, that object requests yet another one, and so on. These chains mean that the client is dependent on navigation along with the class structure. Any changes in these relationships require modifying the client
Total	20

refactoring in which the TeleponeNumber class is separated from the Person class. The reason behind this refactoring is that information and operation on each person's phone should be separated from his/her personal information.

The question is how to preserve the program syntax, semantics, and behavior when applying the extract class refactoring. The class semantics is preserved by delegating the implementation of the methods to the separated class.

Figures 5.16 and 5.17 represent the Person class source code before and after refactoring. The following section describes the use of ANTLR to automate the extract class refactoring.

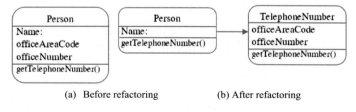

(a) Before refactoring (b) After refactoring

Fig. 5.15 An example of an extract class refactoring

```
public class Person {
    private String name;
    private int officeAreaCode;
    private int officeNumber;
    public String getTelephoneNumber() {
        return String.valueOf(this.officeAreaCode) + "-"
                + String.valueOf(this.officeNumber);
    }
}
```

Fig. 5.16 The Person class before refactoring

```
public class Person {
    public TelephoneNumber telephone=new TelephoneNumber();
    private String name;
    public String getTelephoneNumber() {
        return this.telephone.getTelephoneNumber();
    }
}
```

Fig. 5.17 The **Person class after refactoring**

5.7.1   *Class Cohesion*

Extract class refactoring refers to splitting a class into relatively more cohesive classes. A class is cohesive if all the fields in the class are accessed by every method. Classes are preferred to have high cohesion because if everything in a class is closely related, concerns are likely to be well separated.

A class can be broken down into relatively more cohesive classes by clustering its internal dependencies graph. In an internal dependencies graph, nodes represent fields and methods of the class, and edges connect a field node to all the method nodes accessing the field.

Consider a class C as a dependency graph G(V, E) where each node $v \in V$ represents either a method $m_i \in M$ or a field $f_i \in F$. Here, $M = \{m_1, m_2, ..., m_n\}$ and $F = \{f_1, f_2, ..., F_n\}$ are respectively the set of methods and fields of the class C. An edge $(m_i \rightarrow f_j) \in E$, connects a node $m_i \in M$ to $f_i \in F$ provided that the method m_i uses the class field f_i.

For instance, consider the dependency graph for the class, GodClass, in Fig. 5.19. The class has four fields, f1 to f4, and four methods.

The NetworkX library of Python can be used to build and partition the internal dependency graph as a DiGraph. To build the DiGrapg, G, for a class, C, a dictionary data structure, fields, is built with each class field name as the key and a list of the methods using the field as value.

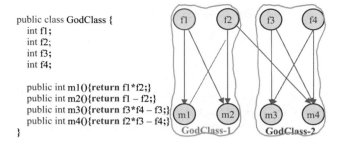

```
public class GodClass {
    int f1;
    int f2;
    int f3;
    int f4;

    public int m1(){return f1*f2;}
    public int m2(){return f1 – f2;}
    public int m3(){return f3*f4 – f3;}
    public int m4(){return f2*f3 – f4;}
}
```

Fig. 5.18 A class and its field-method dependency graph

```
def split_class(self):
    # 1- Create a graph
    G = nx.DiGraph()
    for field, methods in self.field_dict.items():
        for method in methods:
            # print('add edge {0} --> {1}'.format(field, method))
            G.add_node(method[0], method_name=method[0])
            G.add_edge(field, method[0])
    print('---------\nExtracted classes:')
    visualization.graph_visualization.draw(g=G)
    # Generate weakly connected components of G
    S = [G.subgraph(c).copy() for c in nx.weakly_connected_components(G)]
```

Fig. 5.19 A function to construct a class internal dependency graph

For instance, the fields dictionary content for GodClass in Fig. 5.18, is as follows:

$$\text{Fields: } \{\text{'f1'}: [\text{'m1'}, \text{'m2'}],$$
$$\text{'f2'}: [\text{'m1'}, \text{'m2'}, \text{'m4'}],$$
$$\text{'f3'}: [\text{'m3'}, \text{'m4'}],$$
$$\text{'f4'}: [\text{'m3'}, \text{'m4'}], \}$$

The split_class method in Fig. 5.19 uses the fields dictionary to generate the dependency graph as a NetworkX DiGraph. The point is that when having a NetWorkX DiGraph it will be possible to apply any function in the rich library of NetWorkX on the graph.

The question is how to construct the 'fields' dictionary from a class's fields and their dependent methods.

5.7.2 *Extracting Fields and Methods*

Figure 5.20 represents the main body of a program to refactor a Java class in a file, input.java, using the Extract Class refactoring. The program uses ANTLR to create a lexer, parser, listener, parse tree, and parse tree walker for the given class, input.java.

In step 6 of the main function, in Fig. 5.20, an instance of the listener class, ExtractClassRefactoringListener, is created. The parameters passed to the initializer class are the name of the class to be refactored and the token stream extracted from the class. Figure 5.21 represents the class constructor.

The tree walker starts with the root of the parse tree, which is always the starting symbol of the grammar. As the walker encounters a class declaration, it invokes the corresponding method in the ExtractClassRefactoringListener class. When looking in the Java9 grammar for the class declaration, the following rule is found:

classDeclaration
    : normalClassDeclaration #classDeclaration1
    | enumDeclaration #classDeclaration2
    ;

Therefore, considering the grammar rules, the following method is added to the ExtractClassRefactoringListener class. The method checks the class name with the

```python
def main(args):
    # Step 1: Load input source into a stream
    stream = FileStream(args.file, encoding='utf8')
    # Step 2: Create an instance of AssignmentStLexer
    lexer = Java9_v2Lexer(stream)
    # Step 3: Convert the input source into a list of tokens
    token_stream = CommonTokenStream(lexer)
    # Step 4: Create an instance of the AssignmentStParser
    parser = Java9_v2Parser(token_stream)
    parser.getTokenStream()
    # Step 5: Create a parse tree
    parse_tree = parser.compilationUnit()
    # Step 6: Create an instance of AssignmentStListener
    my_listener = ExtractClassRefactoringListener(
                common_token_stream=token_stream, class_identifier='Input')
    # Step 7: Create a walker object to traverse the parse tree in a depth-first order
    walker = ParseTreeWalker()
    # Step 8: Walk through the parse tree
    walker.walk(t=parse_tree, listener=my_listener)
if __name__ == '__main__':
    argparser = argparse.ArgumentParser()
    argparser.add_argument(
        '-n', '--file',
        help='Input source', default=r'input.java')
    args = argparser.parse_args()
    main(args)
```

Fig. 5.20 The main function of a Python program for extract class refactoring

```
class ExtractClassRefactoringListener(Java9_v2Listener):
    def __init__(self, common_token_stream: CommonTokenStream = None,
            class_identifier: str = None):
        self.enter_class = False  # set to true when entering the class to be refactored
        self.token_stream = common_token_stream
        self.class_identifier = class_identifier # name of the class to be refactored
        # Move all the tokens in the source code in a buffer, token_stream_rewriter.
        if common_token_stream is not None:
            self.token_stream_rewriter =    \
                                (common_token_stream)
        else:
            raise TypeError('common_token_stream is None')
        self.field_dict = {}  # the fields dictionary described in the previous section
        self.method_name = []  #holds the names of the class methods
        self.method_no = 0      # The method number in the order of its appearance
```

Fig. 5.21 The initiator of the listener class derived from the listener class created by ANTL

name of the class to be refactored. The "ctx" parameter in the listener methods points to the corresponding parse tree nodes. The ctx.getChildren or ctx.getChild methods access the children of the parse tree nod. The children nodes may be accessed directly with getChildren or getChild. Consider the NormalClassDeclaration rule, shown below:

```
normalClassDeclaration
    : classModifier* 'class' identifier typeParameters? superclass?
      superinterfaces? classBody
    ;
    classModifier : annotation | 'public' | 'protected'  | 'private'
                    |'abstract' | 'static' | 'final' | 'strictfp' ;
```

As shown in Fig. 5.22 the class name can be accessed via the ctx.identifier().getText(). Similarly, the class modifier can be accessed via ctx.classModifier().getText().

The 'ctx' object in Fig. 5.22 points to the class declaration node in the parse tree. Therefore, inside 'enterNormalClassDeclaration,' it is possible to access all the children and parents of the corresponding node in the parse tree via the 'ctx.getParent()' and 'ctx.getChild()' method calls. Below is the rule for declaring fields or, in other words, attributes of a class in Java:

```
normalClassDeclaration
  : classModifier* 'class' identifier typeParameters? superclass?
    superinterfaces? classBody
  ;
  classModifier : annotation |      'public'  | 'protected' | 'private'
                        |'abstract' | 'static' | 'final' | 'strictfp' ;
```

As shown in Fig. 5.22 the class name can be accessed via the ctx.identifier().getText(). Similarly, the class modifier can be accessed via ctx.classModifier().getText().

```
# Checks the class name encountered while traversing the parse tree with
# the name of the class to be refactored.
    def enterNormalClassDeclaration(self, ctx: Java9_v2Parser.
                                    NormalClassDeclarationContext):
      if ctx.identifier().getText() != self.class_identifier:
        return
      self.enter_class = True
```

Fig. 5.22 This function is invoked as the tree walker encounters a class declaration

```
fieldDeclaration
            :        fieldModifier* unannType variableDeclaratorList ';'
            ;
fieldModifier
            :        annotation | 'public' | 'protected' |'private'
            |        'static' | 'final' |  'transient' | 'volatile'
            ;
variableDeclaratorList
            :        variableDeclarator (',' variableDeclarator)*
            ;
variableDeclarator
            :        variableDeclaratorId ('=' variableInitializer)?
            ;
variableDeclaratorId
            :        identifier dims?
            ;
```

Based on the above grammar rules, to access a field name via the "ctx" object, the following call sequence is used:

"ctx.variableDeclaratorList().variableDeclarator(i=0).
            variableDeclaratorId().identifier().getText()"

The fieldDeclation subtree for a class 'a' is shown in Fig. 5.23. Considering the parse tree, in Fig. 5.23, the following call sequence can be used to access the type of the field:

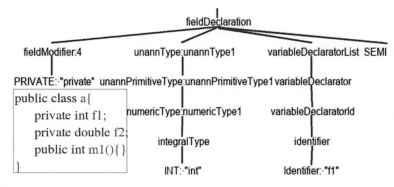

Fig. 5.23 A sample class, a, and its fields declaration subtree

"ctx.UnnType().unannPrimitiveTpe(i=0).
numericType().integerType().getText()"

In the enterFieldDeclaation shown in Fig. 5.24, field names are added as keys to the field_dict dictionary.

The enterMethodDeclaration() method in Fig. 5.25 is invoked each time the tree walker visits a method declaration. Polymorphic methods in a class have the same name. The order of each method in a class can serve as a unique identifier for the method. Therefore, along with the name, the order number of each method in the class is saved in a list, method_name.

The aim is to provide a dictionary, field_dict, to relate each class field/attribute as the key with the methods using the field as the value associated with the key.

```
# Adds attributes of the target class to a dictionary, field_dict.
def enterFieldDeclaration(self, ctx: Java9_v2Parser.FieldDeclarationContext):
    if not self.enter_class:
        return
    field_id = ctx.variableDeclaratorList().variableDeclarator(i=0).
                        variableDeclaratorId().identifier().getText()
    self.field_dict[field_id] = []
```

Fig. 5.24 A method to handle class attributes

```
# Enter a parse tree produced by Java9_v2Parser#methodDeclaration.
def enterMethodDeclaration(self, ctx: Java9_v2Parser.MethodDeclarationContext):
    if not self.enter_class:
        return
    m = []
    m_name = ctx.methodHeader().methodDeclarator().identifier().getText()
    self.method_no = self.method_no + 1
    m.append(m_name)
    m.append(self.method_no)
    self.method_name.append(m)
```

Fig. 5.25 This method inserts the name and the order number of class methods in a list

The exitIdentifier method in Fig. 5.26 is invoked each time a variable is visited. The method ensures that the visited variable is inside a method. If so, it checks whether the variable is a class field. In this case, it adds the method name to the list of methods using the variable.

In the end, the exitNormalClassDeclaration method function is invoked. As shown in below, the exitNormalClassDeclaration method invokes the split_class method, shown in Fig. 5.19, to create the class's internal dependency graph.

```
# Exit a parse tree produced by Java9_v2Parser#normalClassDeclaration.
def exitNormalClassDeclaration(self, ctx: Java9_v2Parser.
                                     NormalClassDeclarationContext):
    self.enter_class = False
    self.split_class()
    self.field_dict = {}
    self.method_name = []
    self.method_no = 0
```

Fig. 5.26 This method is invoked each time a variable is visited

```
# Exit a parse tree produced by Java9_v2Parser#identifier.
def exitIdentifier(self, ctx: Java9_v2Parser.IdentifierContext):
    if not self.enter_class:
        return
    if self.method_no == 0:
        return
    current_method = self.method_name[-1]
    variable_name = ctx.getText()
    if variable_name not in self.field_dict:
        return
    if not current_method in self.field_dict[variable_name]:
        self.field_dict[variable_name].append(current_method)
```

Fig. 5.27 Class diagrams illustrate the Move Method refactoring

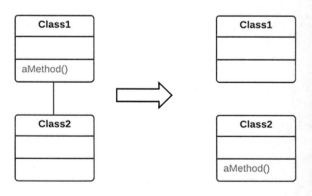

5.8 Move Method

The Move Method refactoring is an immediate solution to fix the Feature Envy smell. This smell occurs when a method in its current class frequently accesses features (field and method) in another class, increasing the coupling between the two classes. The coupling is reduced by moving the method to the class with relatively more interactions with the method. The Move Method refactoring reduces the coupling between the components and enhances cohesion. Figure 5.27 shows the class diagram in which aMethod in class1 is moved to class2.

The move method refactoring program receives the source and destination classes and the method's name to be moved from the source to the destination class. The program delegates the method's task to the destination class by:

1. Make a copy of the method and add it to the destination class.
2. Replace the method's body in the source class with a call to the copied method.

In Fig. 5.28, aMethod in Class1 is not accessed by any other class methods. On the other hand, it is invoked by the methods in Class2.

In Fig. 5.29, a copy of Method is added to the method members of Class2. The aMethod body in Class1 is replaced with Object2.aMethod(), where Object2 is an instance of Class2.

```
package source_package;                      package target_package;
public class Class1 {                        import source_package.Class1;
    public void Method(){ ➜ public void aMethod(){
    class1.Method(); ➜ class1.aMethod();
```

Fig. 5.28 TestMethod In Class2 invokes Method in Class1

```
public void Method(){ ➜ public void aMethod(){
Obj2.Method() }; ➜ Obj2.aMethod() };
Method();} ➜ aMethod();}
public void Method(){ ➜ public void aMethod(){
```

Fig. 5.29 After refactoring, Class2 is independent of Class1

5.8.1   Move Method Refactoring Implementation

Three listeners, CutMethodListener, PasteMethodListener, and DelegateMethodListener classes, are developed to implement the move method refactoring. The CuntMethodListener removes the method to be moved (target method) from the source class containing the method. The listener adds an instance of the target class as an attribute to the source class. Before removing the method, its source code is copied into a buffer. The PasteMethodListener class copies the buffer's content into the target class. This way, the method is moved from the source class to the target class.

If the target method is entirely removed from the source class, then every access to the method should be modified. However, as shown in Fig. 5.29, the method body in the source class is replaced with the method's call in the target class. The DelegateMethod listener delegates a given method to another object.

(1)   **CutMethodListener**

A listener class, **CutMethodListener**, removes the method to be moved from the source class and creates an instance of the destination class in the source class. Below is the initiator of the listener:

```
class CutMethodListener(JavaParserLabeledListener):
    # this listener cuts the method to be removed from the source code
            #.... __init___
            def __init__(self, method_name: str,   # method to be removed
                         target_class: str,  # Move method target
                         source _class: st,  # Move method source
                         rewriter: TokenStreamRewriter,
                is_static = False):
    self.method_name = method_name   # name of the method to be moved
    self.target_class = target_class # class receiving the method
    self.target_class = source_class # class including the method
    self.rewriter = rewriter        # TokenStreamRewriter
    self.is_member = False          # is the method visited
    self.do_delete = False          # Do not remove the method
    self.is_static = is_static      # Method to be removed is static
    self.instance_name = "Target_class_instance"
    self.import_statement = False if not target_class.rsplit('.')[:-1]== \
                            source_class.rsplit('.')[:-1])else True
```

The CutMethodListener initiator takes the source and target class names and the method's name to be moved as input. It is supposed that a different program distinguishes the move method refactoring opportunity. After applying the move method refactoring, the source class needs to access an instance of the target class. Therefore, if the target class is in another package, it should be imported at the top of the source class file to prevent compilation errors. The exitPackageDeclaration method of CutMethodListener adds the import statement after the package declaration. Below is the exitPackageDeclaration method.

```
# Imports the destination class name into the source class
def exitPackageDeclaration(self, ctx:
            JavaParserLabeled.PackageDeclarationContext):
    if self.import_statement:
        self.rewriter.insertAfterToken(token=ctx.stop,
            text=self.import_statement,
            program_name=self.rewriter.DEFAULT_PROGRAM_NAME
        )
        self.import_statement = None
```

The target class is expected to be already imported since it is assumed that the target method, or in other words, the method to be moved, frequently accesses the destination class members due to the feature envy code smell. On the other hand, the destination and source classes may be in the same package. In this case, the import_statement variable is set to False. Below is the syntax of the class declaration in Java8.g4.

The grammar in Java8.g4 is a bit different from Java9.g4. The main reason is to speed up the parsers built based on grammar. According o the grammar in Fig. 5.30 classBodyDeclaration could be either a fieldDeclaration or a methodDeclaration.

```
classDeclaration
  : CLASS IDENTIFIER typeParameters?
    (EXTENDS typeType)?
    (IMPLEMENTS typeList)?
    classBody
  ;
classBody
  : '{' classBodyDeclaration* '}'
  ;
classBodyDeclaration
  : ';' #classBodyDeclaration0
  | STATIC? block #classBodyDeclaration1
  | modifier* memberDeclaration #classBodyDeclaration2
  ;
memberDeclaration
  : methodDeclaration #memberDeclaration0
  | fieldDeclaration #memberDeclaration2
  | ...
```

Fig. 5.30 Class declaration rule in Java8.g4 **grammar**

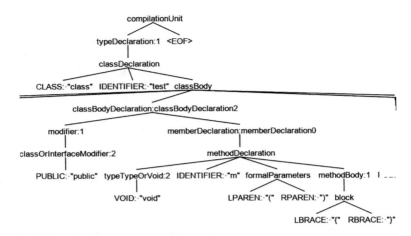

Fig. 5.31 Subtree of the parse tree rooted at class declaration

Figure 5.31 illustrates the subtree of the parse tree rooted at the "classDeclaration." The classBodyDeclartion node in the parse tree is decomposed into a modifier
followed by a "methodDeclaration."

A Boolean variable is_member specifies whether or not the parse tree walker has
visited the "methodDeclaration" on its way done the tree. If so, the method's body
is saved in the method_text variable when exiting the classBodyDeclartion2 node.

While walking up the tree, by the time the tree walker encounters a method or
field declaration node, it invokes the exitClassBody method. If the method to be
moved is found, the value of the variable 'do_delete' will be True. In this case, the
following method finds the address of the first and last tokens of the method to be
removed and saves it in the 'start' and 'stop' variables. The whole method is copied
in the self.method_text variable. Then the method is replaced with an instance of the
target_class, as an attribute of the source_class.

```
def exitClassBodyDeclaration2(self, ctx:
            JavaParserLabeled.ClassBodyDeclaration2Context):
    if self.do_delete: # if the method is found in the source class
        # 1. Copy the method to be moved to the method_text variable
        self.method_text = self.rewriter.getText(
            program_name=self.rewriter.DEFAULT_PROGRAM_NAME,
            start=ctx.start.tokenIndex,
            stop=ctx.stop.tokenIndex
        )
        # 2. Replaces the body of the method to be moved with:
        if self.is_static:
            # 2.1. public static target_class target_class_instance = new target_class
            replace_text = f"public static {self.target_class} {self.instance_name} \
                    = new {self.target_class}();"
        else:
            # 2.2. public static target_class target_class_instance = new target_class
            replace_text = f"public {self.class_name} {self.instance_name} \
                    = new {self.class_name}();"
        #2.3 replace the body of the method with an instance of the target class
        self.rewriter.replace( \
                program_name =self.rewriter.DEFAULT_PROGRAM_NAME,\
                from_idx = start, \
                to_idx = stop, \
                text=replace_text )
        self.do_delete = False
```

(2)   PasteMethodListener

The PasteMethodListener class copies the method to be removed (the target method) to the target class. In the case the source and target classes are in two different packages, the required import instruction is added at the top of the target class package.

As described above, before removing the target method from the source class, its text was saved in the method_text variable. Also, the required import statements were saved in the imports_variable. Below is the initiator of the PasteMethodListener class.

```python
class PasteMethodListener(JavaParserLabeledListener):
    def __init__(self, method_text: str, method_map: dict,
                 imports: str, source_class: str,
                 rewriter: TokenStreamRewriter):
        self.method_text = method_text  # the target method source code
        self.method_map = method_map
        self.source_class = source_class  # class holding the target method
        self.imports = imports            # import instructions tobe inserted
        self.rewriter = rewriter          # the token stream rewriter
        self.fields = None
        self.has_package = False
        self.has_empty_cons = False
```

The following method adds the required import statements to the source and target classes.

```python
    def exitPackageDeclaration(self, ctx:
                       JavaParserLabeled.PackageDeclarationContext):
        if self.has_package and self.imports:
            self.rewriter.insertAfter(
                index=ctx.stop.tokenIndex,
                text="\n" + self.imports,
                program_name=self.rewriter.DEFAULT_PROGRAM_NAME
            )
```

The following method inserts the target method from the self.method_text into the body of the target class:

```python
    def enterClassBody(self, ctx: JavaParserLabeled.ClassBodyContext):
        self.rewriter.insertAfterToken(
            token=ctx.start,
            text="\n" + self.method_text + "\n",
            program_name=self.rewriter.DEFAULT_PROGRAM_NAME
        )
```

5.9 Summary

Automatic refactoring is the main concern of this chapter. This chapter describes how to use the ANTLR listener to develop software tools to automate refactoring. While traversing the parse tree in depth-first order, action can be taken each time a node is visited to detect the refactoring opportunities and refactor the code accordingly. Refactoring should be considered as a part of compiler optimizations. The optimizer should automatically apply any refactoring required by the programmer. In addition to syntax and semantics analysis, smell analysis should also be an integral part of source code analysis. The optimizer may refactor the source code to remove the detected smells. The compiler may report the source code's quality attributes by computing the source code metrics commonly used to measure the quality. Depending on the metrics the programmer is interested in, different sequences of refactorings may be applied to improve the metrics and, consequently, the desired source code quality attribute.

Researchers have recently studied the improvement of different quality attributes through refactoring [22, 23]. One of the critical quality attributes directly affecting software testing is testability, discussed in Chap. 1. Fowler claims that refactoring helps him find bugs [2]. We believe that testability improvement by applying appropriate refactoring helps find bugs. Apparently, when applying a test-first approach to developing software, the testability quality attribute should be improved in the first position to reduce the cost of software development by reducing the cost of tests.

However, a survey in 2007 revealed that despite 90% of the developers having access to refactoring tools, only 40% of the developers could rely on automatic refactoring, and the rest preferred to refactor manually [1]. Most developers prefer manual refactoring despite integrating refactoring tools closely with Integrated Development Environments (IDEs). A reason to avoid refactoring tools is the lack of trust in the automatic code changes performed by refactoring tools [24, 25]. Recent studies revealed that some refactoring tools are sometimes faulty, implying that automatic code changes potentially introduce defects to the developer's codebase [26]. The difficulty is that when refactoring a source code element, such as a method or a class, all the other parts of the code that somehow depend on the refactored part of the code may require modifications. For instance, the move method refactoring entails modifying the objects accessing the method. Symbol tables can be used to keep track of the methods.

Refactorings may be targeted at different quality factors such as testability, reusability, functionality, understandability, flexibility, and extendability. Refactoring is a type of program transformation that preserves the program's behavior. It aims to improve the program's internal structure without changing its external behavior. This way, the program quality, defined and measured in terms of quality attributes, improves. Ongoing research and development efforts are directed at improving different quality attributes through refactoring [26].

5.10   Exercises

1. Complete the encapsulate field refactoring program to update all references of the encapsulated field in a given Java source code.
2. Find the full implementation of described refactoring operations in https://git hub.com/m-zakeri/CodART and apply each refactoring to the Weka project. The source code of this project is available at https://github.com/Waikato/weka-3.8. Report the applicable refactorings.
3. Write a program to find and convert pure abstract classes to interfaces in Java source codes. Pure abstract classes are classes that only contain abstract methods.
4. Implement the automatic application of pull-up and push-down methods according to instructions by Fowler and Beck [2].
5. Implements decent refactoring opportunity detections for extract class and move method refactorings with ANTLR listener mechanism. Describe your algorithms and compare your result with refactoring opportunities in assignment 1.
5. Change the extract class and move method refactoring programs to apply these refactoring on C# programs. Using the C# grammar, you should generate the C# parser with ANTLR. How many modifications did you commit in our existing ANTLR codes which apply these refactoring to Java programs?
7. There are compiler-based tools similar to the ANTLR that can use to automate refactoring operations. For example, the Roslyn compiler provides the parse tree generation and vising mechanism for.NET platform languages, i.e., C# and Visual Basic, or the CLang front-end compiler can be used for C and C++ programs. The underlying techniques in these tools are very similar to the ANTLR. Use the Roslyn tool to automate the Extract Class refactoring application for C# programs. Compare the effort required to automate this refactoring and the performance of the developed program with the ANTLR version of the program automating this refactoring operation.

References

1. Opdyke, W.F.: Refactoring object-oriented frameworks. Ph.D. thesis, The University of Illinois at Urbana-Champaign (1992)
2. Fowler, M.: Refactoring: Improving the Design of Existing Code, 2nd edn. Addison-Wesley (2019)
3. Soetens, Q.D.: Formalizing refactorings implemented in Eclipse. MSc thesis, University of Antwerp (2009)
4. Parr, T., Fisher, K.: LL(*): the foundation of the ANTLR parser generator. In: Proceedings of the 32nd ACM SIGPLAN Conference on Programming Language Design and Implementation, pp. 425–436 (2011). https://doi.org/10.1145/1993498.1993548
5. Par, T.: The Definitive ANTLR 4 Reference, 2nd edn. Pragmatic Bookshelf (2013)
6. Parr, T.: ANTLR (ANother Tool for Language Recognition). https://www.antlr.org. Accessed 10 Jan 2022
7. Islam, Q.N.: Quazi Nafiul Islam, 1st edn. Packt Publishing (2015)

8. Mayer, P., Schroeder, A.: Automated multi-language artifact binding and rename refactoring between Java and DSLs used by Java frameworks. In: Lecture Notes in Computer Science Book Series (LNPSE), vol. 8586 (2015)

9. Martin, R.C.: Clean Code: A Handbook of Agile Software Craftsmanship, vol. 38, no. 6. Prentice-Hall (2009). ISBN: 9-780-13235-088-4

10. Butler, S., Wermelinger, M., Yu, Y., Sharp, H.: Relating identifier naming flaws and code quality: an empirical study. In: 2009 16th Working Conference on Reverse Engineering, pp. 31–35 (2009). https://doi.org/10.1109/WCRE.2009.50

11. Nguyen, S., Phan, H., Le, T., Nguyen, T.N.: Suggesting natural method names to check name consistencies. In: Proceedings of the ACM/IEEE 42nd International Conference on Software Engineering, June, pp. 1372–1384 (2020). https://doi.org/10.1145/3377811.3380926

12. Lanza, M., Marinescu, R.: Object-Oriented Metrics in Practice: Using Software Metrics to Characterize, Evaluate, and Improve the Design of Object-Oriented Systems. Springer-Verlag Berlin Heidelberg (2006)

13. SciTools: Understand. https://www.scitools.com/ (2020). Accessed 11 Sept 2022

14. Rahman, M.S., Roy, C.K.: A change-type based empirical study on the stability of cloned code. In: Proceedings of the 2014 14th International Working Conference on Source Code Analysis and Manipulation (SCAM 2014), pp. 31–40 (2014). https://doi.org/10.1109/SCAM.2014.13

15. Lavoie, T., Mérineau, M., Merlo, E., Potvin, P.: A case study of TTCN-3 test scripts clone analysis in an industrial telecommunication setting. Inf. Softw. Technol. **87**, 32–45 (2017). https://doi.org/10.1016/j.infsof.2017.01.008

16. Roopam, Singh, G.: To enhance the code clone detection algorithm by using a hybrid approach for detection of code clones. In: Proceedings of the 2017 International Conference on Intelligent Computing and Control Systems (ICICCS 2017), vol. 2018-Janua, pp. 192–198 (2017). https://doi.org/10.1109/ICCONS.2017.8250708

17. Dong, W., Feng, Z., Wei, H., Luo, H.: A novel code stylometry-based code clone detection strategy. In: 2020 International Wireless Communications and Mobile Computing (IWCMC), June, pp. 1516–1521 (2020). https://doi.org/10.1109/IWCMC48107.2020.9148302

18. Kalysch, A., Protsenko, M., Milisterfer, O., Müller, T.: Tackling androids native library malware with robust, efficient and accurate similarity measures. In: ACM International Conference Proceeding Series (2018). https://doi.org/10.1145/3230833.3232828

19. GitHub - tech-srl/code2seq: code for the model presented in the paper: 'code2seq: generating sequences from structured representations of code'

20. Pedregosa, F., et al.: Scikit-learn: machine learning in Python. J. Mach. Learn. Res. **12**, 2825–2830 (2011)

21. Mens, T., Tourwe, T.: A survey of software refactoring. IEEE Trans. Softw. Eng. **30**, 126–139 (2004). https://doi.org/10.1109/TSE.2004.1265817

22. Mkaouer, M.W., Kessentini, M., Bechikh, S., et al.: On the use of many quality attributes for software refactoring: a many-objective search-based software engineering approach. Empir. Softw. Eng. **21**, 2503–2545 (2016). https://doi.org/10.1007/s10664-015-9414-4

23. Mohan, M., Greer, D.: Using a many-objective approach to investigate automated refactoring. Inf. Softw. Technol. **112**, 83–101 (2019). https://doi.org/10.1016/j.infsof.2019.04.009

24. Vakilian, M., et al.: Use, disuse, and misuse of automated refactorings. In: Proceedings of the 34th International Conference on Software Engineering (ICSE '12), Zurich, Switzerland, pp. 233–243. IEEE Press (2012)

25. Murphy-Hill, E., et al.: How we refactor, and how we know it. In: Proceedings of the 31st International Conference on Software Engineering (ICSE '09), Washington, DC, USA, pp. 287–297. IEEE Computer Society (2009)

26. Daniel, B., et al.: Automated testing of refactoring engines. In: Proceedings of the 6th Joint Meeting of the European Software Engineering Conference and the ACM SIGSOFT Symposium on the Foundations of Software Engineering (ESEC-FSE '07), Dubrovnik, Croatia, pp. 185–194. ACM (2007)

27. Suryanarayana, G., Samarthyam, G., Sharma, T.: Refactoring for Software Design Smells: Managing Technical Debt. Morgan Kaufmann Publishing Co., (2014)

Part II
Automatic Fault Localization

Sometimes finding a fault in a program is like finding a needle in a haystack. However, the main idea behind automatic fault localization is straightforward. The faulty element is the one that is covered by failing test cases more frequently than the passing ones. The difficulty is the lack of some commonly used software testing tools that act as a barrier to experimenting and developing algorithms for fault localization. The difficulty is to the extent that researchers and scholars repeatedly develop the same software tools before they can implement their algorithm. The software tools include source code instrumentation, slicing, and control flow graph extraction.

Let us break this barrier once and for all. This Part of the book puts an end to this dilemma by showing the scholars and researchers how to use a compiler generator to instrument source code, create control flow graphs, identify prime paths, and finally slice source code. On top of these tools, a software tool, Diagnoser, is offered to facilitate experimenting with and developing new fault localization algorithms. Diagnoser accepts a source code and a test suite to test the source code as input. It runs the code with each test case and creates a diagnosis matrix. Each column in the diagnosis represents statements covered by a test case and the test result as failing or passing. The last columns represent the suspiciousness scores computed for each statement. Each row of the diagnosis matrix represents which test cases have covered a particular statement. In addition, the suspiciousness score for each statement using different formulas is shown.

Conducting the programmer toward the right code region will enable them to find the relevant fault more quickly. Faulty regions often lie in the discrepancies between a faulty execution path and its nearest passing neighbor. Two approaches based on cross entropy and graph matching are offered to determine the discrepancies.

Chapter 6
Graph Coverage

6.1 Introduction

Suppose you want to test a function to ensure it works correctly. The first question you ask yourself could be:

How many test cases are enough?

This is a good question. But it should be corrected as follow:

How many influential test cases are required?

A clear test case is a test case that examines elements of the code that are not examined by any other test cases. This chapter deals with the answers to this question. In addition to answering the question, this chapter shows the reader how to develop their software tool to measure test data adequacy.

Testing is, in a sense, a sort of code analysis. The analysis is facilitated by summarizing the structure of functions/methods into a graph representation called a control flow graph. Graph coverage addresses some test adequacy metrics used to evaluate the effectiveness of tests with a given test data test.

This chapter begins with a brief description of control flow and program dependency graphs. Section 6.2 first raises the question of how to ensure the adequacy of tests. Then, it answers the question by describing the merits and pitfalls of different coverage metrics.

Many books, articles, and reports provide the reader with information about a control flow graph and graph coverage criteria. The missing part is how to develop a software tool that builds a CFG, how to save the CFG, and how to use it to compute different test adequacy criteria for a given test set. This section is concerned with a detailed description of graph coverage criteria. However, the knowledge gained in this section will be useless unless the reader gets the idea of how to build a software tool to extract a CFG from source code. The main question answered in this chapter is how to develop software tools to extract control flow graphs from a function/method body. Section 6.3 describes the application of Python libraries to extract CFGs from

© The Author(s), under exclusive license to Springer Nature Switzerland AG 2023   249
S. Parsa, *Software Testing Automation*,
https://doi.org/10.1007/978-3-031-22057-9_6

Python functions. Section 6.4 describes using ANTLR Listener and Visitor classes to extract control flow graphs. The chapter ends up with a summary of the materials covered.

6.2   Control Flow Graphs

A control flow graph or, in short, CFG is an abstraction of the program as a graph. It can be used to evaluate test data sets regarding the coverage of the graph nodes and edges achieved when running the program with the test data set. For instance, consider the control flow graph for the method listed in Fig. 6.1.

There are five basic blocks in the CFG, shown in Fig. 6.1. Each basic block, i, consists of a sequence of one or more statements $<Si_1, Si_2, ..., Si_n>$ where:

- Si_1 is either the first statement of the method, getNeighbours, or the target of a branching statement, and
- Si_n is either the function's last statement, F, a branching statement, or the statement before the target of a jump.

For instance, in Fig. 6.1, basic block number 3 includes statements 8 to 10, and basic block number 4 includes statement 11. Moreover, there are two particular empty nodes, Start and End in V. Every basic block or node in V is reachable from Start and has a path ending with a fake basic block, labeled end.

As another example, consider the CFG shown in Fig. 6.2. In this CFG, instead of the statement numbers, the statements themselves are copied into the basic blocks.

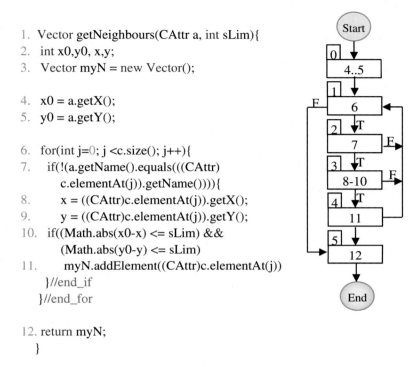

```
1.  Vector getNeighbours(CAttr a, int sLim){
2.  int x0,y0, x,y;
3.  Vector myN = new Vector();

4.  x0 = a.getX();
5.  y0 = a.getY();

6.  for(int j=0; j <c.size(); j++){
7.    if(!(a.getName().equals(((CAttr)
         c.elementAt(j)).getName())))){
8.      x = ((CAttr)c.elementAt(j)).getX();
9.      y = ((CAttr)c.elementAt(j)).getY();
10.   if((Math.abs(x0-x) <= sLim) &&
         (Math.abs(y0-y) <= sLim)
11.       myN.addElement((CAttr)c.elementAt(j))
       }//end_if
     }//end_for

12. return myN;
    }
```

Fig. 6.1 A sample Java method and its corresponding control flow graph[1]

Control flow graphs facilitate the analysis and understanding of a program behavior by summarizing and abstracting each method/function of the program as a graph, illustrating the branching of the execution flow, dependent on the conditions determining the outgoing branch to follow.

As shown in Fig. 6.2, a CFG illustrates all possible paths that might be traversed during the execution of a program. Statements often execute in the order they appear in the source code from top to bottom. Control flow statements, on the other hand, break up the execution flow by decision-making, looping, and branching, thus enabling particular blocks of code to be executed only when certain conditions are met. In effect, a control flow graph represents a combination of *basic control flow graphs*, each representing a control flow statement.

6.2.1   Basic Control Flow Graphs

This section describes how basic control flow graphs are constructed for different types of statements. A CFG includes a single entry and a single exit node. Control

[1] Download the source code for CFG extractor for Java from: https://doi.org/10.6084/m9.figshare.21285915

1. void selectionSort(int arr[], int n)
2. {
3.   int i, j, min_idx;
4.   for (i = 0; i < n - 1; i++)
5.   {
6.   min_idx = i;
7.   for (j = i + 1; j < n; j++)
8.   if (arr[j] < arr[min_idx])
9.   min_idx = j;
10.   swap(&arr[min_idx], &arr[i]);
11. }
12. }

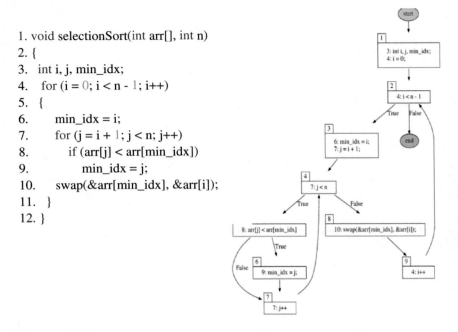

Fig. 6.2 A CFG including the source code

flow enters the CFG at the entry node and leaves the CFG at the exit node. Below, it is shown how to draw a control flow graph for different types of statements, including assignment, if, switch, for, while, do-while, and finally, exceptions in C++.

1. **Assignment statements**

Below is a control flow graph for an assignment statement:

a = b + c;

2. **'If' statements**

The control flow graph for a typical if statement comprises a decision node and two basic blocks representing the then-part and else-part of the if statement. For instance, consider the following if statement:

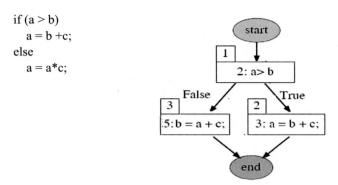

```
if (a > b)
    a = b +c;
else
    a = a*c;
```

3. Switch statements

Both switch and if-else statements are selection statements. Selection statements test an expression of choice amongst different alternatives. An if-else statement has two alternatives, while a switch statement may have several alternatives, each defined as a case part. For instance, consider the following switch statement:

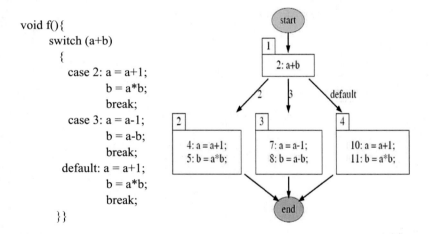

```
void f(){
    switch (a+b)
    {
    case 2: a = a+1;
            b = a*b;
            break;
    case 3: a = a-1;
            b = a-b;
            break;
    default: a = a+1;
            b = a*b;
            break;
    }}
```

4. For statements

The entry point to a basic block is either the target of a jump or after a jump. For instance, consider a 'for' statement in C++. The loop counters are assigned a different value on each iteration of the loop body. Therefore, after each iteration of the loop body, there should be a jump to a basic block modifying the loop counters and then jumping to the basic block, including the loop termination condition. Below is an example of a simple 'for' statement and its corresponding CFG:

```
for( i = 10, j -40;
    i+j < 55; i = i+5, j = j+i)
    a = i*b + j*c;
```

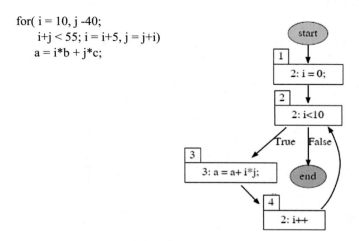

5.   While statements

The condition of a while statement resides in a distinct basic block for two reasons. Firstly, the condition is the target of the jump at the end of the while loop body. Secondly, depending on the condition value, the control either moves to the first basic block of the loop or exits the loop. Below is a sample while statement and its corresponding CFG:

```
void f(){
    while (i < j)
        i++;
}
```

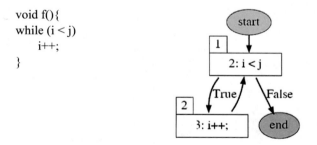

6.   do-while statements

The condition of a do-while statement is checked at the bottom of the loop body. Therefore, based on the original concept of basic blocks, the whole body of the do-while construct should be kept in a single basic block. However, due to specific difficulties discussed in this section, It would be better to construct CFG for do-while statements, as shown below:

```
void foo_1()
  { x = 0;
    do
      {
        y = f(x,y);
        x = x + 1;
      } while( x<y );
    Println(y);
  }
```

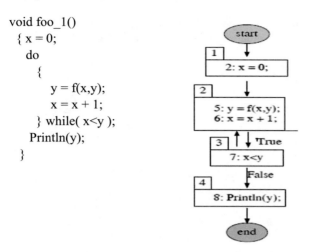

As shown above, the condition of the do-while statement is assigned to a distinct basic block. The reason is that it includes a jump to its previous basic block. At first glance, it seems that the whole body of the do-while statement should be assigned to the same basic block. However, if so, the control flow will not be clearly shown by the CFG. For instance, suppose there is a "continue" statement inside the do-while statement. As shown below, the impact of the "continue" on the control flow can be clearly shown since the condition of the do-while resides in a distinct basic block:

```
1.    void foo_2()
2.    { x = 0;
3.      do
4.        {
5.          y = f(x,y);
6.          if(x>y) continue;
7.          x = x + 1;
8.        } while( x<y );
9.      println(y);
10.   }
```

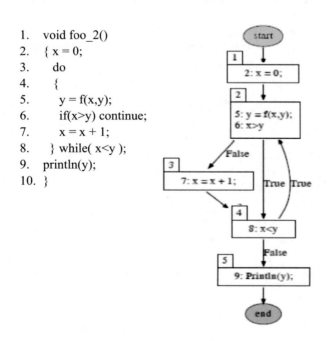

As described in Ref. [1] by Amman and Offutt, the whole do-while loop should be represented as a single basic block. As a result, the CFG for the functions foo_1 will be as follows

```
void foo_1()
  { x = 0;
    do
    {
        y = f(x,y);
        x = x + 1;
    } while( x<y );
    Println(y);
  }
```

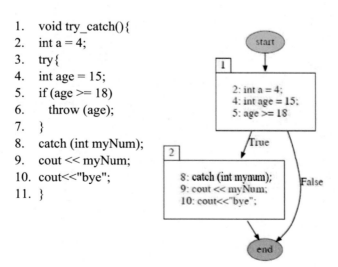

7. Try-block

The final programming construct is exception handling, which in C++ uses the try-catch statement. Below is the CFG for the function try_catch, function an exceptions, called by the program (Exception). The edge from the node labeled 2 to node 3 reflects the programmer-raised exceptions. A try-block, like other statements, can appear anywhere a statement can appear.

```
1.    void try_catch(){
2.    int a = 4;
3.    try{
4.        int age = 15;
5.        if (age >= 18)
6.            throw (age);
7.    }
8.    catch (int myNum);
9.    cout << myNum;
10.   cout<<"bye";
11.  }
```

A control flow graph represents all possible execution paths of a program. Whenever running a program with a test data set, if the test fails, then the bug should be sought inside the basic blocks covered by the test execution paths. The more paths a test data set covers, the more adequate the test data set is. One way to evaluate the adequacy of a test suite for testing a given source code is to apply graph coverage criteria.

6.3 Graph Coverage Criteria

Graph coverage addresses a set of metrics to evaluate the effectiveness of a test suite based on the extent to which the test suite covers a graph representation of the code under test. *Control flow* and *data flow graphs* are two abstract representations of code used for the automated evaluation of coverage metrics.

Data flow coverage criteria evaluate test data based on whether they cover the chain of statements assigning values to variables and the statements using the values. In this way, the tester may trace the propagation of buggy values across the code. Data flow coverage criteria are described in Chap. 9, where I describe how to extract data dependency graphs.

Control flow coverage criteria concern the graph's structure, including nodes, edges, and specific sub-paths. An execution path, including some source code statements, is examined when executing a code with specific test data. The statement coverage metric determines the percentage of statements of the program or blocks in the CFG of the program examined by the test data.

6.3.1 Statement Coverage

It is desired to get the code right before the release. The question is, how much test is adequate to ensure that the code is correct? An immediate answer is to provide test data covering all the statements of the code. For instance, there are 11 lines, including 11 statements in Fig. 6.3. An appropriate test suite should execute each statement at least once. This type of coverage is called *statement* coverage.

Generally, the coverage level of a test set, T, for a set of test requirements, TR, is the percentage of the Number of test requirements satisfied by T divided by the size of TR. Similarly, statement coverage is the percentage of statements a test data set covers. For instance, the statement coverage obtained by running the function "foo" in Fig. 6.3 with the test data set {t1, t2, t3} in Table 6.1 is:

$$\text{Statement coverage} = \frac{\text{No. statements covered by tests}}{\text{Total no. statements}} * 100 = \frac{10}{11} * 100 \approx 90\%$$

```
1.   void foo(int x, int y){
2.     w = 0;
3.     if(x > y)
4.       if(x+y<40) //should be:(x+y<30)
5.          W+ = 1;
6.       else  w+ = 2;
7.     else w+ = 3;
8.     if(x>2*y)
9.        w += 5;
10.    else w = 4;
11.  printf("w = %d", w);
12. }
```

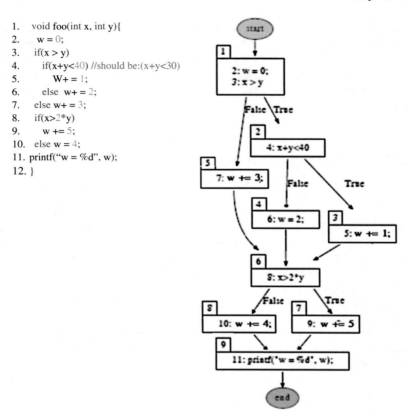

Fig. 6.3 Statement 4 should be if($x + y < 30$)

Table 6.1 The Foo function test results

	Test case			Statements covered										Run
T	x	y	w	2	3	4	5	6	7	8	9	10	11	State
t1	5	50	7	•	•				•	•		•	•	Pass
t2	58	38	6	•	•	•		•		•		•	•	Pass
t3	51	4	7	•	•	•		•		•	•		•	Pass
Statements: (11)				•	•	•		•	•	•	•	•	•	
Statement coverage: 10/11*100 = 90%														

In Table 6.1, the column headed St. indicates the Number of times the test cases have covered each statement. The test suite consists of three test cases t1, t2, and t3. These test cases, all together, cover ten statements.

A test adequacy metric must serve both as an assessment and measurement of the quality of a test data set. A suitable objective would be to add test data covering every source statement. Nevertheless, statement coverage does not adequately consider that many statements (and bugs) involve branching and decision-making. For instance, the test data "t1: a = 2" covers the following control flow graph nodes. However, this test data does not cover the false branch of the if statement. If the value of a is zero, then 1/a is infinity. In general, branch coverage subsumes node or statement coverage.

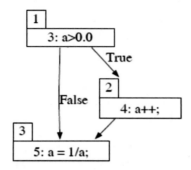

6.3.2 Branch Coverage

A branching statement allows the execution flow to jump to a different program part. *Branch coverage* is a test data adequacy metric to evaluate whether the tests cover all the branches of a codebase. A good rule of thumb is that a branch condition needs to be true and false at least once during testing. Branch coverage is computed as the number of branches covered to the total number of branches. For instance, the branch coverage obtained by running the function "foo" in Fig. 6.3 with the test data set {t1, t2, t3} in Table 6.2 is:

$$\text{Branch coverage} = \frac{\text{total no branches covered by tests}}{\text{total no. branches}} * 100 = \frac{11}{13} * 100 \approx 85\%$$

In Table 6.2, the column headed Br. indicates the Number of branches covered by each test case. The test suite consists of three test cases t1, t2, and t3. These test cases, all together, cover 11 statements.

As shown in Table 6.2, branch coverage is, in fact, the percentage of the CFG edges covered. That is why branch coverage is also known as edge coverage.

Table 6.2 Decision coverage provided by the three test cases t1, t2, and t3

	Test case			Statements covered										Total	
T	x	y	w	2	3	4	5	6	7	8	9	10	11	St.	Br.
t1	5	50	7	•	F				•	F		•	•	6	6
t2	58	38	6	•	T	F		•		F		•	•	7	7
t3	51	4	7	•	T	F		•		T	•		•	7	7
Statements: (11)				•	•	•		•	•	•	•	•	•	10	
Branches: (13)				1	2	1		1	1	2	1	1	1		11

6.3.3   Decision Coverage

Another metric called *decision coverage* considers branching conditions or predicates. Branch coverage considers the coverage of both conditional and unconditional branches, whereas decision coverage only measures the coverage of conditional branches. It is the percentage of decision outcomes exercised by a test suite.

For instance, in Table 6.2, the decisions on lines 3 and 8 evaluate as true or false. The test set covers only the false outcome of the decision at line 4. Therefore, the decision coverage is 5/6*100%, about 83%.

A decision is a predicate. Therefore, no wonder you come across the jargon *predicate coverage* instead of *decision coverage*. It is worth mentioning that branch/edge coverage is the percentage of edges covered by the test set, whereas predicate coverage is the percentage of conditional edges.

A predicate is a boolean expression that can be defined as a combination of non-boolean expressions as follows:

predicate: boolean;
boolean: boolean LogicalOperator clause | clause;
clause: non-boolean RelationalOperator non-boolean | boolean-variable;
non-boolean: expression;
LogicalOperator: '&&' | '||' | '!';
RelationalOperator: '<' | '<=' | '==' | '!=' | '>' | '>=';

The above grammar defines a predicate as a boolean. A boolean itself is either a clause or a boolean joined with a clause by a logical operator. A clause does not contain any logical operator. Sometimes a test requirement could be to cover all the clauses in predicates.

6.3.4 Clause Coverage

Clause or condition coverage is the percentage of clauses covered with a test set. As described above, a clause is either a boolean expression or variable that does not contain any logical operator. For instance, the predicate "a < b || a > c" includes two clauses "a < b" and "a > c". Clause coverage ensures both "a < b" and "a > c" are evaluated as true and false, whereas predicate coverage considers the whole expression "a < b || a > c".

Clause coverage does not generally subsume predicate coverage and vice versa. In general, ensuring each clause is evaluated as true and false does not guarantee that the overall predicate is also evaluated with true and false outcomes. For instance, consider the following tests:

$$T1 : (a, b, c) = (2, 3, 4) => a < b : 2 < 3 = \text{true}$$
$$a > c : 2 > 4 = \text{false}$$
$$T1 : (a, b, c) = (4, 1, 2) => a < b : 4 < 1 = \text{false}$$
$$a > c : 4 > 2 = \text{true}$$

6.3.5 Path Coverage

The question is, how much test data is adequate? Test effectiveness apparently depends on the number of execution paths the test suite covers. An adequate test suite should cover all possible source code execution paths. Nevertheless, the difficulty is that the number of execution paths could be very large. For instance, consider the function f in Fig. 6.4. Depending on the decision made in line 4, the number of different execution paths can be up to 2^n.

Therefore, to evaluate the adequacy of test data based on the coverage of the paths, only those paths in which there are no repeated subpaths or internal loops are considered. Such paths are called *simple paths* [2]. Simple paths have no internal loops, although the entire path itself may be a loop. In Fig. 6.5, the Python function computes all the simple paths in a control flow graph saved as a NetworkX [3] dot file. Below is a dot file representation of the CFG in Fig. 6.4:

```
digraph {
s;
1;
2;
3;
```

1.   int f(int n)
2.   {   int i, a =0, b=0;
3.   while(i<n)
4.   if(i/3) a+=i;
5.   else b+=i;
6.   return a+b;
7.   }

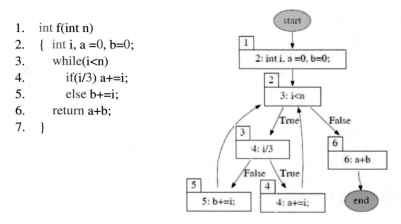

Fig. 6.4 A function f and its corresponding CFG

```
# finds all the simple paths in a CFG represented as a Netwokx graph, g.
def simple_paths(cfg):
    simple_paths_list = []
    for i in cfg.nodes():
        for j in cfg.nodes():
            simple_path = list(nx.all_simple_paths(cfg, i, j))
            for k in simple_path:
                if k not in simple_paths_list and k != []:
                    simple_paths_list.append(k)
    for i in list(nx.simple_cycles(cfg)):
        item = deque(i)
        for j in range(len(i)):
            item.rotate(1)
            x = list(item)
            simple_paths_list.append([*x, x[0]])
    return simple_paths_list
```

Fig. 6.5 A function to extract simple paths from a CFG

```
4;
5;
6;
e;
s -> 1;
1 -> 2;
2 -> 3;
```

```
3 -> 4;
3 -> 5;
4 -> 2;
5 -> 2;
2 -> 6;
6 -> e;
}
```

The simple_path function, shown in Fig. 6.5, computes simple paths in a given CFG. The function invokes the all_simple_paths method of the Python Networkx package to compute simple paths in the CFG. The NetworkX module is imported as 'nx':

$$\text{import networkx as nx}$$

It is worth mentioning that 'nx' is the most common alias for NetworkX.

The above function, collect_simple_paths(), uses "all_simple_paths" method to generate a simple path between the nodes i and j. For instance, the following simple paths were generated when running the above Python function to generate simple paths for the control flow graph in Fig. 6.4.

As shown in Table 6.3, initially, the set of simple paths of length zero, which are the nodes of the CFG, is collected. Afterward, each path of length zero apart from the end node of the CFG is extended to a path of length one. The process continues with extending each simple path to a new simple path of length, say n to a path of length n + 1, provided that the last node of the path is not the end-node of the CFG and the node added to the path is not already in the path.

It is observed that some of the simple paths in the above list are subpaths of the others. For instance:

$$[1, 2] \subset [1, 2, 3] \subset [1, 2, 3, 5]$$

Therefore, if the test requirement is to cover [1,2,3,5], for sure, the test data covering [1,2,3,5] covers [1,2] and [1,2,3] as well. Therefore, those simple paths not covering any other path are kept. Such paths are called *prime paths*.

$$[1, 2, 6, e], [1, 2, 3, 4], [1, 2, 3, 5]$$
$$[2, 3, 4], [1, 2, 3, 5]$$

Table 6.3 The process of generating simple paths

	Prime paths (shaded cells) ⊆ Simple paths					CFG
s	S,1	S,1,2	S,1,2,3	S,1,2,3,4		
				S,1,2,3,5		
				S,1,2,3,6	S,1,2,3,6,e	
1	1,2	1,2,6	1,2,6,e			
		1,2,3	1,2,3,4			
			1,2,3,5			
2	2,3	2,3,4	2,3,4,2			
		2,3,5	2,3,5,2			
	2,6	2,6,e				
3	3,4	3,4,2	3,4,2,6	3,4,2,6,e		
			3,4,2,3			
3	3,5	3,5,2	3,5,2,6	3,5,2,6,e		
			3,5,2,3			
4	4,2	4,2,6	4,2,6,e			
		4,2,3	4,2,3,5			
			4,2,3,4			
5	5,2	5,2,6	5,2,6,e			
		5,2,3	5,2,3,4			
			5,2,3,5			
6	6,e					
e						

6.3.6   Prime Paths

A prime path is a simple path, which does not appear as a subpath of any other simple path, and as a simple path, the last and first nodes may be identical [4]. Therefore, a prime path is a simple one that cannot be extended by adding new nodes. All the shaded cells in Table 6.3 are prime paths.

Below, a function, findPrimePath, is presented to compute prime paths in a given control flow graph (CFG). The CFG is read from a Networkx dot file []. In line 3, all the nodes of the CFG are copied as paths of length zero into a list, epandable_paths. The function findSimplePaths, invoked in line 6, returns all the expandable paths into a list, expandable_path. An expandable path is not a prime path. The list of prime paths is copied into the prime_paths parameter. The list is sorted and copied into a list, final_prime_paths.

```
# Finds prime paths in a CFG as a NetworkX DiGraph format
1.   def findPrimePaths(cfg):
2.       final_prime_paths = list()
3.       # extendable_path holds a list of basic block numbers
4.       extendable_path = [(n, ) for n in cfg['nodes']]
5.       prime_paths = list()
6.       # Find simple paths in the CFG
7.       findSimplePath(CFG, extendable_paths, prime_paths)
8.       prime_paths = sorted(prime_paths, key=lambda a:
9.                   (len(a), a), reverse=True)
10.      for i in primePaths:
11.          if len(i) != 1:
12.              final_prime_paths.append(list(i))
13.      return final_prime_paths
```

The findSimplePaths function computes extendable_paths and prime_paths. An extendable path is not a prime_path because one or more CFG nodes can extend it. Below is the function find_simple_paths:

```
1.   def findSimplePath(cfg, extendable_path, paths):
2.       """extenf all extendable paths apart from the prime paths."""
3.       paths.extend(filter(lambda p: isPrimePath(p, cfg), extendable _path))
4.       """Filter out prime paths from extendable paths."""
5.       extendable_paths = filter(lambda p: extensible(p, cfg),
6.                                 extendable_paths)
7.       newExPaths = list()
8.       for p in extendable_path:
9.           for nxx in cfg['edges'][p[-1]]:
10.              if nxx not in p or nxx == p[0]:
11.                  newExPaths.append(p + (nxx, ))
12.      if len(newExPaths) > 0:
13.          findSimplePath(cfg, newExPaths, paths)
```

The extensible function determines whether a given path is extensible. An extensible path is not a prime path. A path is extensible if it is neither a prime path nor right-extensible. For instance, the simple path 1,2,3 in Table 6.3 can be extended to 1,2,3,4 and 1,2,3,5.

```
1.  def extensible(path, cfg):
2.     # A path is extendable if it is neither prime nor right-extensible.
3.     if isPrimePath(path, cfg) or ! rightExtensible(path, cfg):
4.        return False
5.     else:
6.        return True
```

If the length of a path is greater than one and its first and last nodes are identical, the path is a prime path. For instance, in Table 6.3, path 2,3,4,2 is prime because the first and last nodes are the same. The alternative condition for a prime path is neither left-extensible nor right-extensible.

```
1.  def isPrimePath(path, cfg):
2.     """Determine whether or not a path is prime """
3.     if len(path) >= 2 and path[0] == path[-1]:
4.        return True
5.     elif !(leftExtensible(path, cfg) or rightExtensible(path, cfg)):
6.        return True
7.     else:
8.  return False
```

A path is not left-extensible if none of its nodes apart from the last node is an immediate parent of its first node. In contrast, a path is left-extensible if at least one of the immediate parents of its first node is not in the path. For instance, consider the simple path "3,4,2,6,e" in the CFG, shown in Table 6.3. Node 2 is the parent of node 3 and is in the path. Therefore, the path is not left-extensible.

```
1.  def leftExtensible (path, cfg):
2.     """
3.     1. Compute parent_nodes as a list of source nodes of edges
4.        whose sink is the first node of the given path.
5.     """
6.     parents_first_node = filter(lambda n: path[0] in cfg['edges'][n],
7.                                 cfg['nodes'])
8.     for n in parents_first_node:
9.        if n in path and n != path[-1]
10.          return False
11.    return True
```

A path is right-extensible if its last node is not connected to any CFG nodes in the path, apart from the first node.

```
1.  def rightExtensible(path, cfg):
2.      """
3.      The last node is not connected to any node in the path except the
4.      first node
5.      """
6.      nodes_connected_to_last_node = cfg['edges'][path[-1]]
7.      for n in nodes_connected_to_last_node:
8.          if n in path and n != path[0]:
9.              return True
10.     return False
```

6.3.7   Basis Paths

The number of prime paths can be exponential depending on the number of branches in a method. Complete tests of all prime paths may be resource-intensive. It would be more efficient to use linearly independent paths or basis paths [5]. A basis path is a complete path that is not a subpath of any other unique path. A path from the start node to the end node is called a complete or a test path. Test paths represent complete executions of the code under test. A basis path is a linearly independent test path. A path is linearly independent if it introduces new nodes. The basis paths for the function f in Fig. 6.3 are as follows:

$$1 : (1, 2, 3, 4, 2, 6)$$
$$2 : (1, 2, 3, 5, 2, 6)$$
$$3 : (1, 2, 6)$$

Note that all basis baths are test paths, starting with the first node and ending with the end node of the CFG. The Number of the basis paths is equal to the cyclomatic complexity of the CFG representation of the code under test.

$$\text{no. prime paths} = \text{no.edges} - \text{no.nodes} + 2 * \text{no.connected\_components}$$
$$= 9 - 8 + 2 * 1 = 3$$

McCabe [6] developed cyclomatic complexity to measure the reliability and stability of a source code function. McCabe recommended that the cyclomatic complexity of methods is better off not exceeding 10. The recommendation is to refactor the methods into smaller ones in this case. McCabe introduced the concept of linearly-independent paths. Cyclomatic complexity is equal to the number of linearly-independent paths through a function. Functions with lower cyclomatic complexity are easier to understand and less risky to modify.

Test cases satisfying the basis path coverage criterion indeed execute each "statement" at least once and every "condition" twice (once on its true side and once on its false one). However, basis path coverage is weaker than prime path coverage, and choosing weaker criterion risks overlooking some faults.

The following section describes developing code to extract CFG for a given function/method. Beforehand, let us see how to apply control graphs for software testing.

6.4   Control Flow Graph Extraction

This section provides the reader with tips on developing a software tool to extract CFG from source code. The ANTLR parser generator in the PyCharm IDE environment is used to develop the CFG extractor tool for C++ programs. In general, there are three stages to developing a software tool to extract CFG:

1. Develop a beautifier
2. Develop a graphical software to draw a control flow graph
3. Develop a software tool to extract control flow graphs

Below, the use of the ANTLR visitor [aaa] class to extract a CFG for each function in a given source code is described.

6.4.1   The Main Body of the CFG Extractor Program[2]

Figure 6.6 presents the main function of a Python program, CFGextractor, to extract a control flow graph from each function in a given source code. It takes as input a source file, "input source," and outputs a CFG for each method in the source file. CFGextractor uses the ANTLR parser generator [] to generate lexer, parser, and visitor classes for the CPP14 grammar rules []. The generated lexer, CPP14_v2Lexer, and CPP14_v2Parser, create a parse tree, parse_tree, for the given source, Input-Source. In addition to the parser and lexer classes, ANTLR generates a visitor class, CPP14_v2Visitor.

The visitor class contains a distinct method for each nonterminal. The methods are invoked while traversing the parse tree. In step 7 of the main function in Fig. 6.10, the parse tree traversal starts with the root node of the parse tree. The visit method of the ANTLR class, ParseTreeVisitor, visits a node's children and returns a user-defined result of the operation.

---

[2] Download the source code from: https://doi.org/10.6084/m9.figshare.2128461

```
def main(args):
    # Step 1: load the input source into a stream
    stream = FileStream(args.file, encoding="utf8")
    # Step 2: create an instance of CPP14_v2Lexer
    lexer = CPP14_v2Lexer(stream)
    # Step 3: tokenize the input source into a list of tokens
    token_stream = CommonTokenStream(lexer)
    # Step 4: create an instance of CPP14_v2Parser
    parser = CPP14_v2Parser(token_stream)
    # step 5: create a parse tree
    parse_tree = parser.translationunit()
    # Step 6: create an instance of CFGExtractorVisitor
    cfg_extractor = CFGExtractorVisitor()
    # Step 7: extract the cfg by visiting the parse_tree
    cfg_extractor.visit(parse_tree)
    # Step 8: pass the extracted CFGs to your application
    process_on_cfgs(cfg_extractor.functions)

if __name__ == '__main__':
    arg_parser = argparse.ArgumentParser()
    arg_parser.add_argument(
        '-n', '--file',
        help='Input source')
    args = arg_parser.parse_args()
    main(args)
```

Fig. 6.6 The main body of the CFGExtractor program for C++ functions

6.4.2 CFG Extractor Visitor Class

Whenever the visitor object encounters a nonterminal, it executes the corresponding method. For instance, the visitor object executes the visitFunctiondefinition whenever it visits the "functiondefinition" nonterminal symbol. To extract CFGs from the parse tree, we must override some CPP14_v2Visitor class methods.

Figure 6.17 presents CFGExtractorVisitor as a subclass of CPP14_v2Visitor, generated by ANTLR. The CFGExtractorVisitor class generates a partial CFG for each nonterminal symbol by overriding the corresponding method in the CPP14_v2Visitor class. The methods are named after the nonterminal rules of grammar.

The visitor pattern approach provides a method called visitR for each rule, R, in the grammar. The default behavior of these visit methods is to implement a pre-order traversal of the parse tree. However, the visitor pattern lets users visit the pars tree in any order they want. Each method in Fig. 6.7 returns a sub-graph of the CFG, built for its corresponding grammar rule. The subgraph is represented as a DiGraph (Directed Graph) data structure defined in the NetworkX python library.

Figure 6.8 represents a sample C++ program, test.cpp, and its corresponding CFG, extracted automatically by our CFGExtractor software tool. CFGExtractor is written in Python. It accepts a C++ program as input, extracts CFG for each function, f, and saves the CFG in a file, f.png. Figure 6.8 shows the CFG extracted and depicted by CFGExtractor. CFGExtractor saves the CFG in a dictionary.

CFG extractor uses a dictionary, "functions," to keep the CFG for each function. The dictionary structure for the program in Fig. 6.8 is shown in Fig. 6.9. The dictionary's keys are the signature of the functions as ANTLR context objects. The CFG for each function is kept in a dictionary of the "network.DiGraph" objects. The DiGraphs objects have two properties of nodes and edges.

Saving CFG nodes and edges into two list structures shown in Fig. 6.8, the following function, to_networkx, can be invoked to save the CFG in Networkx DiGraph format. Networkx is a rich library of graph algorithms. Any sort of graph analysis could be applied to the CFG by calling the Networkx functions in this wy.

```
def to_networkx(nodes, edges):
    g = nx.DiGraph()
    g.add_nodes_from(nodes)
    g.add_edges_from(edges)
    return g
```

The function from_netwokx takes a CFG in Networkx DiGrah format and saves it into lists, nodes, and edges.

```
def from_networkx(g):
    nodes = list(g.nodes.data())
    edges = list(g.edges.data()
    return nodes, edges
```

6.4.3 Labeling Rule Alternatives

With ANTLR, you may label rule alternatives. In this way, the ANTLR recognizer generates separate methods for each label. For instance, consider the following grammar, Expression, for expressions:

```
class CFGExtractorVisitor(CPP14_v2Visitor):
    """
    The class includes a method for each nonterminal (i.e., selection, iteration,
    jump and try-catch statements)
    Each method builds a part of a CFG rooted at its corresponding nonterminal.
    The extracted sub-graph is saved using the `networkx` library.
    visit() is the first method of the class, which is invoked initially by the main.
    """

    def __init__(self):
        """
        'functions' is a dictionary to keep each function signature and its CFG reference.
        Each CFG is kept as a `networkx.DiGraph`.
        """
        self.functions = {}

    # Functions definition
    def visitFunctiondefinition(self, ctx: CPP14_v2Parser.FunctiondefinitionContext)

    # Compound statement (statements enclosed in {})
    def visitCompoundstatement(self, ctx: CPP14_v2Parser.CompoundstatementContext)

    # Selection statements
    def visitSelectionstatement1(self, ctx: CPP14_v2Parser.Selectionstatement1Context)
    def visitSelectionstatement2(self, ctx: CPP14_v2Parser.Selectionstatement2Context)
    def visitSelectionstatement3(self, ctx: CPP14_v2Parser.Selectionstatement3Context)

    # Iteration statements
    def visitIterationstatement1(self, ctx: CPP14_v2Parser.Iterationstatement1Context)
    def visitIterationstatement2(self, ctx: CPP14_v2Parser.Iterationstatement2Context)
    def visitIterationstatement3(self, ctx: CPP14_v2Parser.Iterationstatement3Context)

    # Try-catch statement methods
    def visitTryblock(self, ctx: CPP14_v2Parser.TryblockContext)
    def visitHandlerseq(self, ctx: CPP14_v2Parser.HandlerseqContext)
    def visitHandler(self, ctx: CPP14_v2Parser.HandlerContext)
    def visitThrowexpression(self, ctx: CPP14_v2Parser.ThrowexpressionContext)

    # Jump statements
    def visitJumpstatement1(self, ctx: CPP14_v2Parser.Jumpstatement1Context)
    def visitJumpstatement2(self, ctx: CPP14_v2Parser.Jumpstatement2Context)
    def visitJumpstatement3(self, ctx: CPP14_v2Parser.Jumpstatement3Context)

    # Statement sequences
    def visitStatementseq1(self, ctx: CPP14_v2Parser.Statementseq1Context)
    def visitStatementseq2(self, ctx: CPP14_v2Parser.Statementseq2Context)

    # Expression and Declaration statements
    def visitExpressionstatement(self, ctx: CPP14_v2Parser.ExpressionstatementContext)
    def visitDeclarationstatement(self, ctx: CPP14_v2Parser.DeclarationstatementContext)

    # Label statements
    def visitLabeledstatement2(self, ctx: CPP14_v2Parser.Labeledstatement2Context)
    def visitLabeledstatement3(self, ctx: CPP14_v2Parser.Labeledstatement3Context)
```

Fig. 6.7 Visitor class overriding nonterminal symbols affecting control flow

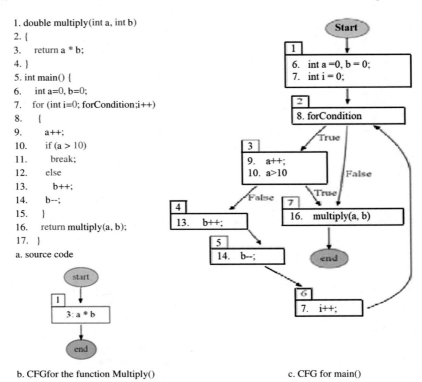

1. double multiply(int a, int b)
2. {
3.   return a * b;
4. }
5. int main() {
6.   int a=0, b=0;
7.   for (int i=0; forCondition;i++)
8.   {
9.   a++;
10.   if (a > 10)
11.   break;
12.   else
13.   b++;
14.   b--;
15.   }
16.   return multiply(a, b);
17. }
a. source code

b. CFGfor the function Multiply()   c. CFG for main()

Fig. 6.8 CFG for the functions of a given program

"nodes": [[1, {"data": [{ "line": 7, "text": "int a = 0, b = 0;" },
                              { "line": 8, "text": "int i = 0;" }]}],
            [2, {"data": [{ "line": 8,   "text": "forCondition" }]}],

            [3, {"data": [{ "line": 10, "text": "a++;" },
                              { "line": 11, "text": "a > 10" }]}],
            [7, {"data": [{ "line": 17, "text": "multiply(a, b)" }]}],
            [5, {"data": [{ "line": 15, "text": "b--;" }]}],
            [6, {"data": [{ "line": 8, "text": "i++" }]}],
            [4, {"data": [{"line": 14, "text": "b++;" }]}],
             [8, {"data": []}]]

"edges": [[1, 2, {}], [2, 3, {"state": "True"}], [2, 7, {"state": "False"}],
            [3, 4, {"state": "False"}], [3, 7, {"state": "True"}],
            [7, 8, {}], [5, 6, {}], [6, 2, {}], [4, 5, {}]]

Fig. 6.9 Dictionary structure used to keep CGGs for the C++ program in Fig. 6.7

Grammar Expression;
exp : exp + exp
| exp * exp
| (exp)
| Number
;
...

Since the grammar name is "Expression," it should be kept in a file named "Expression.g4." Right-click on the grammar file in the PyCharm environment and choose the "Generate ANTLR recognizer" option. The recognizer generates some Python files, including ExpressionLexer.py, ExpressionParser.py, and ExpressionVisitor.py. The ExpressionVisitor.py includes a class ExpressionVisitor as follows:

```
# This class defines a complete visitor for a parse
# tree produced by ExpressionParser.
Class ExpressionVisitor(ParseTreeVisitor):
  # visit this method when encountering an Exp rule.
  def visitExp(self, ctx:ExpressionParser.exp):
    return self.visitChildren(ctx)
```

A unique label for each rule alternative is defined as follows:

Grammar Expression;
exp : exp + exp   # ExpPlusExp
| exp * exp   # ExpTimesExp
| (exp)   # Exp
| Number   # ExpNumbe
;
...

This time the ExpressinVisitor includes a different method for each label. For instance, the following method is created for the rule labeled ExpPlusExp:

```
int inc(int a)
  {
     return a+1;
  }
```

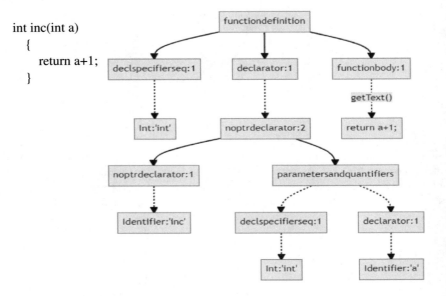

Fig. 6.10   A function, inc, and its corresponding parse tree

6.4.4   Function Definitions

The process of creating control flow graphs begins with the identification of functions in the program text. The production rule defining the syntax of functions is as follows:

    functiondefinition :
        attributespecifierseq? declspecifierseq? declarator virtspecifierseq?
        functionbody
        ;

Based on the above definition of functions, ANTLR generates the following function as a visitor method to access any function defined in the body of C++ programs:

    def visitFunctiondefinition(self, \
                            ctx:CPP14_v2Parser.FunctiondefinitionContext):
        return self.visitChildren(ctx)

The parse tree for each nonterminal, such as function definition, is traversed from the root to the leaves to determine the sequence of method calls to access different parts and elements of the nonterminal. For instance, Fig. 6.10 illustrates the parse tree for a function, inc.

All the methods needed to extract CFGs are copied from the visitor class CPP14_v2Visitor generated by ANTLR into the CFGExtractorVisitor class. To access fields of a function, such as the function name, type, parameters, or declarator, use the parse tree.

In the PyCharm edit pane, Right-click on the functiondefinition rule of the CPP14 grammar and select the "Test Rule functiondefinition" option. The previewer panel opens at the bottom. On the left is a big text box to type some sample code. On the right is a visualization of a parse tree. Beneath the tree, a slider can zoom in and out. For example, Fig. 6.9 represents a sample function, "inc," and its corresponding parse tree. Considering the sub-tree shown in Fig. 6.9, to access the function name, the following call sequence is used:

ctx.declarator().noptrdeclarator().noptrdeclarator().getText()

A function name and its formal parameters can be accessed via the ctx.declarator().getText() call sequence. Figure 6.10 represents the completed body of the visitFuctiondefinition, initially copied from the cpp14_v2Visitor class. The embed_in_function_structure(CFG) method puts a function signature and its CFG into a dictionary called "functions."

The embed_in_function_structure() function appends an ending node to the CFG, and passes the graph to split_on_return() function:

```
def embed_in_function_structure(CFG):
    g = CFG.copy()                          # copy the CFG into the graph g
    end_node = last_node(CFG) + 1           # compute the label of the end node
                                            # of the CFG
    g.add_node(end_node, data=[])           # add the end_node to the CFG
    g.add_edge(last_node(CFG), end_node)    # insert an edge connecting
                                            # the CFG to the end node.
    g = split_on_return(g)                  # Connect returns to end-node
    return solve_null_nodes(g)              # Remove null basic-blocks
```

Next, the split_on_return() function, shown in Fig. 6.12, traverses the CFG and redirects all the basic blocks (nodes), including a "return" statement, to the end node of the CFG. If any expression follows the return statement, add the expression to the basic block.

In Fig. 6.11, the visitor class's visit () function accesses all the nodes of the visitfunctiondefinition subtree in Fig. 6.10 through its input parameter, ctx. The ctx.declarator() method accesses a subtree rooted at the "declaratory" non-terminal symbol. The address of the declarator node in the parse tree is saved as a key in the self.functions dictionary.

The visitor class provides programmers with a method called visit() to select and visit a child of the nonterminal node addressed by the ctx object. When visit() call applies on a parse tree node addressed by a context object, CTX, the visitCTX() method in the visitor class would be invoked. The visitFunctiondefinition() method,

```
def visitFunctiondefinition(self,
                            ctx: CPP14_v2Parser.FunctiondefinitionContext):
    CFG = self.visit(ctx.functionbody())
    self.functions[ctx.declarator()] = embed_in_function_structure(CFG)
```

Fig. 6.11 Embedding the extracted sub-graph into the function pattern

```
def split_on_return(CFG: nx.DiGraph) -> nx.DiGraph:
    """
    Redirects the return statements to the end node of the CFG.
    """
    g = CFG.copy()
    # For each basic block number and statement in it, do
    for bb_number, statements in CFG.nodes(data="data"):
        # For each sub-tree, ctx, whose leaves end up with a statement in statements, do:
        for ctx in statements:
            # If ctx is a nonterminal node which whose immediate child is a return statement
            if is_return(ctx):
                g.remove_edges_from([(bb_number, adj) for adj in
                                     CFG.adj[bb_number]])
                # Connect the basic block to the end-node of the CFG
                g.add_edge(bb_number, last_node(CFG))
                d = statements[:statements.index(ctx)]
                # If the return statement has the following expression
                if ctx.expression():
                    # Append the expression to the enclosing basic block
                    d += [ctx.expression()]
                g.nodes[bb_number]["data"] = d
                break
    return g
```

Fig. 6.12 Connects basic blocks, **including the "return" statement to the end-node**

in Fig. 6.10 makes a visit() call on functionbody. There are four alternative rules for expanding functionbody:

```
Functionbody
    : ctorinitializer? compoundstatement              #functionbody1
    | functiontryblock                                # functionbody2
    | '=' Default ';'                                 #functionbody3
    | '=' Delete ';'                                  #functionbody4
    ;
```

The visit() function selects the appropriate rule by referring to the parse tree shown in Fig. 6.10. Considering the parse tree structure in Fig. 6.10, the visiFunctionbody1() method, generated by the ANTLR recognizer, is invoked:

```
def visitFunctionbody1(self, ctx:CPP14_v2Parser.Functionbody1Context):
    return self.visitChildren(ctx)
```

In the above function, visitChildern(ctx) is a predefined function. It calls the visitor, visitCh, function for each immediate child, ch, of the ctx object until a visitor returns false. The immediate child of functionbody1 node in the parse tree in Fig. 6.10 is compoundStatement. Therefore, the visitCompoundStatement() function is invoked by the visitChildren function.

```
# compoundstatement : '{' statementseq ? '}' :
def visitCompoundstatement(self, ctx:
                            CPP14_v2Parser.CompoundstatementContext):
    return self.visit(ctx.statementseq())
```

The CFG is built while the visitStatementseq() method detects the sequence of statements in the function's body. The statementseq nonterminal symbol and its corresponding visitor method, visitStatementseq(), are described next.

The visitStatementseq() method returns true if all the visitor calls return true; otherwise, it returns false. In the case of the compoundStatement, the visitJumpstatement3() stops these call chains by returning a graph with only one node containing its context. The "visitJumpstatement3" function is invoked whenever a *return* statement is met in the parse tree. The grammar is as follows:

Jumpstatement :   Break ';' #jumpstatement1| Continue ';'   #jumpstatement2
                | Return expression? ';'   #jumpstatement3
                | Return bracedinitlist ';'   #jumpstatement4
                | Goto Identifier ';'   #jumpstatement5 ;

The visitor function for the Jumpstatement3 non-terminal symbol is as follows:

```
def visitJumpstatement3(self, ctx: CPP14_v2Parser.Jumpstatement3Context):
    return build_single_node_graph(ctx)
```

6.4.5   Example

As an example, consider the function, identify_triangle, in Fig. 6.13. The main function of the CFGextractor program invokes the visit method of the CFGExtractorVisitor object on the start symbol, translationunit, of the C++ grammar. The visitTranslationunit() method visits all the children of the start symbol in the parse tree, and

```
1.   int triangle(int a, int b, int c)
2.   {
3.      boolean isATriangle;
4.      if ((a < b + c) && (b < a + c) && (c < a + b))
5.         isATriangle = true;
6.      Else
7.         isATriangle = false;
8.      if (isATriangle)
9.      {
10.        if ((a == b) && (b == c))
11.        {
12.           return EQUILATERAL;
13.           printf("the triangle was a EQUILATERAL");
14.        }
15.        else if ((a != b) && (a != c) && (b != c))
16.        {
17.           return SCALENE;
18.           printf("the triangle was a SCALENE");
19.        }
20.     }
21.     return("invalid");
22.  }
```

Fig. 6.13 Triangle function

each child recursively visits its children as far as a "functiondefinition" node is met and the visitor method, visitFunctionDefinition, shown in Fig. 6.29, is invoked.

The visitFunctionDefinition method invokes the visitFunctionbody1() method which itself invokes the visitCompoundstatement() method. As a result, the CFG shown in Fig. 6.14 is built. Before calling the split_on_return() method, as shown in Fig. 6.13, the "return" statements are ignored.

Once the CFG is completed, the split_on_return method is invoked to detect the basic blocks that include return statements. In Fig. 6.14, the basic block numbers 5 and 7 are connected to basic block number 8 because the return statements are initially ignored. The split_on_return() method corrects the edges and appropriately connects the basic blocks 5 and 7 to the end node of the CFG. The corrected CFG is given in Fig. 6.15. In addition to correcting the edges, all the statements after the return are removed from the basic blocks.

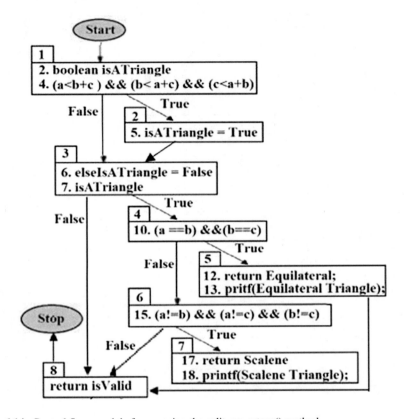

Fig. 6.14 Control flow graph before running the split_on_return() method

6.4.6 Compound Statements

As described above, a function body is a compound statement. For instance, consider the parse tree for a function, Add(), given in Fig. 6.16. Starting with the root node of the parse tree in Fig. 6.16, the visitFunctionbody method invokes the visitCompoundstatement method of the CFGExtractorVisitor class.

C++ defines a sequence of statements embraced in curly brackets as a compound statement:

compoundstatement : '{' statementseq? '}';

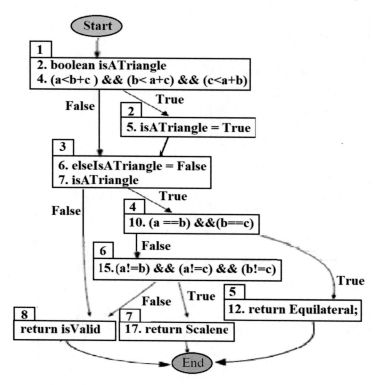

Fig. 6.15 The corrected CFG

```
int add(int a, int b)
{
    int result = a + b;
    return result;
}
```

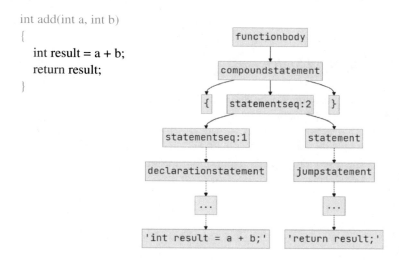

Fig. 6.16 The parse tree for the function add()

The visitCompundstatement of the CFGExtractorVisitor class calls the visit() function to invoke the visitor method of the statementseq nonterminal symbol:

```
def visitCompoundstatement(self, ctx:
                    CPP14_v2Parser.CompoundstatementContext):
    return self.visit(ctx.statementseq())
```

The statementseq nonterminal represents the sequence of statements written in the source code:

```
Statementseq
 : statement                            #statementseq1
 | statementseq statement               #statementseq2
 ;
```

Considering the node labeled compoundstatement in Fig. 6.16, the visitStatemenseq2() method is invoked by the visit() method in visitCompoundstatement():

```
#  statementseq :
#       statement #statementseq1 | statementseq statement #statementseq2;
def visitStatementseq2(self, ctx: CPP14_v2Parser.Statementseq2Context):
    gin1 = self.visit(ctx.statementseq())
    gin2 = self.visit(ctx.statement())
    return merge_graphs(gin1, gin2)
```

In the above method, gin1 and gin2 address subgraphs returned by the visitStatementseq1() and visitStatement() methods. Through the visitStatementseq1() method, considering the node labeled statementseq1 in Fig. 6.16, the visitDeclarationstatement() method is invoked:

```
# buids a CFG with a single basic block
def visitDeclarationstatement(self, ctx:
                    CPP14_v2Parser.DeclarationstatementContext):
    return build_single_node_graph(ctx)
```

The ctx object, in the above function addresses the declarationstatement node in the parse tree. The build_single_node_graph function takes the address as input and invokes the addnode() method of a NetworkX object to build the graph.

```
# buids a CFG with a single basic block
def build_single_node_graph(data=None):
    g = nx.DiGraph()
    g.add_node(0, data=[data] if data else [])
    return g
```

The build_single_node_graph() function builds the following graph for the declarationstatement node in the parse tree:

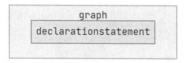

The declarationstatement node is restricted to the:

$$\text{int result} = a + b;$$

When building the final CFG, the subtree rooted at declarationstatement is parsed to reach the actual context object and replace it with the nonterminal symbol declarationstatement in the corresponding basic block.

```
graph
3: int result = a + b;
```

Similarly, a single node graph is built for the jumpstatement symbol in the parse tree. The merge_graphs function takes the two subgraphs of the statemenseq and the jumpstatement as input and merges them in a single node graph as follows:

```
def merge_graphs(gin1: nx.DiGraph, gin2: nx.DiGraph):
    gin2 = shift_node_labels(gin2, len(gin1) - 1)
    gin1_last = last_node(gin1)
    gin2_head = head_node(gin2)
    data = gin1.nodes[gin1_last]["data"] + gin2.nodes[gin2_head]["data"]
    g = compose(gin1, gin2)
    g.nodes[gin1_last]["data"] = data
    return
```

As a result, the following single node graph is built:

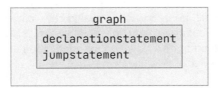

In fact, the above node is a basic block, including the address of the two nodes of the parse tree labeled decarationstatement and jumpstatement that represent two simple statements. After the CFG is completed, the parse tree is traversed, and such simple statements are replaced with the actual statements in the source code.

It is observed that starting with the root node of the parse tree in Fig. 6.16, CFGExtractorVisitor walks down the tree until it reaches a simple statement including expressionstatement, labelstatement jumpstatement and declarationstatement, defined in CPP14 grammar. So far, the sequence of calls is as follow:

visitFunctionbody() => visitCompoundstatement() => visitStatementseq2() =>
visitStatementseq1() **&&** visitStatement()

The visitSequence1() method itself invokes the visitStatement() method:

```
# statementseq1: statement ;
def visitStatementseq1(self, ctx: CPP14_v2Parser.Statementseq1Context):
    return self.visit(ctx.statement())
```

The statement method is also directly invoked by the visitStatementseq2() method. Below is the syntactical definition of statements in C++ grammar:

```
/*Statements*/
statement
  : labeledstatement
  | attributespecifierseq? expressionstatement
  | attributespecifierseq? compoundstatement
  | attributespecifierseq? selectionstatement
  | attributespecifierseq? iterationstatement
  | attributespecifierseq? jumpstatement
  | declarationstatement
  | attributespecifierseq? tryblock
  ;
```

The selection and iteration statements are defined in the following sections.

6.4.7   Selection Statements

A program's flow of control, or control flow, is the order in which instructions, statements, and functions are executed or evaluated. Control flow statements are also referred to as flow control statements. Control flow statements alter, redirect, or control program execution flow based on the application logic. The following are the main categories of control flow statements in C++:

- Selection statements
- Iteration statements
- Jump statements

This section shows how to extract CFG from selection statements, including "if" and "switch."

– CFG for if-then-else structure

if statements are a type of selection statement in C++. The grammar rule of selection statements is presented in Fig. 6.17.

Two different control flow graphs could be depicted for an if statement, depending on whether the else-part is empty. The third type of control flow graph is when there is a return statement in the body of the if statement.

Control flow graphs associate an edge with each possible branching expression in the program and a node with a sequence of statements, a basic block. A sequence of statements executing sequentially, one after the other, is called a basic block. The execution starts with the first statement and continues to the last statement of the block. Therefore, a basic block has only one entry and one exit point.

Our first example language structure is an if statement with an else clause, shown as C++ code followed by the corresponding CFG in Fig. 6.18. The if-else structure results in two basic blocks.

```
selectionstatement
   : If '(' condition ')' statement                  #selectionstatement1
   | If '(' condition ')' statement Else statement   #selectionstatement2
   | Switch '(' condition ')' statement              #selectionstatement3
   ;
```

Fig. 6.17   The CPP14_v2 labeled grammar rule for selection statements

Note that both the statements in the then-part of the "if" statement fall into the same basic block. Basic block number 1 includes the condition, "condition," of the "if" statement. It has two out-edges and is called a decision node. The out edges are called decision edges. Basic block number 3, which has more than one in-edges, is called a junction node.

The visitSelectionStatement2 implements the rule selectionStateent2 in Fig. 6.19. This rule defines the syntax of the if statements having both the then and else parts. Figure 6.20 represents the SelectionStatement2 function. The function is invoked when a visitor's member is called to visit on an if-then-else condition.

If you look at the selectionstatement2 parse tree and grammar, it has seven direct children. The third child, "condition," addresses the root of the subtree restricted to the condition, "condition," of the "if" statement in the function "f" in Fig. 6.18. In this example, the if/then-part and else-part are rooted at the node statement:3 and statement:2, respectively (Fig. 6.19). As shown in Fig. 6.20, the if-part and the else-part are indexed as 0 and 1 and accessed via "ctx.statement(0)" and "ctx.statement(1)" method calls, respectively.

The visitSelectionstatement2() builds two graphs corresponding to the if-part and else-part of the "if" statement by calling the visit() method on their visitor methods. The embed_in_if_else_structure() function takes the two graphs and fits them into an if-then-else CFG rooted at the node, referring to the condition expression of the "if" statement.

1. int f()
2. {
3.   int a = 0, b = 0;
4.   if (condition)
5.   {
6.   a++;
7.   b--;
8.   }
9.   else
10.   a--;
11.   return a + b;
12. }

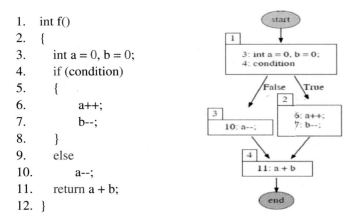

Fig. 6.18 CFG for a function including the if-else structure

selectionstatement : If '(' condition ')' statement Else statement;

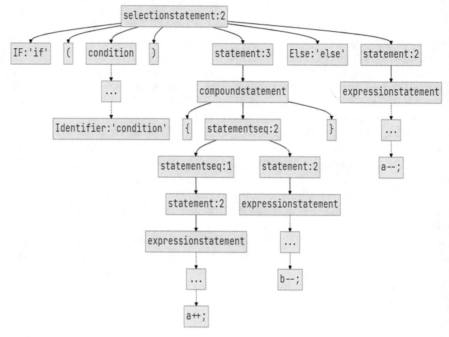

Fig. 6.19 The parse tree of the example if-else code

```
def visitSelectionstatement2(self,
                 ctx: CPP14_v2Parser.Selectionstatement2Context):
    cond = ctx.condition()
    if_body = ctx.statement(0)        # save a reference to the then part in if_body
    else_body = ctx.statement(1)      # save a reference to the then part in if_body
    then_part = self.visit(if_body)       # e.g. invokes visitStatement3()
    else_part = self.visit(else_body)      # e.g. invokes visitStatement2()
    return embed_in_if_else_structure(then_part, else_part, cond)
```

Fig. 6.20 Embeds then part and else part into an if-else structure right away

"""

Fits the then_part and else_part graphs into an if-then-else graph structure
"""

```
def embed_in_if_else_structure(then_part, else_part, condition):
    # Build a grpah including the condition expression of the if satetment
    if_graph, if_graph_head = build_initial_conditional_graph(condition)
    # shift then-part and else-parts basic block numbers to fit into the CFG
    else_part, then_part = shift_if_else_graphs(len(if_graph), else_part, then_part)
    # retrieve then-part and else-part graphs head and last node labels
    then_part_head, then_part_last = get_graph_label_range(then_part)
    else_part_head, else_part_last = get_graph_label_range(else_part)
    # calculate the last node number of the if-then-else graph
    if_graph_last = else_part_last + len(if_graph)
    # Connect the if-then-else basic blocks in a diamond shape
    if_graph.add_edges_from([(if_graph_head, else_part_head, {"state": "False"}),
                (if_graph_head, then_part _head, {"state": "True"}),
                (then_part_last, if_graph_last),
                (else_part_last, if_graph_last)])
    # set the basic junction block like an empty block
    if_grap.nodes[if_grap _last]["data"] = []
    # Connect the if-then-else basic blocks in a diamond shape
    if_grap = compose(if_grap, then_part, else_part)
    return if_grap
```

For instance, the graph built for the "if" statement in Fig. 6.18 is as follows:

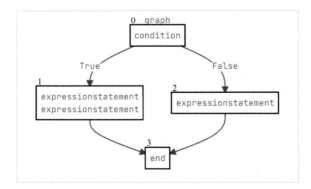

The build_initial_conditional_graph() function builds the head node for the "if" and "while" statements graph:

```
def build_initial_conditional_graph(condition):
    graph = nx.DiGraph()
    graph_head = 0
    graph.add_node(graph_head, data=[condition])
    return graph, graph_head
```

For instance, the graph built for the "condition" of the "if" statement in Fig. 6.18 is as follows:

To insert the graph, if_graph, for an "if" statement into the graph of its enclosing function, g, the number assigned to each basic block in the if_graph should be rearranged to match the numbers assigned to the basic blocks in g. The following method, shift_if_else_graphs(), shifts the basic block numbers of the then-part and else-part of an "if" statement by "n":

```
# shifts the basic block numbers by n
def shift_if_else_graphs(n, then_part, else_part):
    then_part = shift_node_labels(then_part, n)
    else_part = shift_node_labels(else_part, len(then_part) + n)
    return then_part, else_part
```

The get_graph_label_range() function returns the head and the last node number of a given graph.

```
def get_graph_label_range(graph):
    return head_node(graph), last_node(graph)
```

Finally, the compose() function accepts a list of graphs passed as input and combines them into a unified graph.

```
def compose(*graphs) -> nx.DiGraph:
    return reduce(lambda acc, x: nx.compose(acc, x), graphs)
```

– **CFG fragment for if-then structure**

Let us now examine if statements without an else clause. The control graph for the if-else statements includes only three nodes. The tree nodes are covered by a test case, provided that the condition of the if-else statement is true. However, the edges cannot be covered by a single test case. For example, consider the if statement, shown in Fig. 6.21.

```
1.   int f()
2.   {
3.       int a = 0, b = 0;
4.
5.       if (condition)
6.       {
7.           a++;
8.           b--;
9.       }
10.
11.      return a + b;
12.  }
```

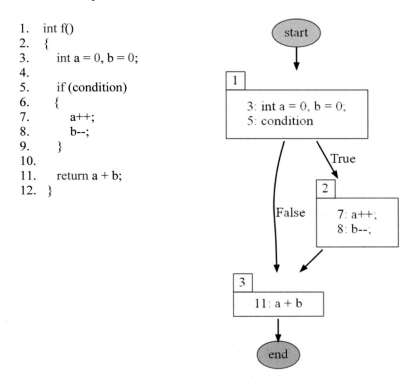

Fig. 6.21 CFG fragment for the if structure without an else clause

The parse tree for the code fragment in Fig. 6.21 is given in Fig. 6.22. Due to the lack of space, the tree is pruned by removing some branches.

Below is the graph pattern for an "if" statement without an "else" is shown. It has a decision node with two out-edges and a junction node with two in-edges.

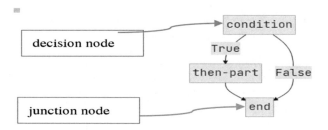

Fig. 6.22 The trimmed
parse tree for the function f()
in Fig. 6.21

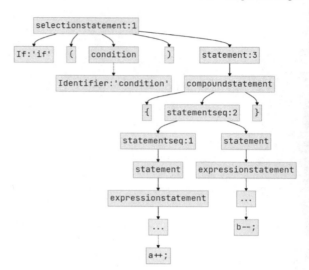

The following function is invoked to build a CFG for a given if-then statement. The
function is similar to the embed_in_if_else_structure() function described above.

```
def embed_in_if_structure(then_part, condition) -> nx.DiGraph:
    # Build a grpah including the condition expression of the if satetment
    decision_graph, decisin_head = build_initial_conditional_graph(condition)
    # shift then-part basic block number to fit into the CFG
    then_part = shift_node_labels(then_part, len(decision_graph))
    # retrieve then-part and else-part graphs head and last node labels
    then_part_head, then_part_last = get_graph_label_range(then_part)
    # calculate the last node number of the if-then-else graph
    decision_graph_last = then_part_last + len(decision_graph)
    # Connect the ifthen basic blocks
    decision_graphadd_edges_from([(decision_graph_head,
                decision_graph_last, {"state": "False"}),
                (decision_graph_head, then_part_head, {"state": "True"}),
                (then_part_last, decision_graph_last)])
    # set the basic junctionblock like an empty block
    decision_graphnodes[decision_graph_last]["data"] = []
    # Connect the ifthen-else basic blocks in a diamond shape
    decision_graph= compose(decision_graph, then_part)
    return decision_graph
```

- **CFG structure for switch statements**

The switch statement is the last selection statement in the grammar. The syntax for
switch statements in C++ is as follows:

selectionstatement
  : If '(' condition ')' statement #selectionstatement1
  | If '(' condition ')' statement Else statement #selectionstatement2
  | Switch '(' condition ')' statement #selectionstatement3
  ;

statement
  : labeledstatement
  | Expressionstatement
  | jumpstatement
  | ... ;

labeledstatement
  : attributespecifierseq? Identifier ':' statement #labeledstatement1
  | attributespecifierseq? Case constantexpression ':' statement #labeledstatement2
  | attributespecifierseq? Default ':' statement #labeledstatement3
  ;

Each case may contain a break statement at the end of its instruction sequence, which would direct the execution flow to the statement after the switch. If there is no jumping statement, such as break and return within a case part, it continues with the following statement. If the switch has a default case, a case is always entered. For instance, consider the CFG for the switch statement shown in Fig. 6.23. The out edges of the basic block number 1, including the switch condition expression, are labeled with the case values.

Figure 6.24 shows the pruned parse tree for the function f shown in Fig. 6.23. The parse tree determines the chain of method calls to access different parts of the switch statement.

The visitor method for a nonterminal symbol node in a parse tree controls the order of visiting the child modes during a preorder walk. Considering the syntactic structure of switch statements, the visitor method to build CFG fragments for switch statements is as follows:

```
#selectionstatement Switch '(' condition ')' statement #selectionstatement3
def visitSelectionstatement3(self,
      ctx: CPP14_v2Parser.Selectionstatement3Context):
   condition = ctx.condition()
   statement_graph = self.visit(ctx.statement())
   return embed_in_switch_structure(statement_graph, condition)
```

The embed_in_switch_structructure function implements the graph pattern for the switch statement, shown in Fig. 6.25. The head of the graph addresses the switch condition expression. Each case part is a graph that points by an arrow link to the end node or the next case part, depending on whether or not it contains a break statement.

```
int f() {
    int switcher;
    switch (switcher)
    {
        case 0:
            c0++;
            break;
        case 1:
            c1++;
            c1--;
        case 2:
            c2++;
            c2--;
            break;
        default:
            dfl++;
    }
    return c0 * c1 + c2;
}
```

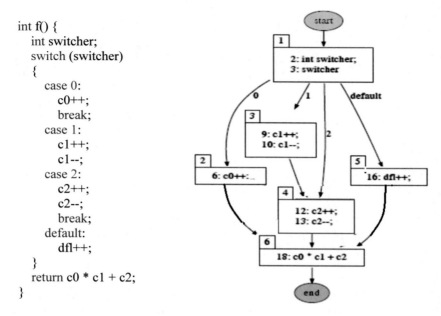

Fig. 6.23 A switch statement and its corresponding CFG

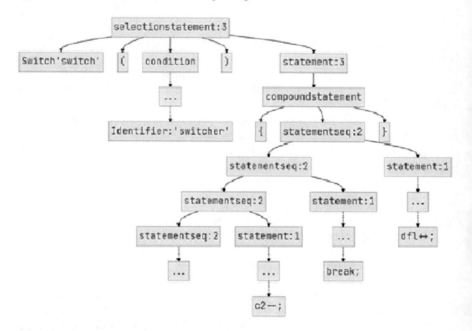

Fig. 6.24 The pruned parse tree for the switch statement in Fig. 6.23

Fig. 6.25 Switch statements
graph pattern

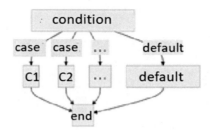

If the default (part of the switch statement) is not present, the default node is removed
from the pattern.

6.4.8   Iteration Statements

A loop, or iteration statement, executes a statement, known as the loop body, repeatedly until the controlling expression is false (0). The C++ grammar for the iteration
statements is as follows:

    iterationstatement
        : While '(' condition ')' statement   #iterationstatement1
        | Do statement While '(' expression ')' ';'   #iterationstatement2
        | For '(' forinitstatement ';' condition? ';' expression? ')' statement
        | For '(' forrangedeclaration ':' forrangeinitializer ')' statement
        ;

– **While statements**

Genuinely, the CFG representation of a "while" loop is almost similar to an if statement without an else clause. The only difference is a back edge connecting the last
node of the "while" body to its head node. Care should be taken about drawing CFG
for an iteration statement, including a jumping statement such as break or continue.
In such cases, the node containing a "break" or "continue" statement should respectively point to the head node or end node of the "while" statement graph. For instance,
consider the control flow graph for a while, including break and continue statements,
in Fig. 6.26.

   Another code fragment, including a simple while statement and its corresponding
control flow graph, is given in Fig. 6.27.

```
int f()
  {
    int a = 0, b = 0;
    while (whileCondition)
    {
       a++;
       if (continueCondition) continue;
       a--;
       if (breakCondition) break;
       b--;
    }
    return a * b;
  }
```

Fig. 6.26 Control flow graph for a while statement

```
1.    int f()
2.    {
3.        int a = 0, b = 0;
4.        while (whileCondition)
5.        {
6.            a++;
7.            b--;
8.        }
9.        return a * b;
10.  }
```

Fig. 6.27 A while statement and its CFG representation

It is observed that both of the non-conditional instructions inside the while body appear in the same basic block. The condition of a "while" statement is the only element in the head node of the "while" graph. Once a "while" condition occurs in a basic block, it is split into a new basic block. The parse tree for the function f() in Fig. 6.27 is illustrated in Fig. 6.28.

The parse tree helps us to write code for the visitor methods. Below is the visitor method, visitIterationstatement1(), for the iterationstatement1 nodes:

```
iterationstatement: While '(' condition ')' statement;     #iterationstatement1
def visitIterationstatement1(
      self, ctx: CPP14_v2Parser.Iterationstatement1Context):
   condition = ctx.condition()
   gin = self.visit(ctx.statement())
   return embed_in_while_structure(gin, condition)
```

iterationstatement : While '(' condition ')' statement;

```
                        iterationstatement:1

While:'while'    (    condition    )              statement:3

                             ...              compoundstatement

          Identifier:'whileCondition'    {   statementseq:2   }

                                       statementseq:1   statement:2

                                          statement:2      ...

                                             ...           b--;

                                             a++;
```

Fig. 6.28 The parse tree for the "while" statement in Fig. 6.28

Below is the pattern graph for the while statements. The pattern is used to write the code for the embed_in_while_structure() function.

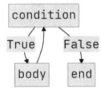

Note that the iteration structures may contain "continue" and "break" jump statements in their bodies. The "break" statement splits its enclosing basic block and points it to the statement after the "while" statement. The "continue" statement does not terminate the while loop and simply points its enclosing basic block to the basic block, including the "while" condition.

```
# Implements the "while" statement pattern graph
def embed_in_while_structure(while_body, condition):
    # Build a graph including the condition node of the while pattern graph
    while_graph, while_graph_cond, while_graph_head = \
                          build_initial_while_graph(condition)
    # Shift the label numbers of the "while" body nodes by the size of the head
    while_graph_body = shift_node_labels(while_body, len(while_graph))
    while_body_head, while_body_last = get_graph_label_range(while_body)
    while_graph_last = while_body_last + 1
    # Add edges to the condition graph
    while_graph.add_edges_from([(while_graph_head, while_graph_cond),
             (while_graph_cond, while_body_head, {"state": "True"}),
             (while_graph_cond, while_graph_last, {"state": "False"}),
             (while_body_last, while_graph_cond)])

    while_graph.nodes[while_graph_last]["data"] = []
    while_graph = compose(while_graph, while_body)
    while_graph = split_on_continue(while_graph, while_cond)
    while_graph = split_on_break(while_graph)
    return while_graph
```

The body of a "while" statement is a statement. The embed_in_while_structure()
function takes the graph of the "while" body as input and builds a CFG for the
"while" statement. At first, it builds a graph for the condition expression of the
"while" statement by calling the build_initial_while_graph() function.

The build_initial_while_graph() function creates the head node and the "condition" node of the CFG structure and copies the "condition" node address from the
parse tree to the "condition" basic block of the CFG.

```
def build_initial_while_graph(condition):
    graph = nx.DiGraph()
    graph_head, graph_cond = 0, 1
    graph.add_nodes_from([(graph_head, {"data": []}),
      (graph_cond, {"data": [condition]})])
    return graph, graph_cond, graph_head
```

The shift_node_labels() function, invoked by embed_in_while_structure, shifts
the basic block numbers of the "while" body graph by the length of the graph of the
condition expression. After calculating the last node's Number and adding the edges
to the graph built for the condition expression, the compose() function is invoked to
combine the resulting graph with the graph of the body of the "while" statement.

The "continue" statement inside the body of a while statement splits the basic block containing the "continue" statement and adds an edge from the basic block to the basic block of the "while" condition.

```
def split_on_continue(graph: nx.DiGraph, continue_return) -> nx.DiGraph:
    return direct_node_to_if(graph, continue_return, is_continue)
```

Similar to the "continue" statement, a "break" statement also splits its enclosing basic block. The difference is that the basic block points to the statement immediately after the "while" statement.

```
def split_on_break(graph: nx.DiGraph) -> nx.DiGraph:
    return redirect_nodes(graph, last_node(graph), is_break)
```

The is_break method checks whether the current context object is a break statement.

```
def is_break(rule: ParserRuleContext) -> bool:
    return rule.start.type == CPP14_v2Lexer.Break
```

The redirect_nodes function looks for a jump statement, jump_statement, in a given graph node. Any node, including the jump statement, is split and connected to a given target node.

```
# Splits and redirects node, including a jump statement to the target node
def redirect_nodes(graph: nx.DiGraph, dtarget_node,
                   jump_statement) -> nx.DiGraph:
    # gin is a networkx graph including a list of nodes: [(label, statements), ...]
    g = graph.copy()
    for label, data in graph.nodes(data="data"):
        # for each statement in data
        for ctx in data:
            # if it is a jump statement? (break, continue, return, or throw)
            if jump_statement(ctx):
                # remove all the out edges to the successor nodes
                g.remove_edges_from([(label, adj) for adj in graph.adj[label]])
                # Connect to the target node
                g.add_edge(label, target)
                # remove statements after the jump statement from the  node
                g.nodes[label]["data"] = data[:data.index(ctx)]
                break
    return g
```

```
1.   int f()
2.   {
3.       int a = 0, b = 0;
4.       Do
5.       {
6.           a++;
7.           b--;
8.       } while (whileCondition);
9.       return a / b;
10.  }
```

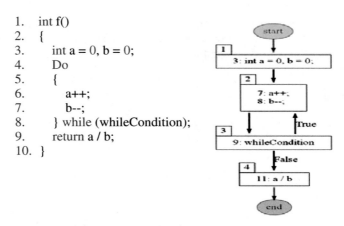

Fig. 6.29 A do-while statement and its control flow graph

– **do-while statements**

A do-while loop is similar to a while loop, except that the loop condition is after the loop's body. It is also known as an exit-controlled loop. A do-while loop executes at least once because its termination condition is checked after the loop body. A graphical representation of all possible runtime control-flow paths for a do-while statement is shown in Fig. 6.29.

The do-while condition resides in basic block number 2 because it is not a jump target. However, as shown in Fig. 6.30, the condition is moved to a distinct basic block because of the if statement.

6.5   Control Flow Graphs for Python Functions[3]

This section provides the reader tips on developing a software tool to extract CFG from Python source code. For this purpose, we use the AST module of Python [7]. AST stands for Abstract Syntax Tree. The AST module helps Python applications process trees of the Python abstract syntax grammar. The abstract syntax itself might change with each Python release; this module helps determine programmatically what the current grammar looks like. In general, there are two stages to developing a software tool to extract CFGs:

1. Develop graphical software to draw a control flow graph
2. Develop software to extract control flow from Python code.

Below, we describe our program to understand how to extract CFGs from Python source code.

---

[3] Download the source code from: https://doi.org/10.6084/m9.figshare.2128461

1.   **int** do_while_st(**int** am **int** b)
2.   {
3.   do
4.   {
5.   a++;
6.   b--;
7.   if(b>a) **continue**;
8.   a = a+2;
9.   } **while** (whileCondition);
10.   **return** a / b;
11. }

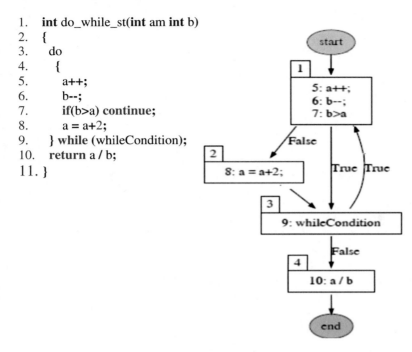

Fig. 6.30 Control flow graph for a do-while statement

6.5.1   What is an AST?

Abstract syntax tree or, in short, AST is a compressed representation of the parse tree. It can be generated by eliminating non-terminal symbols from the parse tree. ASTs are also called syntax trees. For instance, consider the following grammar representing the structure of an expression, exp:

Figure 6.31 shows the parse tree for a given assignment statement. The parse tree includes both non-terminal and terminal symbols. As shown in Fig. 6.32, after eliminating all the nonterminals, the parse tree converts into an abstract syntax tree.

Figure 6.33 illustrates the AST for an if statement as a trinary operator, including three operands. The block operator is an n-ary operator where n is the number of statements in the compound statement, represented as a block. A function name is also an operator whose operands are the function call's actual parameters.

Compilers generate ASTs as intermediate codes because they can be easily converted to assembly code. ASTs are used heavily in program transformation and static analysis tools. Refactoring, smell detection, and reverse engineering techniques are sorts of static analysis. In this section, ASTs are used to extract CFGs from the source code of the Python functions and methods.

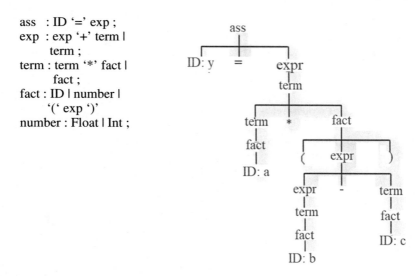

```
ass  : ID '=' exp ;
exp  : exp '+' term |
       term ;
term : term '*' fact |
       fact ;
fact : ID | number |
       '(' exp ')'
number : Float | Int ;
```

Fig. 6.31 Parse tree for the assignment statement "y = a* (b − c)"

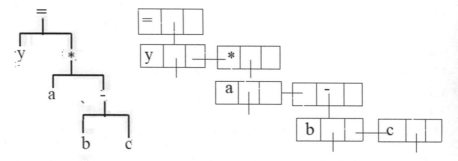

(1) Ast for y=a*(b-c)   (2) binary tree representation of AST

Fig. 6.32 Two AST representations for the statement in Fig. 6.31

6.5.2 Using AST Class in Python

As described in the previous section, each language construct is represented by a different CFG fragment. An AST, a condensed representation of a parse tree, is useful for detecting and analyzing language constructs. Therefore, AST representations of functions facilitate building CFGs for Python functions. Each language constructs, such as if-then-else, for, compound statement, and try-catch, is represented by a subtree rooted at an operator indicating the type of the construct. For instance, in Fig. 6.33, the 'if,' 'block,' and 'f' operators indicate the root of the subtrees for the if, compound, and function call statements, respectively.

if a > b + c
    then a := a + 1
    else begin

        a := f(a*b, c, 5);

        a := a + b;
end;

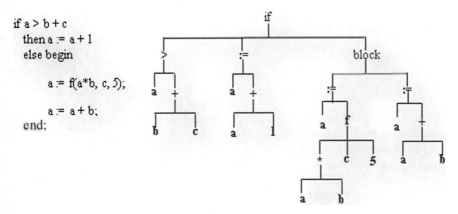

Fig. 6.33 AST representation of an if statement as a trinary operator

AST representation facilitates the automated generation of CFGs for the functions. Hence, before anything else, the ast.Parse method is invoked to create an AST tree for a given function as follows:

```
source_file = 'input_function.py'
# 1. Open input file
with open(source_file, 'r') as src_file:
    srouce = src_file.read() + "\n" + "End"
# 2. Parse the tree
ast_root = ast.parse(source, mode='exec')
```

The above code fragment reads a Python function into a buffer called 'source.' The ast.Parse method takes 'source' as a string and converts it into an AST. The parameter 'mode' determines the compile mode of the code. In this case, it is an exec, which takes a code block with statements, classes, functions, etc. The ast.Parse() method returns the AST root's address, stored in ast_root. The visit(ast_root) method call traverses the AST in order.

A CFG fragment can be generated for each type of statement while traversing the AST. The visit(ast_root) starts walking the AST from the tree's root, 'ast_tree.' While visiting a node labeled 'm', it calls the corresponding method 'visit_m.' if the method does not exist, it proceeds with the next child of the node in the AST.

```
# 2. Parse the tree
ast_root = ast.parse(source, mode='exec')
# 3. Build the CFG
cfg = CFGVisitor().visit(ast_root)
```

The class CFGVisitor extends from the parent class, ast.NodeVisitor. The reason for defining CFGVisitor is to override the predefined methods visit_R in

ast.NodeVisitor to build a CFG fragment for each non-terminal symbol R. For
instance, the visit_break() method is rewritten as follows:

```
def visit_Break(self, node):
    assert len(self.loop_stack), " break not inside a loop"
    self.add_edge(self.curr_block.number, self.loop_stack[-1].number, ast.Break())
```

Note that child nodes with a custom visitor method will not be visited unless the
visitor calls generic_visit() or visits them.

As traversing the abstract syntax tree when visiting a break statement node, the
visitor invokes the function visit_break to operate on the node. Here, the visit_break
function invokes the add_edge method to add an edge from the current block number,
cur_block.number, including the break statement, to the basic block immediately
after the loop body.

```
def add_edge(self, from_block_number: int, to_block_number: int,
             condition=None) -> BasicBlock:
    self.cfg.blocks[from_ block_number].next.append(to_block_number)
    self.cfg.blocks[to_ block_number].prev.append(from_ block_number)
    self.cfg.edges[(from_ block_number, to_ block_number)] = condition
    return self.cfg.blocks[to_ block_number]
```

For instance, Fig. 6.34 illustrates a CFG including a break statement inside a
while loop. Figure 6.34a represents the graph before the break statement is met. The
complete CFG is shown in Fig. 6.34b.

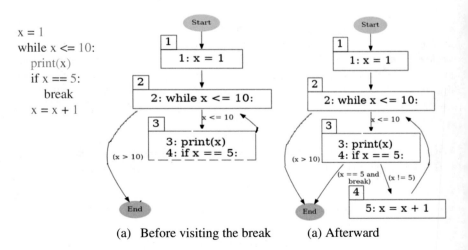

(a) Before visiting the break   (a) Afterward

Fig. 6.34 A break statement inside a while loop

6.5.3 Automatic Construction of CFGs for Python Programs

Figure 6.35 presents the main function of a Python program, CFGextractor, to extract a CFG from a Python function. It takes as input a source file, "filename," and outputs a CFG representing the flow of control between the statements in each function within the source code. CFGextractor uses the AST visitor classes for the Python grammar rules [8].

The visitor class contains a distinct method for each non-terminal. The methods are invoked while traversing the abstract syntax tree (AST). The AST is built by the method ast.parse(). The ast.parse() method tokenizes the source code into tokens based on Python lexical rules and then builds an abstract syntax tree (AST) upon the tokenized code. The resulting abstract syntax tree, "ast_root," is passed as a parameter to the build method of the CFGVisitor class, described in the next section, to build the CFG.

6.5.4 The Visitor Class

The CFGVisitor class, shown in Fig. 6.36, overrides the ast.NodeVisitor methods to build CFGs for different program constructs. The build method of the CFGVisitor class is called on AST by the main function in Fig. 6.35. While walking through the abstract syntax tree, the nodeVisitor class calls a function, visit(), for every node it meets.

```
if __name__ == '__main__':
    filename = 'code.py'
    # 1. Open input file
    with open(filename, 'r') as src_file:
        src = src_file.read() + "\n" + "End"
    # 1. Build abstrat syntax tree
    ast_root = ast.parse(src, mode='exec')
    #2- Generate CFG
    cfg = CFGVisitor().build(filename, ast_root)
    #3- Save the generated CFG as an image in a pdf file
    cfg.show()
```

Fig. 6.35 The main body of the CFGExtractor program for Python functions

The visit() method, by default, calls the method self.visit_classname, where the suffix classname is replaced with any class node name such as "if," "while," "for," "pass," "return," "break," or "continue," depending on the AST node name met. For instance, if a node labeled "if" is visited while traversing the AST, the method named visit_if() is invoked.

The build method of the **CFGVisitor** class takes the file's name, including the source of the Python function and the root address of the AST, built for the function, as input. The method returns the CFG for the function. It initially creates an instance, self.cfg, of the class CFG. Then invokes ast.visit(ast_root) to traverse the AST. The default behavior of the visit methods is to implement a traversal of the abstract syntax tree. Each of the methods in Fig. 6.36 returns a sub-graph of the CFG, built for its corresponding grammar rule.

Figure 6.37 represents a sample Python program and its corresponding CFG, extracted automatically by our CFGExtractor software tool. CFGExtractor is written in Python. It accepts a Python program as input, builds a CFG for it, and saves the CFG in a file, 'output.pdf.' Figure 6.37 shows the CFG extracted and depicted by CFGExtractor:

CFGextractor builds a separate CFG for each function it meets in a given Python program. It uses a dictionary called functions. This dictionary stores the function's name as the key and an object of a class called CFG as the value. The object address the CFG built for the function. Two other dictionaries are blocks and edges. The *blocks* dictionary keeps basic-block numbers as the key and the statements contained in the basic block as the value. The key of the edges dictionary is a pair of basic block numbers connected through an edge, and its value is the edge's label. The class is listed below:

```
Class CFG:
    # Stores CFG nodes and basic blocks in two dictionaries
    def __init__(self, function_name: str):
        self.function_name: str = function_name

        #The BasicBlock object holds the details of each basic block
            # 1. Fake nodes including start and end nodes
            self.start: Fake_nodes[BasicBlock] = None
            # 2. The function name for which the CFG will be built.
            self.functions: Dict[str, CFG] = {}
            # 3. CFG basic blocks
            self.blocks: Dict[int, BasicBlock] = {}
            # 4. CFG edges
            self.edges: Dict[Tuple[int, int], Type[ast.AST]] = {}
    # Saves CFGs in DiGraph format.
```

```python
class CFGVisitor(ast.NodeVisitor):
    """
    Control flow graph builder.
    A control flow graph builder is an ast.NodeVisitor that can walkthrough
    a program's AST and iteratively build the corresponding CFG.
    """
    # Invert the operation in an AST node object (get its negation).
    invertComparators: Dict[Type[ast.AST], Type[ast.AST]] = \
        {ast.Eq: ast.NotEq, ast.NotEq: ast.Eq, ast.Lt: ast.GtE,
         ast.LtE: ast.Gt,
         ast.Gt: ast.LtE, ast.GtE: ast.Lt, ast.Is: ast.IsNot,
         ast.IsNot: ast.Is, ast.In: ast.NotIn, ast.NotIn: ast.In}
    def __init__(self):
        super().__init__()
        # the loop stack keeps the basic block after the body of a loop
        self.loop_stack: List[BasicBlock] = []
        # the loop stack keeps the body of the loop being processed
        self.curr_loop_guard_stack: List[BasicBlock] = []
        # ifexp = if there is an if statement within a loop
        self.ifExp = False
    # Build a CFG from an AST
    def build(self, filename: str, ast_root: Type[ast.AST]) -> CFG:
        # 1- initialize 3 dictionaries to store the CFG name, nodes, and edges
        self.cfg = CFG(filename)
        # 2- create a new basic block
        self.curr_block = self.new_block()
        # 3- Use the newly created block as the start node
        self.cfg.start = self.curr_block
        # 4- Start traversing the AST from its root node
        self.visit(ast_root)
        # 4- remove empty blocks created for the else-parts of 'if' and 'for' statements
        self.remove_empty_blocks(self.cfg.start)
        return
    # Create a new block with a new id
    def new_block(self) -> BasicBlock
    # Add a statement to a block
    def add_stmt(self, block: BasicBlock, stmt: Type[ast.AST]) -> None
    # Link between blocks in a control flow graph
    def add_edge(self, frm_id: int, to_id: int, condition=None) -> BasicBlock
    # Create a new block for a loop's guard if the current block is not empty. Links the
    # current block to the new loop guard
```

Fig. 6.36 The visitor class to be completed to extract CFGs

```
        def add_loop_block(self) -> BasicBlock:
        # ---------- AST Node visitor methods ---------- #
            def visit_Assign(self, node): ...
            def visit_Expr(self, node): ...
        # Iteration statements
            def visit_For(self, node): ...
            def visit_While(self, node): ...
        # Selection statements
            def visit_If(self, node): ...
            def visit_IfExp(self, node): ...
        # compact syntax
            def visit_Lambda(self, node): ...
            def visit_ListComp(self, node): ...
            def visit_DictComp(self, node): ...
        # Jump statements
            def visit_Return(self, node): ...
            def visit_Continue(self, node): ...
            def visit_Break(self, node): ...
        # Try-except statement methods
            def visit_Try(self, node): ...
            def visit_Raise(self, node): ...
```

Fig. 6.36   (continued)

```
    def _traverse(self, block: BasicBlock, visited: Set[int]=set(), calls: bool=True,\
            is_verbose: bool = True) -> None:
        if block.number not in visited:
            visited.add(block.number)
            content = block.fill_with_sts(is_verbose)
            self.graph.make_node(str(block.number), content)
            for next_block_number in block.next:
                from_node = str(block.number)
                to_node = str(next_block_number)
                edge = block. edge_to(next_block_number)
                if ("break" or "continue") not in edge:
                    self.graph.edge(from_node, to_node, label="True")
                    elif 'None' in edge:
                        self.graph.edge(from_node, to_node, label="")
                    elif "break" in edge:
                        self.graph.edge(from_node, to_node, label="break")
                    elif "continue" in edge:
                        self.graph.edge(from_node, to_node, label="continue")
                    else:
                        self.graph.edge(from_node, to_node, label="False")
        self.graph.render("filename.zip")
```

```
def max(a, b, c):
    if a > b and a > c:
        print(f'{a} is maximum among all')
    elif b > a and b > c:
        print(f'{b} is maximum among all')
    else:
        print(f'{c} is maximum among all')
```

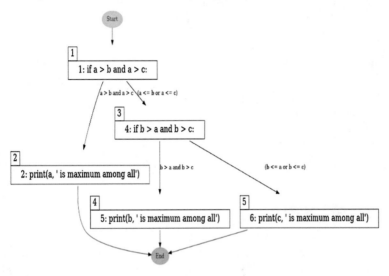

Fig. 6.37 CFG built by CFGExtractor for the function max

The dictionary data structures holding the CFG for the program in Fig. 6.37 is shown below.

"functions": {'max': <__main__.CFG object at 0x7f242701e9a0>}
"blocks" : {1: <model.BasicBlock object at 0x7f115fd756a0>,
          2: <model.BasicBlock object at 0x7f115fd757c0>,
          3: <model.BasicBlock object at 0x7f115fd757f0>,
          4: <model.BasicBlock object at 0x7f115fd75880>,
          5: <model.BasicBlock object at 0x7f115fd75a30>,
          6: <model.BasicBlock object at 0x7f115fd75a90>,
          7: <model.BasicBlock object at 0x7f115fd75af0>}
"edges": {(1, 3): <_ast.BoolOp object at 0x7f115fd37220>,
          (1, 4): <_ast.BoolOp object at 0x7f115fd75940>,
          (4, 6): <_ast.BoolOp object at 0x7f115f643fa0>,
          (4, 7): <_ast.BoolOp object at 0x7f115fd75b80>,
          (3, 2): None, (7, 2): None, (6, 2): None}

The CFG class includes two methods, _traverse, and _show. The following method, _show, invokes the _traverse method to walk through the CFG, and label the edges. The _show saves the CFG in the DiGraph format. The Digraph is a Python built-in class to store the NetworkX and Graphviz graphs.

```
# Converts a NetworkX to graphviz graph using the Digraph format
def _show(self, fmt: str = 'pdf', is_verbose: bool = True) -> gv.dot.Digraph
    # Creates a graph labeled name, with format = fmt & named cluster_name
    self.graph = gv.Digraph(name='cluster_' + self.name, format=fmt,
                    graph_attr={'label': self.name})
    self.graph.attr('node', shape='none')
    # Add a start node clolored green and shaped oval
    self.graph.node('Start', _attributes=
                                {'color': '#aaffaa', 'style': 'filled', 'shape': 'oval'})
    # Connect the start node to the first basic block
    self.graph.edge('Start', '1')
    self._traverse(self.start, is_verbose=is_verbose)
    return self.graph
```

6.5.5 If Statements

The visit_if function is an example of the visitor functions of the CFGVisitor class. If, while traversing the AST, the visitor visits a node representing an 'if' operator, it invokes the visit_if method of the class CFGVisitor. The visit_if function first adds the if statement condition to the end of the current basic block. Then, it creates a new basic block to add the if body. Another block is created to include statements after the if-else.

```
def visit_If(self, node):
    # Add the if-statement condition at the end of the current basic block.
    self.add_stmt(self.curr_block, node)
    # Create a block for the code after the if-else.
    afterif_block = self.new_block()
    # Create a new block for the then-part of the if statement.
    if_block = self.add_edge(self.curr_block.number,
                             self.new_block().number, node.test)
    # Create a new block for the else part if there is an else part
    if node.orelse:
        self.curr_block = self.add_edge(self.curr_block.bid,
                            self.new_block().bid, self.invert(node.test))
        # Visit the children in the else's body to populate the block.
        self.populate_body(node.orelse, afterif_block.bid)
    else:
        self.add_edge(self.curr_block.bid, afterif_block.bid, self.invert(node.test))
    # Visit children to populate the if block.
    self.curr_block = if_block
    self.populate_body(node.body, afterif_block.bid)
    # Continue building the CFG  the after-if block.
    self.curr_block = afterif_block
```

In the continuation, examples of using CFGExtractor to extract CFG for different types of "if" statements are given.

1. **Simple if statement**

A Simple If statement only has one condition to check.

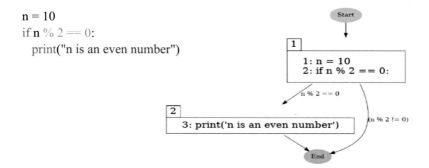

2. **if-else statements**

The if-else statement evaluates the condition and executes the if body provided that the condition is True, but if the condition is False, then the body of else is executed.

```
n = 5
if n % 2 == 0:
    print("n is even")
else:
    print("n is odd")
```

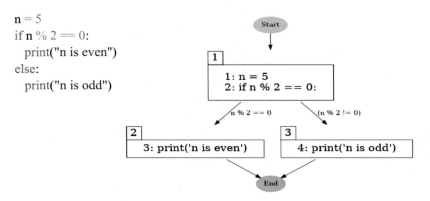

3. Nested if statements

If statements are written in a syntax that is either simple or nested. Nested if statements are an if statement inside another if statement.

```
a = 5
b = 10
c = 15
if a > b:
    if a > c:
        print("a value is big")
    else:
        print("c value is big")
elif b > c:
    print("b value is big")
else:
    print("c is big")
```

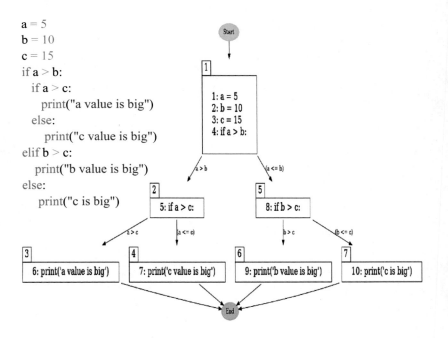

4. if-elif-else statements

There are times when several conditions coexist. As a solution, Python allows adding elif clauses before and after an if statement.

```
x = 15
y = 12
if x == y:
    print("Both are Equal")
elif x > y:
    print("x is greater than y")
else:
    print("x is smaller than y")
```

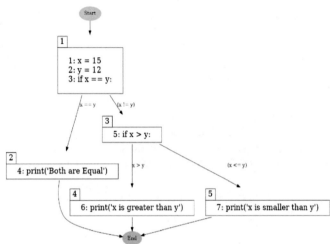

6.5.6 Iteration Statements

The loop statements support the iteration mechanism in Python. This section presents a sample for and a while statement and their CFG, extracted by CFGExtracor.

1. **For loops**

A for loop iterates over a sequence that is either a list, tuple, dictionary or set. We can execute a set of statements once for each item in a list, tuple, or dictionary:

```
lst = [1, 2, 3, 4, 5]
for i in range(len(lst)):
    print(lst[i], end = " ")

for j in range(0,10):
    print(j, end = " ")
```

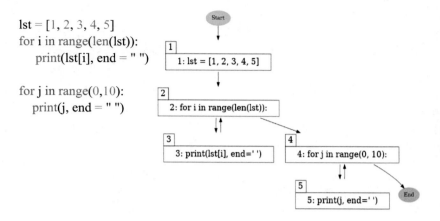

2. while loops

In Python, "while" loops are used to execute a block of statements repeatedly until a given condition is satisfied. Then, the expression is rechecked, and if it is still valid, the body is executed again. This continues until the expression becomes false.

```
m = 5
i = 0
while i < m:
    print(i, end = " ")
    i = i + 1
```

```
m = 5
i = 0
while i < m:
    print(i, end = " ")
    i = i + 1
```

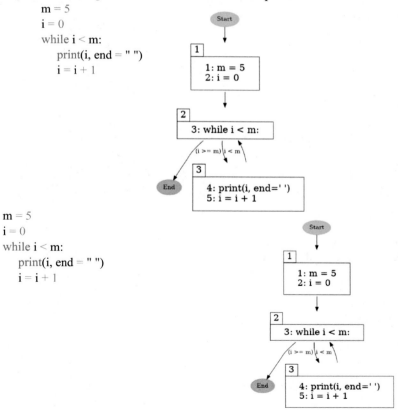

3. Jump statements

Jump statements in python alter the flow of loops by skipping a part of a loop or terminating a loop. Jump Statements in Python are:

- Break Statement
- Continue Statement
- Pass Statement

Break statements terminate their enclosing loop iterations and move the control to the statement immediately after the loop body:

```
x=1
while x<=10:
    print(x)
    if x==5:
        break
    x=x+1
```

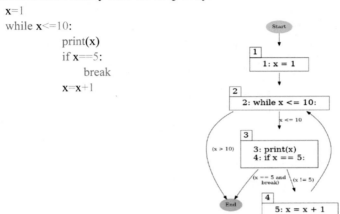

Continue statements move the control flow to the next iteration by ending the current iteration and returning the control to the loop condition.

```
x=0
while x<=10:
    x=x+1
    if x==5:
        continue
    print(x)
```

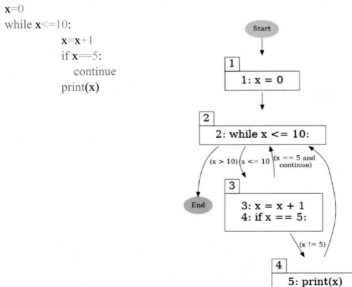

4. **Try Except Statement**

Try and Except statements handle errors in Python code. The try block checks the code inside the try block at runtime to ensure no error. The except block executes whenever the program encounters some error in the preceding try block.

    try:

            k = 5 // 0

            # raises divide by zero exception.

            print(**k**)

    # handles zerodivision exception

    except ZeroDivisionError:

            print("Can't divide by zero")

    finally:

            # this block is always executed

            # regardless of exception generation.

            print('This is always executed')

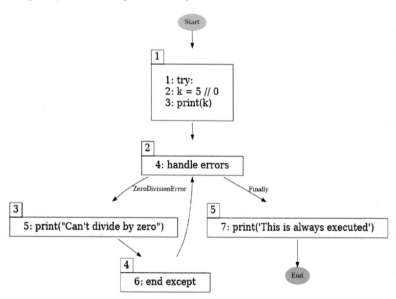

6.6 **Exercises**

1. Does clause coverage subsume predicate coverage or vice versa?
2. What is the difference between branch coverage and predicate coverage? Does branch coverage subsume predicate coverage or vice versa?

3. Compute simple paths, prime paths, and basis paths for the triangle method in
    the Triangle class, shown below:

```
1.    class Triangle
2.    {
3.       public:
4.          static const int OUT_OF_RANGE = -2;
5.          static const int INVALID = -1;
6.          static const int SCALENE = 0;
7.          static const int ISOSELES = 1;
8.          static const int EQUILATERAL = 2;
9.          static int triangle(int a, int b, int c)
10.      {
11.         boolean c1, c2, c3, isATriangle;
12.         // Step 1: Validate Input
13.         c1 = (1 <= a) && (a <= 200);
14.         c2 = (1 <= b) && (b <= 200);
15.         c3 = (1 <= c) && (c <= 200);
16.         int triangleType = INVALID;
17.         if (!c1 || !c2 || !c3)
18.             triangleType = OUT_OF_RANGE;
19.         else
20.           {
21.             // Step 2: Is A Triangle?
22.             if ((a < b + c) && (b < a + c) && (c < a + b))
23.                 isATriangle = true;
24.             else
25.                 isATriangle = false;
26.             // Step 3: Determine Triangle Type
27.             if (isATriangle)
28.                 {
29.                   if ((a == b) && (b == c))
30.                       triangleType = EQUILATERAL;
31.                   else if ((a != b) && (a != c) && (b != c))
32.                           triangleType = SCALENE;
33.                       else
34.                               triangleType = ISOSELES;
35.                 }
36.             else
37.                 triangleType = INVALID;
38.           }
39.         return triangleType;
40.      }
41.   };
42.
```

References

1. Ammann, P., Offutt, J.: An Introduction to Software Testing, 2nd edn. Amazon (2017)
2. Bushmais, A.H.: Graph based unit testing. Master Thesis, The University of Texas at Austin (2011)
3. Sedgewick, R.: Algorithms in C, Part 5: Graph Algorithms, 3rd edn. Addison Wesley Professional (2001)
4. Dwarakanath, A., Jankiti, A.: Minimum number of test paths for prime path and other structural coverage criteria. In: 26th IFIP International Conference on Testing Software and Systems (ICTSS), Sept 2014, Madrid, Spain, pp. 63–79. https://doi.org/10.1007/978-3-662-44857-1_5. hal-01405275f
5. Jiang, S., Zhang, Y., Yi, D.: Test data generation approach for basis path coverage. ACM SIGSOFT Softw. Eng. Notes **37**(3), 1–7 (2012)
6. McCabe, T.: A complexity measure. IEEE Trans. Softw. Eng. (4), 308–320 (1976). https://doi.org/10.1109/tse.1976.233837
7. Woodruff, B., Taylor, J., Meyer, C., Lai, J., Zeng, R.: LibCST Documentation. 2 Jan 2022
8. Robinson, P.J.: MyPyTutor: an interactive tutorial system for Python. In: Proceedings of the Thirteenth Australasian Computing Education Conference (ACE 2011), Perth, Australia (2011)

Chapter 7
Spectrum-Based Fault Localization

7.1 Introduction

Fault localization is among the most costly tasks in software development [1]. It can be a very cumbersome and time-consuming task. As a software system grows in size, the cost of fault localization grows steeply. Developers usually decide on debugging locations using their experience and intuition; thus, the decision is prone to error. Statistical fault localization techniques tend to make the decision systematically and based on the influence of the program elements on the execution results [2]. The influence is measured by contrasting the involvement of each element, e.g., statement, in the failing and passing executions of the program under debug.

This chapter provides an empirical application of a spectrum-based technique to locate suspicious statements in a sample faulty source code. Spectrum-based techniques use the execution traces of the faulty code to compute the possibility of each statement being faulty based on the frequency of its involvement in the program executions with correct and incorrect results. The execution traces of a given source code can be captured by inserting probes into the source code at the branching instructions. Such a trace is just a long sequence of statements encountered while the code is running. This chapter gives insight into how to automatically develop your own code to instrument a given source code.

© The Author(s), under exclusive license to Springer Nature Switzerland AG 2023   317
S. Parsa, *Software Testing Automation*,
https://doi.org/10.1007/978-3-031-22057-9_7

Fault localization is sometimes like finding a needle in haystack

7.2   Spectrum-Based Fault Localization in Practice

Despite the simplicity, spectrum-based fault localization (SBFL) techniques are very efficient [1]. This section uses a motivating example to empirically demonstrate the applicability and effectiveness of spectrum-based fault localization (SBFL) techniques to compute the suspiciousness score of the statements by comparing their association with failing and passing executions of the program under debug.

Generally, examples make things easier to understand. Therefore, to begin with, instead of giving definitions, I present a faulty sample code to illustrate how spectrum-based techniques work in practice. Afterward, I show the reader how to use the execution traces to locate suspicious elements.

7.2.1   A Motivating Example

Consider the function in Fig. 7.1a with a seeded fault in statement S9. The faulty statement, S9,should be 'Max = a' instead of 'Max = b'. The question is how to locate the faulty statement automatically.

A faulty code and its corresponding control flow graph (CFGcontains five branching conditions or predicates at s5, s8, s10, s12, and S14 (Fig. 7.1). Therefore, there are $2^5 = 32$ different execution paths in the function. Each execution path is represented as a sequence of predicate values and locations visited during the program execution:

Execution-Path := Sequence of (location, value)

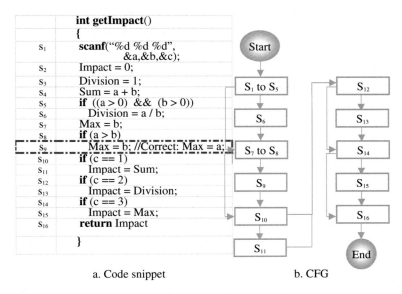

	int getImpact()
	{
S_1	scanf("%d %d %d", &a,&b,&c);
S_2	Impact = 0;
S_3	Division = 1;
S_4	Sum = a + b;
S_5	if ((a > 0) && (b > 0))
S_6	Division = a / b;
S_7	Max = b;
S_8	if (a > b)
S_9	Max = b; //Correct: Max = a;
S_{10}	if (c == 1)
S_{11}	Impact = Sum;
S_{12}	if (c == 2)
S_{13}	Impact = Division;
S_{14}	if (c == 3)
S_{15}	Impact = Max;
S_{16}	return Impact
	}

a. Code snippet   b. CFG

Fig. 7.1 A faulty code and its corresponding control flow graph (CFG)

For instance, in Fig. 7.1a:

$$Ex\text{-}Path\text{-}k := <(S5, \; Value(S5)), \; (S8, \; Value(S8)),$$

$$(S10, \; Value(S10)),(S12, \; Value(S12)),$$

$$(S14, \; Value(S14))>$$

where Value(Si) is either true or false depending on the values of the input data.
Running the getImpact() function with the following test case:

$$\text{"}T1: (a = 4, \; b = 1, \; c = 3, \; expected \; result = 4)\text{"}$$

the execution path will be:

$$Ex\text{-}Path\text{-}1 := <(S5, \; true), \; (S8, \; true), \; (S10, \; false),$$

$$(S12, \; false), \; (S14, \; true)>$$

This function's run is labeled as "failing" because the expected return value is
4 while the actual value returned by the function is 1. Table 7.1 shows statement
coverage for 12 test cases, T1 to T12, executed on the running example in Fig. 7.1. As
described in Chap. 10, the most practical technique for automatic test data generation
is to use genetic algorithms to generate test data, covering execution paths with
minimal overlapping.

Table 7.1 Execution traces and execution results

S #.	Test case #	Coverage											
		T1	T2	T3	T4	T5	T6	T7	T8	T9	T10	T11	T12
		a:4	a:7	a:6	a:2	a:3	a:9	a:8	a:9	a:-6	a:7	a:6	a:-8
		b:1	b:6	b:3	b:1	b:2	b:7	b:-3	b:-2	b:8	b:6	b:8	b:9
		c:3	c:3	c:3	c:3	c:3	c:3	c:2	c:2	c:3	c:1	c:3	c:3
S_1	**Scanf**(…, a,b,c);	*	*	*	*	*	*	*	*	*	*	*	*
S_2	result = 0;	*	*	*	*	*	*	*	*	*	*	*	*
S_3	Div = 1;	*	*	*	*	*	*	*	*	*	*	*	*
S_4	Sum = a + b;	*	*	*	*	*	*	*	*	*	*	*	*
S_5	**if** ((a > 0) && (b > 0))	*	*	*	*	*	*	*	*	*	*	*	*
S_6	Div = a / b;	*	*	*	*	*	*				*	*	
S_7	Max = b;	*	*	*	*	*	*	*	*	*	*	*	*
S_8	**if** (a > b)	*	*	*	*	*	*	*	*	*	*	*	*
S_9	Max=b; //Correct: Max =a;	*	*	*	*	*	*	*	*		*		
S_{10}	**if** (c == 1)	*	*	*	*	*	*	*	*	*	*	*	*
S_{11}	result=Sum;										*		
S_{12}	**if** (c == 2)	*	*	*	*	*	*	*	*	*	*	*	*
S_{13}	result = Div;							*	*				
S_{14}	**if** (c == 3)	*	*	*	*	*	*	*	*	*	*	*	*
S_{15}	result=Max;	*	*	*	*	*	*			*		*	*
S_{16}	**return** result	*	*	*	*	*	*	*	*	*	*	*	*
Execution Result:		1	1	1	1	1	1	0	0	0	0	0	0

Source code instrumentation makes it possible to keep a log of a program execution path by inserting instructions to record the locations where execution flow deviates from proceeding with the instruction located at the following memory address.

A program may keep a log of its execution trace provided it is instrumented by inserting instructions to record the value and location of predicates visited at run time. A compiler generator could be best applied to automatically develop software tools to instrument programs. Section 7.2.2 offers a C# method that automatically uses ANTLR to instrument C# programs.

7.2.2 Fault Localization

Coverage-based fault localization techniques, also known as statistical or spectrum-based, are the most popular and studied methods for automated debugging. Such techniques gather statistics from execution traces to determine the fault suspiciousness score of the statements in the program under test [5, 6]. For instance, in Fig. 7.1, there are six failing tests, using the test cases T1 to T6, and six passing tests, using T7 to T12. It is worth mentioning that a passing test is when the program appears to work correctly. An asterisk in row R and column C of Table 7.1 indicates the execution of the statement at the front of row number R with the test data, TC, at the top of column number C.

Apparently, if a statement is observed in failing executions more than the passing ones, it is more fault suspicious. Ochiai [1] is a spectrum-based fault localization method that uses the coverage statistics of a test suite run to determine fault suspicious statements. It defines the suspiciousness of a statement as follows:

$$
\begin{aligned}
&Suspiciousness(aStatement) \\
&= \frac{failed(aStatement)}{\sqrt{totalFailed \times (passed(aStatement) + failed(aStatement))}}
\end{aligned}
$$

where:

- passed(aStatement): Number of passed tests covering aStatement
- failed(aStatement): Number of failed tests covering aStatement t
- totalPassed: The number of passing tests
- totalFailed: The number of failing tests

Tarantula is another spectrum-based method that uses code coverage information obtained from a test suite run to help developers localize faults [2]. Taran tula defines the suspiciousness score of a statement, aStatement, as follows:

$$
\begin{aligned}
&Suspiciousness(aStatement) \\
&= \frac{Failed(aStatement)/totalFailed}{failed(aStateent)/total\,Failed + Passed(aStatement)/totalPassed}
\end{aligned}
$$

As the input, both the Tarantula and Ochiai methods accept a test suite and a program under test [3]. As the output, if there is at least one failing test, it ranks the statements based on a measure of suspiciousness. Tarantula and Ochiai instrument the program under test to collect statement coverage and calculate a measure of suspiciousness for each statement based on the coverage information and each test run's pass/fail result.

Table 7.2 Tarantula and Ochiai results

S_4	Sum = a + b;	6	6	0.5	0.71
S_5	**if** ((a > 0) && (b > 0))	6	6	0.5	0.71
S_6	Div = a / b;	2	6	0.75	0.87
S_7	Max = b;	6	6	0.5	0.71
S_8	**if** (a > b)	6	6	0.5	0.71
S_9	Max=b; //Correct: Max =a;	3	6	0.67	0.81
S_{10}	**if** (c = = 1)	6	6	0.5	0.71
S_{11}	result = Sum;	1	0	0.0	0.00
S_{12}	**if** (c = = 2)	6	6	0.5	0.71
S_{13}	result = Div;	2	0	0.0	0.00
S_{14}	**if** (c = = 3)	6	6	0.5	0.71
S_{15}	result = Max;	3	6	0.67	0.81
S_{16}	**return** result	6	6	0.5	0.71
Dynamic Analysis Results:					

Table 7.2 represents the suspiciousness score calculated for the statements of the getImpact() function, according to the Tarantula and Ochiai metrics.

It is observed that even the most known methods, Tarantula and Ochiai, cannot detect the root cause of the failure. These methods should not account for statements with no direct or indirect impact on the expected result in executing the code under test. In Table 7.3, all the statements directly or indirectly affect the return value, and the result is marked with a ☑ symbol.

A slice is a program statement sequence that potentially affects the computation of variable 'V' at statement 'S'. The dynamic slice is a sequence of statements that affect the value of a variable, v, at a statement, S, at runtime [2, 3]. Table 7.3 shows dynamic slices for the variable result in statement S16.

Fault localization is a search over the space of program statements for the faulty statement. Slicing provides a suitable way of shifting out irrelevant code. By restricting execution paths to slices for the return value, the search space to locate the root cause of the failure is significantly reduced. As described in Chap. 8, compiler optimization techniques can be applied to obtain slices.

Table 7.3 Backward dynamic slices for the return value

#.	int getImpact()	T1 a:4 b:1 c:3	T7 a:8 b:-3 c:2	T8 a:9 b:-2 c:2	T9 a:-6 b:8 c:3	T10 a:7 b:6 c:1	T11 a:6 b:8 c:3	T12 a:-8 b:9 c:3
S1	scanf("%d %d %d",a,b,c);	☑	☑	☑	☑	☑	☑	☑
S2	result = 0;	*	*	*	*	*	*	*
S3	Div = 1;	*	☑	☑	*	*	*	*
S4	Sum = a + b;	*	*	*	*	☑	*	*
S5	if ((a > 0) && (b > 0))	*	*	*	*	*	*	*
S6	Div = a / b;	*				*	*	
S7	Max = b;	*	*	*	☑	*	☑	☑
S8	if (a > b)	☑	*	*	*	*	*	*
S9	Max=b; //Correct: Max =a;	☑	*	*		*		
S10	if (c == 1)	*	*	*	*	☑	*	*
S11	result = Sum;					☑		
S12	if (c == 2)	*	☑	☑	*	*	*	*
S13	result = Div;		☑	☑				
S14	if (c == 3)	☑	*	*	☑	*	☑	☑
S15	result = Max;	☑			☑		☑	☑
S16	return result	☑	☑	☑	☑	☑	☑	☑
	Execution Result (0=Successful/ 1=Failed)	1	0	0	0	0	0	0

7.3 Source Code Instrumentation[1]

Typically, dynamic analysis of runtime behavior encompasses instrumentation of the program under investigation with extra instructions to record and keep track of its execution [7]. Instrumentation is the technique of inserting extra code and probes into a program to gather information about its runtime behavior and coverage [3]. This technique is called instrumentation because it is focused on probing and gathering code information. The gathered information for each run of the instrumented program, P, with input data set, di, is a sequence of predicate or branching condition values, $C_i = < C_{i1}, C_{i2}, \ldots, C_{ik} >$.

Suppose di is the complete input data set for testing a program, P. When running P with di, the program traverses a specific program path, Pi. Pi is the sequence of statements, including repeated statements, caused by loops. The entry point determines pi in the program, P, and the sequence of decisions ($C_i = (C_{i1}, C_{i2}, \ldots)$ made at branching points.

Below, in Fig. 7.2, the instrumented version of the getImpact() function is listed.

---

[1] Download the source code from: https://doi.org/10.6084/m9.figshare.2128461

	int getImpact()
	{
S_{ins}	Fprintf("logfile", "\nEnter: getImpact\n");
S_1	**scanf**("%d %d %d", &a,&b,&c);
S_2	Impact = 0;
S_3	Division = 1;
S_4	Sum = a + b;
S_5	**if** ((a > 0) && (b > 0))
S_{ins}	{ Fprintf("logFile", "P1 ");
S_6	Division = a / b; }
S_7	Max = b;
S_8	**if** (a > b)
S_{ins}	{ Fprintf("logFile", "P2 ");
S_9	Max = b; //Correct: Max = a; }
S_{10}	**if** (c == 1)
S_{ins}	{ Fprintf("logFile", "P3 ");
S_{11}	Impact = Sum; }
S_{12}	**if** (c == 2)
S_{ins}	{ Fprintf("logfile", "P4 ");
S_{13}	Impact = Division;
S_{14}	**if** (c == 3)
S_{ins}	{ Fprintf("logfile", "P5 ");
S_{15}	Impact = Max; }
S_{ins}	Fprintf("\n Exit: getImpact( )\n");
S_{16}	**return** Impact
	}

Fig. 7.2 The instrumented version of the getMethod() function

7.3.1 Automatic Instrumentation

This section offers a simple Python program to instrument C++ programs. The program uses the ANTLR parser generator to create a lexical analyzer, CPP14Lexer, and CPP14Parser in Python. The CPP grammar, CPP14.g4, for ANLR is available to download from:

https://github.com/antlr/grammars-v4/tree/master/cpp

The parser and lexical analyzer for CPP programs could be generated by running the following commands:

java -jar./antlr-4.8-complete.jar -Dlanguage = Python3 CPP14.g4

As a result of running the command, the following files are generated:

1. CPP14Lexer.py: including the source of a class, CPP14Lexer.

2. CPP14Parser.py: including the source of a class, CPP14Parser.
3. CPP14Listener.py: including the source of a class, CPP14Listener.

In addition to the above three files, four other files are generated that are used by ANTLR. All these files should be copied into the folder where the instrumentation program, CppInstrumenter resides. Also, to access the ANTLR4 runtime library within a Python project, MyProject, in the PyChram environment, select:

File>Setting>MyProject>Project interpreter>+ (add)

Search for "ANTLR4" in the popped-up window, select "ANTLR4 runtime Python3" option, and click on the "install package" button. The main body of CppInstrmenter.py is shown in Fig. 7.3.

In step 7 of the Python code, listed in Fig. 7.3, a parse tree walker is created to walk through the parse tree generated in step 5. In step 8, the walker, 'walker', goes through each node of the parsTree, 't = parse_tree,' and invokes the methods of the InstrumentationListener class in Fig. 7.5 to insert probes into appropriate locations within the given source file.

ANTLR generates a ParseTreeListener subclass specific to the given grammar with entering and exiting methods for each nonterminal symbol. For instance, Fig. 7.4 lists all the programming constructs that, when executed, may deviate the execution flow and should be instrumented to show the deviation.

```
# Step 1: Load input source into a buffer
stream = FileStream('gcd.cc', encoding='utf8')
# Step 2: Create an instance of CPP14Lexer
lexer = CPP14Lexer(stream)
# Step 3: Convert the input source into a list of tokens
token_stream = CommonTokenStream(lexer)
# Step 4: Create an instance of the CPP14Parser
parser = CPP14Parser(token_stream)
# Step 5: Create a parse tree
parse_tree = parser.translationunit()
# Step 6: Create an instance of InstrumentationListener
instrument_listener = InstrumentationListener(token_stream)
# Step 7: Create a walker object to traverse parse_tree (DFS)
walker = ParseTreeWalker()
# Step 8: Walkthrough parse_tree & apply listener methods to their
#         corresponding nodes
walker.walk(listener=instrument_listener, t=parse_tree)
# Step 9: Get the instrumented code
new_source_code =
instrument_listener.token_stream_rewriter.getDefaultText()
print(new_source_code)
sourceFile = open('gcd_instrument.cpp', 'w')
print(new_source_code, file = sourceFile)
sourceFile.close()
```

Fig. 7.3 The main body of the CppInstrumenter.py program

statement
: selectionstatement | iterationstatement | jumpstatement | tryblock | ...
functionbody
: compoundstatement | functiontryblock | '=' Default ';' | '=' Delete ';'
selectionstatement
: If '(' condition ')' statement | If '(' condition ')' statement Else statement
| Switch '(' condition ')' statement
iterationstatement
: While '(' condition ')' statement
| Do statement While '(' expression ')' ';'
| For '(' forinitstatement condition? ';' expression? ')' statement
| For '(' forrangedeclaration ':' forrangeinitializer ')' statement
jumpstatement
: Break ';' | Continue ';' | Return expression? ';' | Return bracedinitlist ';' |
Goto Identifier ';'

Fig. 7.4 List of instrumental constructs in C++

For each nonterminal symbol, such as a statement, ANTLR generates two methods: enterStatement and exitStatement(). The programmer could override these methods. For instance, in Fig. 7.5, the enterStatement, and enterFunctionBody, override their corresponding methods generated by ANTLR. While walking through the parse tree and visiting the tree nodes, the treeWalker, described above, invokes the overridden methods whenever it comes acrosses their corresponding nonterminal nodes. In Fig. 7.5, the class instrumentationListener is listed. This class instruments C++ programs.

The class instrumentationListener has six methods. The __init__ constructor method moves all the tokens of the C++ program into a buffer, token_stream_rewriter. The second method, enterTranslationunit, inserts an instruction for creating a log file, 'logfile', into the program to be instrumented. The other three methods insert probes into the C++program. In Fig. 7.6, a C++ program used as input and its instrumented version provided by the CppInstrumenter, is shown.

7.4 Summary

In software testing, the code under test is executed with different inputs. Output is validated to see whether the code executes as expected. Correct output means a successful or passing test case, and unexpected output represents a failed test case. A test case is an input data set and its expected output. Program Spectra is the execution information of program components in passed and failed test cases. Spectrum-based

```
from antlr4 import *
from CPP14Lexer import CPP14Lexer
from CPP14Parser import CPP14Parser
from CPP14Listener import CPP14Listener
from antlr4.TokenStreamRewriter import TokenStreamRewriter
class InstrumentationListener(CPP14Listener):
    def __init__(self, tokenized_source_code: CommonTokenStream):
        self.branch_number = 0
        self.has_return_statement = False
        if tokenized_source_code is not None:
        # Move all the tokens in the source code in a buffer, token_stream_rewriter.
            self.token_stream_rewriter = TokenStreamRewriter( \
                                                    tokenized_source_code)
        else:
            raise Exception(" common_token_stream is None ")

    # Create and open a log file to save execution traces.
    def enterTranslationunit(self, ctx: CPP14Parser.TranslationunitContext):
        new_code = '\n #include <fstream> \n std::ofstream \
                    logFile("log_file.txt"); \n\n'
        self.token_stream_rewriter.insertBeforeIndex(ctx.start.tokenIndex, \
                                                    new_code)
    # DFS traversal of a branching condition subtree, rooted at ctx, and
    # insert a prob

    def enterStatement(self, ctx: CPP14Parser.StatementContext):
        if isinstance(ctx.parentCtx, (CPP14Parser.SelectionstatementContext,
            CPP14Parser.IterationstatementContext)) or \
            isinstance(ctx.children[0], CPP14Parser.JumpstatementContext):
        # if there is a compound statement after the branchning condition:
            if isinstance(ctx.children[0], \
                    CPP14Parser.CompoundstatementContext):
            self.branch_number += 1
            new_code = '\n logFile<< "p'+ str(self.branch_number)+ '" << endl; \n'
            self.token_stream_rewriter.insertAfter(\
                                            ctx.start.tokenIndex, new_code)
```

Fig. 7.5 A Python class for instrumenting C++ programs

```
    # if there is only one statement after the branching condition
    #                                    then creates a block.
         elif not isinstance(ctx.children[0],
                     (CPP14Parser.SelectionstatementContext,
                     CPP14Parser.IterationstatementContext)):
             self.branch_number += 1
             new_code = '{'
             new_code += '\n logFile << "p' + str(self.branch_number) \
                     + '" << endl; \n'
             new_code += ctx.getText()
             new_code += '\n}'
             self.token_stream_rewriter.replaceRange(ctx.start.tokenIndex,
                     ctx.stop.tokenIndex, new_code)


    def enterFunctionbody(self, ctx: CPP14Parser.FunctionbodyContext):
        #Insert a prob at the beginning of the function.
        self.branch_number += 1
        new_code = '\n logFile << "p' + str(self.branch_number) + '" << endl;\n'
        self.token_stream_rewriter.insertAfter(ctx.start.tokenIndex, new_code)
```

Fig. 7.5 (continued)

fault localization (SBFL) is a debugging technique that uses execution spectra to locate faults. Execution spectra are collected by instrumenting the code under test.

7.5   Exercises

1. Five of the best-studied SBFL techniques [4] are given below. In the following let, let *total_failed* be the number of failed test cases and failed(s) be the number of those that executed statement s (similarly for total_passed and passed(s)):

   1. Ochiai (s) $= \frac{failed(s)}{\sqrt{totalFailed \times (passeds) + failed(s))}}$
   2. Tarantula (s) $= \frac{Failed(s)/totalFailed}{failed(s)/totalFailed + Passed(s)/totalPassed}$
   3. Op2(s) $= failed(s) - \frac{passed(s)}{totalPassed+1}$
   4. Barinel (s) $= 1 - \frac{passed(s)}{Passed(s)+failed(s)}$
   5. DStar (s) $= \frac{failed(s)^2}{Passed(s)+(totalfailed-failed(s))}$

   In Table 7.2 and Table 7.4, the Tarantula and Ochiai metrics are used to compute the suspiciousness score for each statement. Add three more columns to each of these tables to show the suspiciousness scores computed by the other three metrics.

Fig. 7.6 A sample CPP program and its instrumented version

```cpp
#include <stdio.h>
#include <iostream.h>

int main()
{
    int num1, num2, i, gcd;
    std::cout << "Enter two integers: ";
    std::cin >> num1 >> num2;
    for(i=1; i <= num1 && i <= num2; ++i)
        // Checks if i is a factor of both the integers
        if(num1%i==0 && num2%i==0)
                        gcd = i;
    std::cout << "G.C.D is " << gcd << std::endl;
    return 0;
}
```

a. Original code

```cpp
#include <stdio.h>
#include <iostream>
#include <fstream>
std::ofstream logFile("log_file.txt");

int main()
{
    logFile << "p1" << std::endl;
    int num1, num2, i, gcd;
    std::cout << "Enter two integers: ";
    std::cin >> num1 >> num2;
    for(i=1; i <= num1 && i <= num2; ++i)
    {
        logFile << "p2" << std::endl;
        // Checks if i is a factor of both integers
        if(num1%i==0 && num2%i==0)
        {
            logFile << "p3" << std::endl;
            gcd=i;
        }
    }
    std::cout << "G.C.D is " << gcd << std::endl;
    logFile << "p4" << std::endl;
    return 0;
}
```

b. Instrumented by CppInstrumenter

Table 7.4 Suspiciousness score calculated for each of the statements

	totalPassed = 6	totalFailed = 6			
Stmt.#	int getImpact()	passed	failed	Tarantula	Ochiai
S_1	scanf("%d %d %d",a,b,c);	6	6	0.50	0.71
S_2	result = 0;	0	0	0.00	0.00
S_3	Div = 1;	2	0	0.00	0.00
S_4	Sum = a + b;	1	0	0.00	0.00
S_5	if ((a > 0) && (b > 0))	0	0	0.00	0.00
S_6	Div = a / b;	0	0	0.00	0.00
S_7	Max = b;	3	0	0.00	0.00
S_8	if (a > b)	0	6	1.00	1.00
S_9	Max=b; //Correct: Max =a;	0	6	1.00	1.00
S_{10}	if (c == 1)	1	0	0.00	0.00
S_{11}	result = Sum;	1	0	0.00	0.00
S_{12}	if (c == 2)	2	0	0.00	0.71
S_{13}	result = Div;	2	0	0.00	0.00

2. The effectiveness of the backward slicing technique for debugging depends on the size of the slice. Improve the effectiveness by examining only the subset of the statements in the slice that reside in a suspicious basic block. The suspicious score, *Score*, for a basic block, BB_i, is computed as follows [4]:

$$Score(BB_i) = \frac{|failed(BB_i) - passed(BB_i)|}{no.test\_cases} * \frac{failed(BB_i)}{passed(BB_i)}$$

Use equations 1 to 5 in Exercise 1 to compute the suspiciousness scores for the statement in the most suspicious basic block of the InsertionSort function, given below:

int **insertionSort(** int *arr, int n, int flag1, int flag2){ int key,j;	Coverage											
	T1	T2	T3	T4	T5	T6	T7	T8	T9	T10	T11	T12
S1 **for**(int i=1; i<n; i++){	1	1	1	1	1	1	1	1	1	1	1	1
S2 key = arr[i];	1	1	1	1	1	1	1	1	1	1	1	1
S3 **if**(flag1%2 == 0){	1	1	1	1	1	1	1	1	1	1	1	1
S4 j= i-2;//correct: j=i-1;}	1	1	1	1	1	1	1	1	1	1	1	1
S5 **If**(flag2 > flag1){	1	1	1	1	1	1	1	1	1	1	1	1
S6 j = i-1;}	1	1	0	1	1	0	0	0	1	1	0	0
S7 **if**(flag1>0){	1	1	1	1	1	1	1	1	1	1	1	1
S8 **while**(j>=0&&arr[j]>key){	1	1	1	1	0	1	1	1	1	1	0	1
S9 arr[j+1] = arr[j];	1	1	1	1	0	1	1	1	1	1	0	1
S10 j = j-1;}}	1	1	1	1	0	1	1	1	1	1	0	1
S11 arr[j+1]=key;}	1	1	1	1	1	1	1	1	1	1	1	1
S12 **return** arr;}	1	1	1	1	1	1	1	1	1	1	1	1
Execution Result (P=Passed ,F=failed)	P	P	F	P	P	F	F	F	P	P	F	F

The test cases T1 to T12 are as follows:

	arr[]	n	flag1	flag2
T1	{15,12,13,5,7} // T	5	2	51
T2	{95,-60,42,62,17,24,-5,255,104,-500} // T	10	18	30
T3	{-16,15,14,10,8,11,-1,10,105} // F	9	4	3
T4	{5,-5,0} // T	3	8	11
T5	{-3,-2,-1,0,1,2} // T	6	12	17
T6	{26,44,-18,25,85,17,20,-15,-3,18,64,74,19,12}	14	10	6
T7	{66,2,49,3} // F	4	16	0
T8	{76,34,89,23,54,72,94} // F	7	14	3
T9	{6,43,72,53,67,23,65,68} // T	8	6	15
T10	{54,4,22,28,20} // T	5	12	18
T11	{13,-6,34} // F	3	20	10
T12	{21,75,19,54,-65} // F	5	36	7

3. Give an example of a faulty method with multiple return statements and describe how to slice the code to locate the faulty statement.
4. How can we apply spectrum-based techniques to locate the root cause of failure in a program whose execution result is always erroneous?
5. Give an example of a faulty code, including "for" or "while" loops in which statements are executed many times, and then show how to apply Tarantula or Ochiai formulas to compute their probability of being faulty.
6. Write a program that accepts a Java project as input and uses ANTLR to instrument the project.

References

1. Neelofar.: Spectrum Based Fault Localization Using Machine Learning Technique. Ph.D. thesis, School of Computing and Information Systems, the University of Melbourne (2017)
2. Parsa, S., Vahidi-Asl, M., Zareie, F.: Statistical based slicing method for prioritizing program fault relevant statements. Comput. Inf. **34**(4) (2015)
3. Zareie, F., Parsa, S.: A non-parametric statistical debugging technique with the aid of program slicing. Int.J. Inf. Eng. Electr. Bus. **2**, 8–14 (2013)
4. Wong, W.E., Gao, R., Li, Y., Abreu, R., Wotawa, F.: A survey of software fault localization. IEEE Trans. Softw. Eng. (TSE) (2016)
5. Parsa, S., Zareie, F., Vahidi-Asl, M.: Fuzzy clustering the backward dynamic slices of programs to identify the origins of failure. In: SEA 2011, LNCS 6630, pp. 352–363 (2011). Springer-Verlag Berlin Heidelberg
6. Parsa, S., Asadi-Aghbolaghi, M., Vahidi-Asl, M.: Statistical debugging using a hierarchical model of correlated predicates. In: Proceedings of the 3rd International Conference on Artificial Intelligence and Computational Intelligence (AICI'11), Taiyuan, China, September 24–25 (2011)
7. Isaacs, R., Sigelman, B.: Distributed Tracing in Practice Instrumenting, Analyzing, and Debugging Microservices. Published by O'Reilly Media (2020)

Chapter 8
Fault Localization Tools: Diagnosis Matrix and Slicing

8.1 Introduction

Manual debugging is a costly task. Automatic fault localization techniques mitigate the burden of manual debugging. The most prevalent diagnosis technique is spectrum-based fault localization (SBFL). It collects statistical information at runtime to diagnose faults. Runtime information, including execution spectra, is conventionally saved in diagnosis (fault localization matrices).

The execution spectra for different test cases and test results for each code under test are tabulated in a diagnosis matrix. The matrix columns, representing the statements covered by each test data, are used to compute the suspiciousness scores for each statement in the program under test. This chapter describes the implementation details of a software tool, Diagnoser, to generate a diagnosis matrix for a given Python function automatically.

As described in the last chapter, faults propagate from the root cause of the failure through data and control dependencies to the location where a faulty result is observed. Data flow testing techniques focus on the movement of the values all over the code to determine the test paths. The extent to which various test data covering the test paths is evaluated is based on the data flow coverage metrics. Backward dynamic slices identify fault propagation paths as sequences of statements affecting the faulty results [2]. This chapter presents an algorithm to compute slices.

A slice is, in fact, a part of the program dependency graph. A dependency graph is a labeled directed graph in which each node represents a program statement and edges represent control and data dependencies. Definition-use and use-definition algorithms and their implementation in Python are given in Sect. 8.6. Section 8.8 offers a straightforward algorithm to compute control dependencies.

© The Author(s), under exclusive license to Springer Nature Switzerland AG 2023
S. Parsa, *Software Testing Automation*,
https://doi.org/10.1007/978-3-031-22057-9_8

8.2 Diagnosis Matrix[1]

Diagnosis matrices keep data for applying spectrum-based fault localization formulas to compute the fault suspiciousness of the statements [1]. Traditionally, log traces for test case execution are kept in a diagnosis matrix, along with the test case numbers and fault suspiciousness scores for each program statement.

This section provides practical advice on developing a software tool, Diagnoser, to construct a decision matrix for a given function. The input to Diagnoser is a Python function and test cases to run the function and evaluate its return value. The output Diagnoser is an excel file representing the diagnosis matrix. For instance, Fig. 8.1 represents a function alpha4 to normalize an array of integers.

The alpha4 function, shown in Fig. 8.1, has two input parameters, array, and type. The first parameter is an array of integers to be normalized by the function alpha4. The second input parameter, type, controls the normalization type to be applied. The test cases used to test the alpha4 function are as follows:

```
# Test suite = list of [Input parameters : array: list, type:int, exec. Res. = p/f]
test_suite =
[[[4,4,8,9,10],1,"f"],[[4,5,8,1,7],0,"p"],[[1,6,4,3,2],1,"p"],[[3,3,4,3,3],1,"t"],[[8,7,6,5,4],1,"f"],[[7,7,7,3,7],1,"p"],[[7,8,9,4,2],0,"p"]]
```

The test suite, test_suite, includes seven test cases. Each test case itself is a list of [(array), type, "fail/pass"]. The diagnosis results provided by Diagnoser are shown in Fig. 8.2.

Running the Diagnoser program with the faulty function alpha4 and test_suite as input, the program outputs the diagnosis matrix shown in Fig. 8.2. The first column of the matrix includes the statement numbers. The coverage of each statement by a test case is shown by the numbers 1 and 0. The last row shows the test results as failing or passing, and the last two columns represent the suspiciousness scores computed for each statement by the Ociai and Tarantula formulas.

The question is how to access the state of a function during its execution. Interpreted languages such as Python interrupt the execution of programs to continue translating and subsequently executing the next part of the program. Thus, it is usually relatively easy to control execution and inspect the state of the interpreted languages since that is what the interpreter is doing already. Debuggers are implemented on top of the hooks interrupting execution and accessing the program state.

The Python function sys.settrace() provides a mechanism for such a hook. The function is typically invoked with a tracing function that will be called at every line executed, as in:

sys.settrace(traceit)

The sys.settrace() function sets the hook. The trace hook is modified by passing a callback function, traceit, to sys.settrace(). The traceit function is called during

---

[1] Download the source code from: https://doi.org/10.6084/m9.figshare.21317940

```
    # input function
1.  def alpha4(array: list, type: int):
2.      arrayLenght = len(array)
3.      is_sorted = True
4.      sum = 0
5.      sum_square = 0
6.      Average = 0
7.      arr_norm_sd = list()
8.      arr_norm = list()
9.      for i in range(0, arrayLenght):
10.         arr_norm_sd.append(array[i])
11.         arr_norm.append(array[i])
12.     for i in range(0, len(arr_norm_sd) - 1):
13.         if arr_norm_sd[i] > arr_norm_sd[i + 1]:
14.             is_sorted = False
15.             Break
16.     if not is_sorted:
17.         for i in range(0, arrayLenght - 1):
18.             for j in range(0, arrayLenght - i -1):
19.                 if (arr_norm[j] > arr_norm[j + 1]):
20.                     temp = arr_norm[j]
21.                     arr_norm[j] = arr_norm[j + 1]
22.                     arr_norm[j + 1] = temp
23.     Range = arr_norm[arrayLenght - 1] - arr_norm[0]
24.     for i in range(0, arrayLenght):
25.         sum = sum + arr_norm_sd[i]
26.     avrage = sum / arrayLenght
27.     for number in arr_norm_sd:
28.         sum_square =sum_square+((number-avrage)*(number–avrage))
29.     SD = math.sqrt(sum_square / (arrayLenght - 1))
30.     if type == 0:
31.         for i in (0, arrayLenght - 1):
32.             arr_norm_sd[i]=((array[i]-avrage) / (SD+0.1)) //should be (SD)
33.         return arr_norm_sd
34.     else:
35.         for i in range(0, arrayLenght):
36.             arr_norm[i] = array[i] - array[0]  / ( Range + 1)
37.         return arr_norm
```

Fig. 8.1 A sample function with an error in line 32

several events, one of which is the "line" event. Before the Python interpreter executes a line, it calls the trace function, traceit, with the current frame to get "f_lineno", the line number about to be run.

The cornerstone of the Diagnoser software tool is the traceit function. As shown in Fig. 8.3, the traceit function takes three arguments frame, event, and arg. The *frame* parameter allows access to the current location and variables:

#	T1	T2	T3	T4	T5	T6	T7	tarantula	Ochiai
2.	1	1	1	1	1	1	1	0.5	0.534522
3.	1	1	1	1	1	1	1	0.5	0.534522
4.	1	1	1	1	1	1	1	0.5	0.534522
5.	1	1	1	1	1	1	1	0.5	0.534522
6.	1	1	1	1	1	1	1	0.5	0.534522
7.	1	1	1	1	1	1	1	0.5	0.534522
8.	1	1	1	1	1	1	1	0.5	0.534522
9.	1	1	1	1	1	1	1	0.5	0.534522
10.	1	1	1	1	1	1	1	0.5	0.534522
11.	1	1	1	1	1	1	1	0.5	0.534522
12.	1	1	1	1	1	1	1	0.5	0.534522
13.	1	1	1	1	1	1	1	0.5	0.534522
14.	1	1	1	1	0	1	1	0.333333	0.288675
15.	1	1	1	1	0	1	1	0.333333	0.288675
16.	1	1	1	1	1	1	1	0.5	0.534522
17.	1	1	1	1	0	1	1	0.333333	0.288675
18.	1	1	1	1	0	1	1	0.333333	0.288675
19.	1	1	1	1	0	1	1	0.333333	0.288675
20.	0	1	1	1	0	1	1	0	0

Fig. 8.2 A diagnosis matrix generated by Diagnoser for the function alpha4

- **frame.f_code**: addresses the currently executed code,
- **frame.f_code.co_name**: the function name,
- **frame.f_lineno**: the current line number,
- **frame.f_locals:** the current local variables and arguments.

21.	0	1	1	1	0	1	1	0	0
22.	0	1	1	1	0	1	1	0	0
23.	1	1	1	1	0	1	1	0.333333	0.288675
24.	1	1	1	1	0	1	1	0.333333	0.288675
25.	1	1	1	1	0	1	1	0.333333	0.288675
26.	1	1	1	1	0	1	1	0.333333	0.288675
27.	1	1	1	1	0	1	1	0.333333	0.288675
28.	1	1	1	1	0	1	1	0.333333	0.288675
29.	1	1	1	1	0	1	1	0.333333	0.288675
30.	1	1	1	1	0	1	1	0.333333	0.288675
31.	0	1	0	0	0	0	1	0	0
32.	0	1	0	0	0	0	1	0	0
33.	0	1	0	0	0	0	1	0	0
34.	0	0	0	0	0	0	0	0	0
35.	1	0	1	1	0	1	0	0.454545	0.353553
36.	1	0	1	1	0	1	0	0.454545	0.353553
37.	1	0	1	1	0	1	0	0.454545	0.353553
res	0	1	1	1	0	1	1	0: Failing && 1: Passing	

Fig. 8.2 (continued)

The *event* parameter is a string naming the type of notification. The event string takes the following values:

-   **"call"**: A "call" event is raised immediately before a function call,
-   **"line"**: A "line" event is raised before a line is executed,
-   **"return"**: A "return" event is raised immediately before a function returns. In this case, the arg parameter takes the return value.
-   **"exception"**: An "exception" event is raised after an exception occurs. In this case, the arg parameter takes the tuple:
(exception, value, traceback),
-   **"c_call"**: A "c_call" event raises immediately before a C function is called. In this case the parameter arg takes the C object,
-   **"c_return"**: A "c_return" event raises after a C function returns.
-   **"c_exception"**: A "c_exception" event raises after a C function throws an error.

The *traceit* function, in Fig. 8.3, simply appends the line number of the statement to be executed after a global list named execution_trace. The total number of times each line executes is kept in a global array, linenos_frequency.

```
# Keeps records of execution traces
def traceit(frame, event, arg):
    # if event = the following line is executed
    if event == "line":
        # Get the line no.
        lineno = frame.f_lineno
        # if the line no. is not already observed, add it to the list
        if lineno not in execution_trace:
            execution_trace.append(lineno)
        if lineno in linenos_sum:
            linenos_frequency[lineno] +=1
        else:
            linenos_ frequency [lineno] = 1
    return traceit
```

Fig. 8.3 The traceit function is triggered by the "line" event

Below is the main body of a Python program to generate a diagnosis matrix for a given Python function. The function under test, function_under_test, is called with each test case. After each call, the execution trace and the test result as failing, "f", or passing, "p", are appended to a list called test_case_trace.

All the parameters required to compute the Ochiai and Tarantula suspiciousness score for each statement are computed in the main body of the Diagnoser in Fig. 8.4. The functions to compute the suspiciousness score are presented in Fig. 8.5.

As described in the last chapter, when looking for the root cause of the failure, the effectiveness of tests can be highly improved by solely considering those statements that directly or indirectly impact the results. Slicing is a technique used to shrink the search space to the statements affecting a particular output value, a slicing criterion designated by the tester. Therefore, when generating test data to examine the results provided at a particular location in the code, test data should be generated to cover the static backward slice for that location.

For instance, consider the function f shown in Fig. 8.6. Since there is only one return statement in the given source code, all the feasible paths of the function are included in the backward static slice of the Impact variable in statement number 18.

As shown in Fig. 8.7, the test case T1 is a failing test case. The output of the function getImpact is its return value in statement S16. The backward dynamic slice (BDS) of the test criterion (S16, Impact) is <S18, S17, S16, S11, S10, S3>. These statements, marked with '?' in Fig. 8.7, are all suspected to be the root cause of the failure.

```python
# Main body of the program
# sys.settrace() is used to implement debuggers, profilers, and coverage tools.
# This is thread-specific and must register the trace using threading.settrace().
sys.settrace(traceit)
for test_case in test_suite:
    expected_result = test_case[2]
    # Run the function under test with a test case and save log
    # trace of the execution in an array named execution_trace
    function_under_test(test_case[0], test_case[1])
    if (expected_result == "p"):       # If the execution is passed
        exec_result.append("1")        # Execution_result = "passed"
        total_passed += 1              # total number of passing executions
        for line_no in execution_trace:        # for each line in the execution trace
            if line_no in linenos_passed:
                linenos_passed[line_no] += 1
            else:
                linenos_passed[line_no] = 1
    else:
        exec_result.append("0")
        total_failed += 1
        for line_no in execution_trace:
            if line_no in linenos_failed:
                linenos_failed[line_no] += 1
            else:
                linenos_failed[line_no] = 1
    test_suite_trace.append(execution_trace)
    for i in linenos_frequency.copy():
        passed_no = 0
        failed_no = 0
        if i in linenos_passed:
            passed_no = linenos_passed[i]
        if i in linenos_failed:
            failed_no = linenos_failed[i]
        Ochiai(i,passed_no,failed_no)
        Tarantula(i,passed_no,failed_no)
        # leave the remaining part to the reader
        ...
        ...
```

Fig. 8.4 The main body of the Diagnoser software tool

```
# Compute fault suspiciousness score using Tarantula
def Tarantula(line_number,passed,failed):
    Tarantula_suspiciousness[line_number] = (failed/total_failed) /
((failed/total_failed) + (passed/total_passed))
def Ochiai(line_number,passed,failed):
    #print(line_number,passed,failed)
    Ochiai_suspiciousness[line_number] = failed /
                            sqrt(total_failed*(passed+failed))
```

Fig. 8.5 Two functions to compute the suspiciousness scores

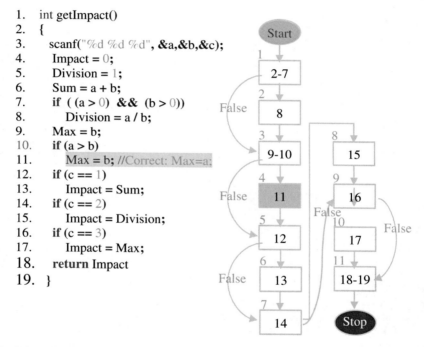

```
1.   int getImpact()
2.   {
3.       scanf("%d %d %d", &a,&b,&c);
4.       Impact = 0;
5.       Division = 1;
6.       Sum = a + b;
7.       if ( (a > 0)  &&  (b > 0))
8.           Division = a / b;
9.       Max = b;
10.      if (a > b)
11.          Max = b; //Correct: Max=a;
12.      if (c == 1)
13.          Impact = Sum;
14.      if (c == 2)
15.          Impact = Division;
16.      if (c == 3)
17.          Impact = Max;
18.      return Impact
19.  }
```

Fig. 8.6 Statement number 11 should be "Max = a;" instead of "Max = b;"

8.3 Slicing

Mark Weiser introduced the concept of slicing in 1981 [3]. Slicing techniques tend to extract those statements from a program that may be influenced by or influence a specific statement of interest, called a slicing criterion. Slices are either forward or backward [4]. A forward slice is a portion of the code influenced by the slicing criterion, whereas a backward slice influences the slice criterion.

8.3.1 Backward Slicing

The computation of a slice is called slicing. The backward slice for the slicing criterion (V, L), where V is a variable used in a statement at the location L of the program P, is the set of statements, St, involved in computing V's value at L. For instance, the backward slice for the slicing criterion <S11, Sum> in Fig. 8.7, is:

1. S11: Since Sum is used in S11.
2. S10: Since S11 is control dependent on S10 or, in other words, the execution of S11 depends on the decision made in S10.
3. S4: Since in S14, the Sum is defined or, in other words, the value of Sum is updated.
4. S1: Since the last value assigned to Sum depends on the values of the 'a' and 'b' variables.

Slices are either computed dynamically at runtime or statically at compile time. When looking for the root cause of a program failure, we can speed up the search by restricting it to those statements that somehow affect the incorrect result. In other words, we should look for the sequence of statements in the execution trace on which the execution result depends. Such a sequence is called a backward dynamic slice (BDS) for the slicing criterion <V, L>, where L is the statement in which an unexpected incorrect value for the variable V is observed. For instance, in Fig. 8.8, the BDS for the variable Impact in line S16 when executing the function getImpact with the test case T1 is:

#.	int getImpact()	T1 a:4 b:1 c:3	T7 a:8 b:-3 c:2	T10 a:7 b:6 c:1	T11 a:6 b:8 c:3
S_1	scanf("%d %d %d",a,b,c);	?	√	√	√
S_2	result = 0;	*	*	*	*
S_3	Div = 1;	*	√	*	*
S_4	Sum = a + b;	*	*	√	*
S_5	if ((a > 0) && (b > 0))	*	*	*	*
S_6	Div = a / b;	*		*	*
S_7	Max = b;	*	*	*	√
S_8	if (a > b)	?	*	*	*
S_9	Max=b; //Correct: Max =a;	?	*	*	
S_{10}	if (c == 1)	*	*	√	*
S_{11}	Impact = Sum;			√	
S_{12}	if (c == 2)	*	√	*	*
S_{13}	Impact = Div;		√		
S_{14}	if (c == 3)	?	*	*	√
S_{15}	Impact = Max;	?			√
S_{16}	return Impact	?	√	√	√
	Execution result (0=Successful/ 1=Failed)	1	0	0	0

Fig. 8.7 Finding faulty path

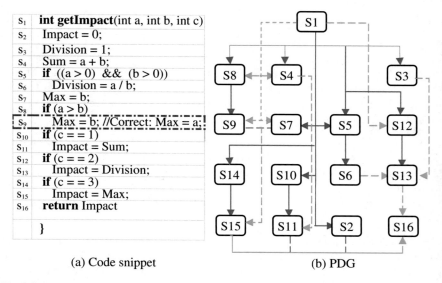

S_1	**int getImpact**(int a, int b, int c)
S_2	Impact = 0;
S_3	Division = 1;
S_4	Sum = a + b;
S_5	**if** ((a > 0) && (b > 0))
S_6	Division = a / b;
S_7	Max = b;
S_8	**if** (a > b)
S_9	Max = b; //Correct: Max = a;
S_{10}	**if** (c == 1)
S_{11}	Impact = Sum;
S_{12}	**if** (c == 2)
S_{13}	Impact = Division;
S_{14}	**if** (c == 3)
S_{15}	Impact = Max;
S_{16}	**return** Impact
	}

(a) Code snippet   (b) PDG

Fig. 8.8 Solid and dotted arrows represent control and data dependencies, respectively.

$$BDS(Impact, S16, T1) :< S16, S15, S14, S8, S9, S1 >$$

The backward dynamic slice or, in short, the BDS for a variable at a specific location, S_n, in a program execution trace is a sequence of statements:

$$S = < S_n, S_{n-1}, \ldots, S_1 >$$

where statement S_i is either control or data-dependent on S_{i-1}.

A statement, T, is data dependent on another statement, D, if at least a value used at T is defined or computed in D. On the other hand, T is control dependent on D if its execution depends on a decision made at T.

8.3.2 Forward Slicing

A forward slice is a part of the program, including statements somehow affected by the value of a variable at a point of interest. The forward slice for the slicing criterion (V, L), where V is a variable defined in a statement at the location L of the program P, is the set of statements, St, involved in computing V's value at L. For instance, in Fig. 8.8, the forward slice for the slicing criterion <S1, a> is:

< S1, S4, S5, S6, S8, S11, S13, S15, S16 >

Because:

1. S1: The variable 'a' is defined in S1.
2. S4, S4, S6, S8: The value of 'a', defined in S1, is used in S4, S5, S6, and S8.
3. S11: The variable Sum used in S11 depends on the value of 'a' in S8.

   A slice is a sub-graph of the program dependency graph (PDG) that starts or ends
with the slicing criterion, depending on whether it is a forward or backward slice.
To write a slicing tool, you need to learn about PDGs and how to develop a software
tool to extract a PDG from a given source code.

8.4   Program Dependency Graphs

A program dependency graph (PDG) is a graph (V, E) where V is the program
statements and E represents control and data dependencies between the statements
[9]. A statement S_i is control-dependent on a statement S_j if and only if the execution
of S_i depends on the decision made in S_j, whereas S_i is data-dependent on S_j if S_i
uses a value computed in S_j. For example, consider the program dependency graph
illustrated in Fig. 8.8.

Figure 8.8b illustrates the program dependency graph for the getImpact method. Solid edges represent data dependencies, and dashed edges indicate data dependencies. For instance, statement S11 is control-dependent on S10 and data-dependent on S4, whereas S4 is itself data-dependent on S1. The question is how to extract a dependency graph from the context of a given source code. As shown in Fig. 8.8b, a PDG combines a data dependency and a control dependency graph. The following section describes how to build a data dependency graph.

8.4.1   Slicing Algorithm

The essential idea of slicing in software testing is to focus on the statements that directly or indirectly impact an erroneous result [4, 5]. Faults propagate across slices and propagate via any statement somehow affected by the faulty value. Backward and forward slices can be extracted from the PDG [3].

The forward slice for a given testing criterion, addressing a node n, is a connected subgraph of the program dependency graph rooted at n. The subgraph is traversed to compute the forward static slice for the variable defined in the statement addressed by 'n'. The same procedure can be applied to compute static backward slices if all the dependencies' directions are inversed.

For instance, the forward slice for the variable 'fac' at line 7 and the backward slice for 'fac' at line 9 of the code fragment in Fig. 8.9 are as follows:

As another example, let us obtain the forward slice of the variable 'i' at line 8 of the C function given in Fig. 8.9. The forward slice is computed while traversing the subtree of the PDG rooted at node 8. The traverse starts with the outgoing edges from node 8 forward and adds nodes 5, 6, attached to the outgoing edges to the slice. All the nodes added to the slice are marked to avoid repeated nodes in the slice. The same process is performed for the added nodes until all the subtree nodes are traversed. As a result, the forward slice calculated for the variable i at line 8 contains the lines of code numbered 8, 5, 6, 7, 9, and 10. Figure 8.10 represents an algorithm for the forward slicing, slice_forward.

Forward_slice(fac, 6) = (7, 9)
Backward_slice(fac, 9) = (9, 6, 8, 5, 4, 2, 1)

```
1  int SF(int n){
2      int i=1;
3      int sum=0;
4      int fac =1;
5      while (i<=n) {
6          fac = fac * i;
7          sum = sum + fac;
8          i = i+1;
       }
9      printf(" Factorial:%d ", Fac);
10     return sum;
   }
```

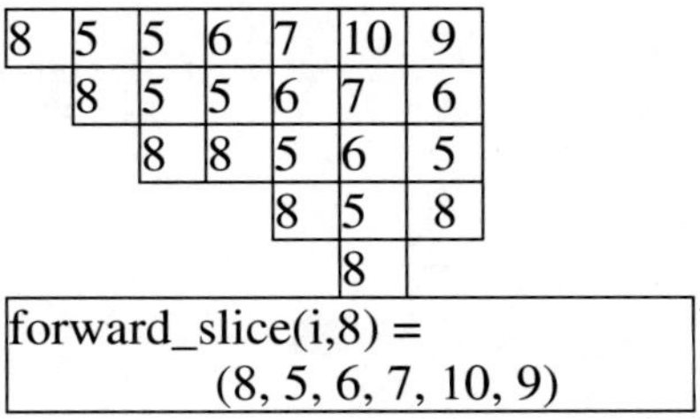

8	5	5	6	7	10	9
8	5	5	6	7	6	
	8	8	5	6	5	
		8	5	8		
		8				

forward_slice(i,8) =
(8, 5, 6, 7, 10, 9)

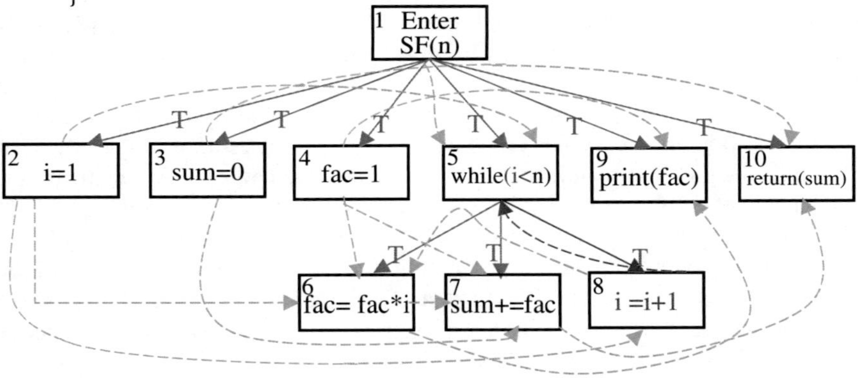

Fig. 8.9 A PDG with a mutual dependency between nodes 5 and 6

Slice_forward is a recursive algorithm that, in fact, uses a hidden stack to traverse a subtree of the PDG, rotted at a given node of the tree. Figure 8.10 represents the trace of the algorithm to compute the forward slice for the definition of fac at line 8 of the C function.

```
Algorithm slice_forward(PDG, node) : SliceForNode
    if not is_visited(node)
        then node.mark = _visisted
             node.slice = [node]
             foreach _node in get_dependent(node) do
                     _node_slice = Slice_forward(PDG, _node)
                     node.slice = node.slice + [ _node_slice]
                 end_foreach
    end_if
    return node.slice
emd_algorithm
```

Fig. 8.10 An algorithm to compute forward slice for a given node

Slices can be easily derived from a PDG. The question is how to derive a PDG from the source code. A PDG is a combination of data flow and a control dependency graph. A data flow/dependency graph is a directed graph, $G(V, E)$, where each vertex $v \in V$ is a statement and each directed edge $e \in E$ has an arrow from a vertex p to q, which means the statement q uses a variable, defined by the statement P. It is worth mentioning that a statement such as "$x = y*z$" uses the variables "y" and "z"; it kills the previous definition of "x" and creates the new one. A statement Si is data-dependent on Sj if Si uses at least a value defined by Sj.

When a value is assigned to a variable, it is said that the variable is defined. In a data dependency graph, the node representing a statement, D, holding the definition of a variable V, is connected to all the nodes, U, using the value of V. Thus, a chain of definitions and uses, def-use chain, for all the variables is created at the source code level. A def-use chain is a set of triples <D, U, V> where D defines the variable V, and U is a statement using the defined value before V is redefined. Here, D is called a reaching definition for the statement U. The following sections describe how to derive these graphs from the source code of the functions.

8.5   Data Flow Analysis

Faults propagate through dependencies to the outputs. Dependencies reveal through data flow analysis [3]. A data flow is a path for data to move from one part of the code to another. Any statement S_i that assigns a value to a variable, V, is a definition of V and any statement Sj that uses the value assigned to V is a use of V. Data flow in a function is represented as definition-use or, in short, du-chains and conversely ud-chains as shown in Fig. 8.11.

A definition-use chain or, in short, du-chain relates a value assigned to a variable, v, in line, n, to the line numbers where the value is used. Conversely, an ud-chain relates the use of a variable to where it is defined. These chains help debuggers to trace fault propagation from one statement to the other and find the root cause of the failure or, in the case of any changes to the code, determine the statements affected by the change.

The question is, what is data flow analysis doing with software testing and fault localization? Data-flow testing examines the history of the definition and use of the variables. The history can be examined statically by looking for specific patterns in source code or dynamically considering the coverage of the data flow graph provided by the test data [7]. Static and dynamic data flow testing are described in Sects. 8.5.1 and 8.5.2, respectively.

When a test fails, typically, the debugging process starts with the statement where the faulty result is observed. The debugger traces backward the execution path and examines all the statements affecting the faulty result as far as the root cause of the failure is found. Apparently, the debugging process will be accelerated if the chain of statements affects the faulty value. Section 8.5.2 describes an algorithm, reaching definitions, and another algorithm, use-def chain, to determine the chain of statements affecting the value of a variable at a given location in the program under test.

In certain situations, the programmer may suspect a statement to cause the test to fail. Def-use chains, described in the next section, can benefit such cases. The recently developed parts of code are often more likely to be fault-prone [4]. Code complexity is a reason for fault-prone and lousy code quality. The question is how to relate the values computed in fault-prone regions to the faulty results.

In fact, the def-use chain and the use-def chain are two examples of test requirements concerning data flow analysis. Search-based techniques attempt to generate test data satisfying one or more test requirements. Data flow coverage concerns.

8.5.1 Static Data Flow Testing

Specific bugs in programs are detected by considering the definition and use of each variable [6, 8]. Compiler optimizers look for specific patterns in data flow graphs to detect anomalies. Examples of such patterns are definitions with no use and definitions killed before being used. Using a variable with no definition is another type of anomaly detected by the compiler optimizers. For instance, consider the code fragment in Fig. 8.12.

In the data flow graph shown in Fig. 8.12, the variable d is used before being defined. The variable a is redefined before being used, and finally, the variable c is defined but not used. Such anomalies can be detected simply by inspecting the data flow graph.

S1.	a = b;		S1.	a = b;
S2.	b = c * d;		S2.	b = c * d;
S3.	e = a + 5		S3.	e = a + 5
S4.	b = a*e;		S4.	b = a*e;
S5.	f = b*2;		S5.	f = b*2;

(a) du-chains link each def to its uses   (b) ud-chains for each use to its defs

Fig. 8.11 Definition-use and use-definition chain

```
      void main(){
        int a, b, c, d;
1.      scanf("%d %d %d", a, b, c);
2.      a = b*d;
3.      if (a > 5)
4.              b++;
5.      printf("a = %d , b = %d", a,b);
6.      exit(0);
      }
```

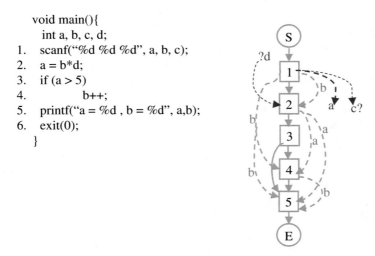

Fig. 8.12   A data flow graph with some anomalies shown with red dashed arrows

8.5.2   *Dynamic Data Flow Testing*

The primary objective of dynamic data flow testing is to uncover possible bugs in data usage during code execution [10]. The test case generation objective is to cover the traces of every definition of a data variable to each of its use and every use and vice versa. The data flow coverage criteria, including **All-uses (AU)**, **All-defs (AD)**, and **All-def-uses**, employed for creating test cases are defined in the continuation.

The value assigned to a variable or, in other words, the definition of a variable may be used in several locations before the definition is killed (the variable is redefined). For instance, consider the following code fragment:

```
      ...
11.   x = a*b;       // 1- (x,11): x is defined in line 11.
12.   if (i>5)
13.       y = x -1;   // 2- (x,11,13): (x,11) is used in line 13.
14.       y = y*x;    // 3- (x,11,14): (x,11) is used in line 14.
15.       x = x + 1;  // 4- ~(x,1): (x,11) is killed in line 15.
16.       z = x -1;   // 5- (x,15,16): (x,15) is used in line 16.
17.   else
18.       a = x+1;    // 6 – (x,1,18): (x,11) is used in line 18.
19.   a = x -1;       //7- ((x,11), (x,15), 19)
```

The *definition* of a variable x at line n is represented as def(x,n). For instance, in the above code fragment, the variable x has two definitions def(x,11) and def(x,15). The two definitions reach line 19. A path is *def-clear* for a variable x if there is no redefinition of x along the path. The definition def(x,n) is *live* as far as x is not redefined. A *du-path* indicates a simple def-clear path between a definition of a variable x and its use. The simple def-clear path between the definition of the variable x at the line number n, def(x,n), and its use at line m, is represented as du-path(x, n, m). For instance, in the above code fragment, the definition of x reaching line 19 includes:

reach(x, 19) = {(x,1), (x,15)}

The reaching definition algorithm, presented in the next section, computes the reach set for each definition.

The *All-uses* (AU) criterion states that there is at least a path between every use of a variable x and its definition. The All-uses coverage criterion requires the test data set to examine all possible paths between the use of def(x,n) in any line number m, du-path(x,n,m), and def(x,n) at any line n. The point is that if there is no path connecting the use of a variable to its definition, then the variable is used before being defined!

A test data set satisfies the *All-def(AD)* criterion provided that for each set S = du-path(x,n,m), there is a test data examining at least a path d in S. If the test data set covers all the members of S then it satisfies *All-du Paths (ADUP)* criterion. In other words, a test data set satisfies the All-du paths coverage criterion provided it examines all the du-paths in the code under test.

The use of a definition in the predicate of a branch statement is a predicate use or "p-use." Any other use is a computation-use or "c-use." For instance, consider the following if statement:

$$\text{if } (\ \underbrace{x > w^*v}_{\text{p-use of x,w and v}}\ )\ y = \underbrace{a * b}_{\text{c-use of a and b}}$$

Some faults may cause wrong execution paths; therefore, the faulty output will not be data-dependent on the faulty value. The *All-p-uses* coverage criterion ensures that the test data set covers at least a du-path(x, n, m) for each def(x,n) used in a predicate in line m. Here, n and m can be any line number in the code under test. Similarly, the *All-c-uses* criterion ensures that if test data covers a c-use of a variable x at line number m, it covers the du-path(x, n, m).

The coverage criterion *All-p-uses/Some c-uses* states that if there is a p-use at line m for a variable def(v,n), then test data should cover du-path(v,n,m); otherwise, the test data should cover du-path(v,n, m') where m' is a line including a c-use of def(v,n). Similarly, another coverage criterion, *All-c-uses/Some p-uses*, ensures that there is a use for each definition. However, if an execution path covers a definition, its use should also be covered by the execution path. The compiler optimizer uses the reaching definition algorithm described in the next section to discover such faults.

8.5.3   *Reaching Definitions*

The reaching definition is a classic method adapted from compiler construction for the analysis and testing of software [11, 12]. A reaching definition for a given statement, U, is an earlier statement, D, whose target variable, v, can reach (be assigned to) U without an intervening assignment. The algorithm ReachingDefinitions, in Fig. 8.13, computes the definitions reaching the beginning and the end of each basic block, B, in two sets In(B) and Out(B), respectively, where:

$$In[B] = \cup \, Out[p] \forall p \in Predecessor(B)$$

Gen(B) includes all the definitions generated in a basic-block B that reaches the end of B. Finally, Kill(B) includes all the definitions killed in B. Out(B) for each basic block, B, is computed as follows:

$$Out[B] = Gen[B] \cup (In[B] \backslash Kill[B])$$

Algorithm ReachingDefinitions

Input:
    CFG: Control flow graph for a given function
Output:
    In(B) and Out(B) for each basic block B $\in$ CFG
Method:
1.   Let Gen(CFG) = φ
2.   for each B $\in$ CFG do
    2.1. Compute Gen(B);
    2.2. Gen(CFG) = Gen(B) $\cup$ Gen(CFG)
3.   for each B $\in$ CFG do
    3.1. Kill(B) = Gen(B) $\cap$ (Gen(CFG)\Gen(B)
    3.2. Out(B) = Gen(B)
    3.3. In(B) = φ
4.   changes = true
5.   while changes do
    5.1. changes = false
    5.2. for each node n in traverse(CFG) do
        5.2.1. IN[n] = $\cup$ OUT[P], where P is an immediate predecessor of n
        5.2.2. OLDOUT = OUT[n]
        5.2.3. OUT[n] =GEN[n] $\cup$ (IN[n] - KILL[n])
        5.2.4. if OUT[n] != OLDOUT
        5.2.5.   then change = true

Fig. 8.13 Applies each flow equation iteratively until the solution stabilizes

Initially, the Gen and Kill sets for each basic block are determined to compute the reaching definitions.

Reaching definitions is a data-flow analysis technique that statically determines which definitions may reach a given point in the code. The reaching definition algorithm in Fig. 8.13 initially computes the Gen and Kill sets for all the basic blocks. The Gen and Kill set for the basic blocks of a sample code fragment are given in Fig. 8.14.

The algorithm uses Gen and Kill sets of each basic block to compute the In and Out sets. It iterates until reaching a stable state so that the value of Out does not alter in two consecutive iterations. Figure 8.15 illustrates the process of computing the In and Out sets for the basic blocks of a given code fragment.

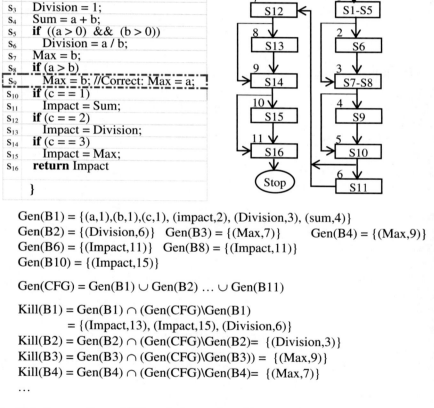

Gen(B1) = {(a,1),(b,1),(c,1), (impact,2), (Division,3), (sum,4)}
Gen(B2) = {(Division,6)}   Gen(B3) = {(Max,7)}   Gen(B4) = {(Max,9)}
Gen(B6) = {(Impact,11)}   Gen(B8) = {(Impact,11)}
Gen(B10) = {(Impact,15)}

Gen(CFG) = Gen(B1) ∪ Gen(B2) ... ∪ Gen(B11)

Kill(B1) = Gen(B1) ∩ (Gen(CFG)\Gen(B1)
         = {(Impact,13), (Impact,15), (Division,6)}
Kill(B2) = Gen(B2) ∩ (Gen(CFG)\Gen(B2)= {(Division,3)}
Kill(B3) = Gen(B3) ∩ (Gen(CFG)\Gen(B3)) = {(Max,9)}
Kill(B4) = Gen(B4) ∩ (Gen(CFG)\Gen(B4)= {(Max,7)}
...

Fig. 8.14 Compute Gen and Kill sets before starting with the algorithm

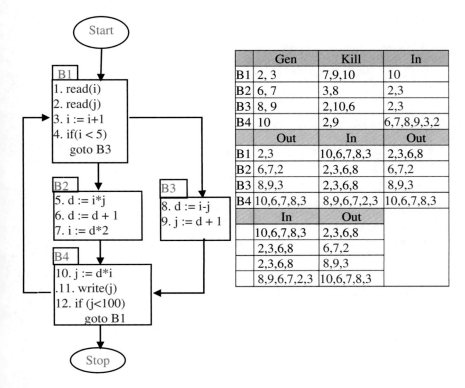

Fig. 8.15 An illustration of the reaching definitions algorithm

8.5.4 Definition-Use Chains

Each definition-use pair associates a definition of a variable with the use of the same variable. The definition-use pair, (d,u), is formed when the value assigned by the definition, d, can reach the point of use, u, without being overwritten by another value. The definition-use pairs form a graph called the data-dependency graph.

For each basic block, B, the reaching definition algorithm computes in (B) as the set of definitions that can reach the leading statement of B. In this way, the use-definition and definition are formed considering the use of the definitions. For instance, in Fig. 8.16, the reaching definition of each statement is shown. The reaching definition for each statement can be computed using the reaching definitions of each previous statement in its enclosing basic block.

Figure 8.17 illustrates the definition-use chain for the variable, 'i.' in Fig. 8.16. The dashed arrows show data dependencies to the variable i.

In Fig. 8.18, an algorithm to compute the definition-use pairs for all the variables in a program is given.

Program Sums
1. read(n);
2. i = 1;   // In = {1:n}
3. sum = 0;   // In = {1:n, 2:i}
4. while(i <= n) do   // In = {1:n, 2:i, 11:i, 3:sum, 5:sum, 8:sum,
                                    //   j:6, j:9 }
5.   sum = 0;   // In = {1:n, 2:i, 11:i, 3:sum, 5:sum, 8:sum,
                                    //   j:6, j:9 }
6.   j = 1;   // In = {1:n, 2:i, 11:i, 5:sum, j:6, j:9 }
7.   while(j <= i) do   // In = {1:n, 2:i, 11:i, 5:sum, 8:sum, j:6, j:9 }
8.   sum = sum + j;   // In = {1:n, 2:i, 11:i, 5:sum, 8:sum, j:6, j:9 }
9.   j = j + 1;   // In = {1:n, 2:i, 11:i, 8:sum, j:6, j:9 }
        endwhile
10.   write(sum, i);   // In = {1:n, 2:i, 11:i, 5:sum, 8:sum, j:6, j:9 }
11. i = i + 1;   // In = {1:n, 2:i, 11:i, 5:sum, 8:sum, j:6, j:9 }
        endwhile
12. write(sum, i);   // In = {1:n, 2:i, 11:i, 3:sum, 5:sum, 8:sum,
                                    //   j:6, j:9 }

end sums

Fig. 8.16 Reaching definition for each statement

Fig. 8.17 Definition use
chain for the variable i

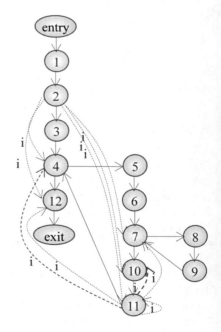

Algorithm Def-Use Chain
Input: A flow graph, CFG, for which the IN sets for reaching definitions have
been computed for each node B.
Output: DUPairs: a set of definition-use pairs.
Method: Visit each node in the control flow graph.
For each node, use upwards exposed uses and reaching definitions
to form definition-use pairs.
1. DUPairs = Null
2. for each node B do
      InS = In(B)
3.   for each statement, S, in B do
4.   for each use, U, in S do
5.   for each definition, D, of U in InS do
6.   DUPairs = DUPairs $\cup$ (D,U)
            endFor
         endFor
7.   For each definition D in S do
8.   Remove all the definitions similar to D from InS
10.   Add D to InS
         endFor
      endfor
   endfor

Fig. 8.18 An algorithm to generate definition-use pairs for a given CFG

A use-definition or, in short, ud-chain for a variable, v, connects the statement, U, using the value of v to all those statements that have assigned the value, U, to v.

8.6 Control Dependency Graphs

Duplicates are avoided through dependencies; however, dependencies increase the risk of fault propagation. Control and data dependencies are identified to diagnose fault propagations [9]. In a control flow graph, the children of a basic block are control dependent on the basic block. If the execution of the statement depends on a condition, then the statement is control dependent on the condition responsible for its execution. For instance, all the statements in the then part and else part of an 'if statement are control dependent on the if statement condition'. The following subsections present three algorithms for the automatic extraction of control dependency graphs.

8.6.1　Dominator and Post-Dominant Trees

In CFG = (V, E, Start, End), node x dominates node y if every path from Start to y passes through x. In fact, x is a bottleneck to reaching y from Start. The entry node of a natural loop, i.e., a loop with a single entry point, dominates all its body nodes in CFG. The formation of the loop is due to the back edges. A directed edge from node A to node S is a back edge if S dominates A. The algorithm Dominant in Fig. 8.19 computes for each node, X, of a CFG the set of nodes, Dominant(X), dominating the node X.

The above algorithm computes Dominant(n) as the set of nodes dominating each node, n, in a given CFG. Figure 8.20 illustrates a CFG and its corresponding dominator tree.

Figure 8.21 represents an algorithm, DominantTree; the algorithm takes as input a control flow graph, CFG, and the set, Dominant(n), including all the nodes dominating the node "n ∈ CFG" and builds a dominators tree, DT, for the CFG.

A node, y, post-dominates a node, x, if every path from x to the Stop node passes through y. After inverting the direction of the arrows and considering the Stop node as the Start node and inversely Start as Stop, the dominant tree algorithms can be

Algorithm Dominant
   Input-　　　ControlFlowGraph : CFG
   Output-　　Dominant : Set of nodes dominating each CFG node

  1.　Dominant(Start) = {Start}
  2.　forall node <> Start in ControlFlowGraph do
         Dominant(node) = All Nodes
    endfor
  3.　Changes = true
  4.　while Changes do
  5.　　Changes = false
  6.　　forall node <> Start in ControlFlowGraph do
  7.　　NewDominant =
         {node} ∪ { ∩ Dominatnt(p), ∀ p ∈ {predecessors(node)}}
  8.　　　　if Dominant(node) <> NewDominant
  9.　　　　then changes = true
      endif
 10.　　　　Dominat(node) = NewDominant
    endfor
  endwhile

Fig. 8.19 An algorithm to compute nodes dominating each node in CFG

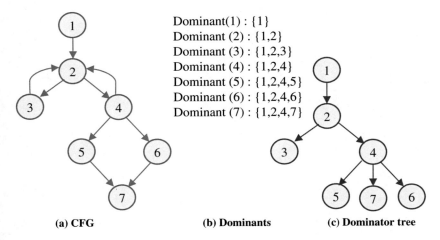

Dominant(1) : {1}
Dominant (2) : {1,2}
Dominant (3) : {1,2,3}
Dominant (4) : {1,2,4}
Dominant (5) : {1,2,4,5}
Dominant (6) : {1,2,4,6}
Dominant (7) : {1,2,4,7}

(a) CFG   (b) Dominants   (c) Dominator tree

Fig. 8.20 A CFG and its corresponding Dominant tree

Algorithm DominatorTree
    Input. A CFG & Dominant(n) for all nodes, n, in CFG
    Output. Dominator tree DT for CFG.
    1. **let** Start be the root of DT;
    2. **put** Start on queue Q;
    3. **for each** node n **in** N **do** D(n) = D(n) – n; **enddo**;
    4. while Q is not empty, do
    5.   m = the next node on Q; remove m from Q;
    6.   for each node n in N such that D(n) is nonempty, do
    7.   if D(n) contains m
    8.   D(n) = D(n) - m;
    9.   if D(n) is now empty
    10.   add n to DT as a child of m;
    11.   add n to Q;
                endif
              endif
            endfor
          endwhile

Fig. 8.21 An algorithm to build a dominator tree for a given CFG

used to construct the post-dominant tree. Figure 8.22 represents a sample control
flow graph and its dominant and post-dominant trees.

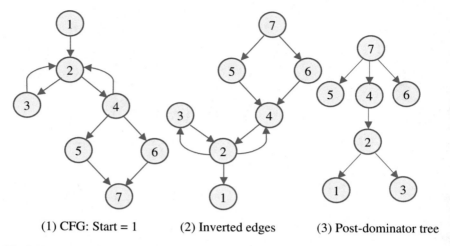

| (1) CFG: Start = 1 | (2) Inverted edges | (3) Post-dominator tree |

Fig. 8.22   Constructing a post-dominant tree from a CFG

8.6.2   *Control Dependency Graph*

Branching conditions allow control of execution jump jumps from one position to another. A Control dependency occurs when a program statement's execution depends on another. Control dependencies are determined by analyzing the control flow graphs.

In a CFG a node y is control dependent on the node x if:

1. there is a path from x to y,
2. y post-dominates all the nodes on the path apart from x.

Since y is not a bottleneck for reaching the Stop from x, then there should be another path to reach Stop from x. For instance, in the CFG fragment shown in Fig. 8.23, node y is control dependent on x.

Fig. 8.23   A decision is
made at node x to execute y

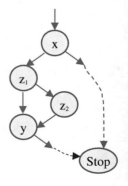

In the continuation of this section, the following operators and terms are used to describe control dependencies:

    (1) X dominates Y :

      ➢ X ∇d Y

    (2) X does not dominate Y :

      ➢ X ∇/d Y

    (3) X post=dominates Y :

      ➢ X ∇p Y

    (4) X does not post-dominates Y :

      ➢ X ∇/p Y

    (5) Y is control dependent on X :
      ➢ Y ∇c X

    (6) Post-dominant tree:
      ➢ PDT

The above term and operations simplify the formal definition of control dependencies. Let nodes X and Y represent two basic blocks in a CFG, then Y is control dependent on X if and only if there is a path in the post-dominant tree.

Y ∇c X $\Leftrightarrow$ $\exists$ path P $\in$ PDT from X to Y / $\forall$ Z $\in$ P $\Rightarrow$ Y ∇p Z $\wedge$ Y ∇/p X

A more detailed description of the algorithm is given below:

    Algorithm parsa-control-dependencies (input: CFG) returns : CDG

      Let PDT = post-dominator-tree(CFG)

      Let S = Set of edges (X→Y) $\in$ CFG / Y ∇/p X

      Forall (X→Y) in S do

        Let W = closest-ancestor(X, Y) in PDT

        $\forall$Z $\in$ Path(W to Y) in PDT) $\wedge$ Z $\neq$ W $\wedge$ Z ∇/p X $\Rightarrow$ Z ∇c X

Figure 8.24 shows how to follow the algorithm to compute control dependencies for a sample CFG.

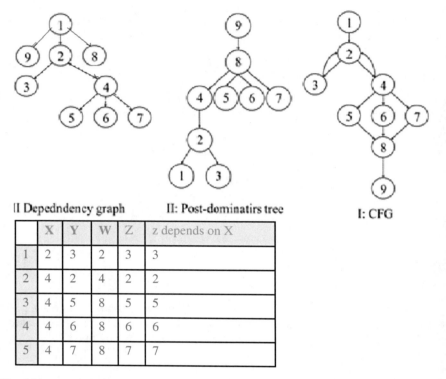

<table>
<tr><td></td><td>X</td><td>Y</td><td>W</td><td>Z</td><td>z depends on X</td></tr>
<tr><td>1</td><td>2</td><td>3</td><td>2</td><td>3</td><td>3</td></tr>
<tr><td>2</td><td>4</td><td>2</td><td>4</td><td>2</td><td>2</td></tr>
<tr><td>3</td><td>4</td><td>5</td><td>8</td><td>5</td><td>5</td></tr>
<tr><td>4</td><td>4</td><td>6</td><td>8</td><td>6</td><td>6</td></tr>
<tr><td>5</td><td>4</td><td>7</td><td>8</td><td>7</td><td>7</td></tr>
</table>

Fig. 8.24 An example to illustrate the parsa-control-dependencies algorithm steps

8.7   Summary

This chapter describes the design and implementation of a software tool, Diagnoser. The Python function sys.settrace() provides a mechanism defining hooks to interrupt execution and access the program state. The input to Diagnoser is a Python function and test cases to run the function and evaluate its return value. Diagnoser uses the evaluation results to compute the fault suspiciousness score for each statement using the Tarantula and Ochiai equations. The accuracy and speed of the fault localization process can be accelerated by computing the backward dynamic slice for the output statements.

Slices are built upon the program dependency graph. Program dependence graphs mainly consist of nodes representing the statements of a program and control and data dependence edges:

- Control dependence between two statement nodes exists if one statement controls the execution of the other.
- Data dependence between two statement nodes exists if a definition of a variable at one statement might reach the usage of the same variable at another statement.

Algorithms to implement data and control dependencies are given in this chapter. Faults propagate through data and control dependencies. Data flow testing is a white-box testing technique examining the data flow concerning the values assigned and used in the code.

Data flow testing discovers improper utilization of data values by examining the test paths of a program covering the locations of definitions and uses of variables in the program. Data flow testing bases the test coverage criterion on how variables are defined and used in the program:

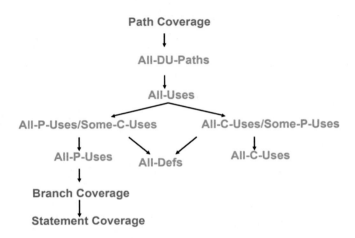

8.8 Exercises

[1] Complete the code given for the Diagnoser software tool. Provide some examples to demonstrate the applicability of the software tool to facilitate spectrum-based fault localization techniques.

[2] Use the traceit function to develop a class to compute branch and statement coverage. For more information, refer to the following address:
     https://www.fuzzingbook.org/beta/html/Coverage.html

[3] Implement a software tool to compute dynamic backward slices for Python functions. Add the tool as a plugin to the Diagnoser to improve its fault localization accuracy and speed.

[4] Provide Diagnoser with the option to accept any number of equations to compute fault suspiciousness score for the program statements.

[5] Write ten failing and passing test cases to locate the fault in the quicksort function given below. Create a diagnosis table to represent the test results and suspiciousness scores.

```
/* QuickSort:
arr[] --> Array to be sorted,
low  --> Starting index,
high  --> Ending index */
void quickSort(int arr[],
            int low, int high)
{
    int i, pivot;
    if (low >= high)
        return;
    //Error:should be pivot= arr[high]
    pivot = arr[high + 1];
    i = (low - 1);
    for (int j = low; j <= high- 1; j++)
    {
        if (arr[j] <= pivot)
{ //Swap

        i++;
        temp = arr[i];
        arr[i] = arr[j]);
        arr[j] = temp;
      }
    }
    temp = arr[i+1];
    arr[i+1] = arr[high]);
    arr[high] = temp;
    quickSort(arr, low, i);
    quickSort(arr, i + 2, high);
    return arr;
}
```

[6] Use chain of 'ud-pairs' to compute dynamic backward slice for the variable 'arr' used in the "return arr" instruction. Notice: start with the "arr instruction."

[7] Consider the following code:

```
              Program Sums
    1.        read(n);
    2.        i = 1;
    3.        sum = 0;
    4.        while( i <= n) do
    5.            sum = 0;
    6.            j = 1;
    7.            while( j < i) do // should be i<= j
    8.                sum = sum + j;
    9.                j = j + 1;
              endwhile
    10.           write(sum, i);
    11.           i = i + 1;
          endwhile
    12.   write(sum, i);
    end sums
```

(a) Draw a CFG for the above function.
(b) Annotate the CFG with reaching definitions for each basic block.
(c) Draw the program dependency graph.
(d) Compose ten test cases for the program and draw a diagnosis table to represent the test traces am test results.

[8] Modify the def-use chain algorithm to compute ud-pairs. Write a Python function to compute du-pairs for each basic block in a CFG. Chapter 6 describes how to implement the control flow graphs, and Chap. 7 shows how to instrument code. The functions you have already written to instrument code and to extract CFG, should be invoked to trace backward the du-pairs to compute the backward dynamic slice for a given slicing criterion.

References

1. Christakis, M., Heizmann, M.: Semantic fault localization and suspiciousness ranking. In: International Conference on Tools and Algorithms for the Construction and Analysis of Systems: TACAS 2019, pp. 226–243
2. Zareie, F., Parsa, S.: A non-parametric statistical debugging technique with the aid of program slicing (NPSS). Int. J. Inf. Eng. Electr. Bus. (2013)
3. Weiser, M.: Program slicing. In: Proceedings of the 5th International Conference on Software Engineering, pp. 439–449. IEEE Computer Society Press, March (1981)
4. Parsa, S., Zareie, F., Vahidi-Asl, M.: Fuzzy clustering the backward dynamic slices of programs to identify the origins of failure. In: SEA 2011, LNCS 6630, pp. 352–363. Springer-Verlag Berlin Heidelberg (2011)
5. Parsa, S., Vahidi-Asl, M., Arabi, S.: Software fault localization using elastic net: a new statistical approach. In: Advances in Software Engineering—Communications in Computer and Information Science, pp. 127–134 (2009). https://doi.org/10.1007/978-3-642-10619-4_16

6. Parsa, S., Vahidi-Asl, M., Arabi Naree, S.: Finding causes of software failure using ridge regression and association rule generation method. In: Proceedings of the 9th ACIS International Conference on Software Engineering, Artificial Intelligence, Networking, and Parallel/Distributed Computing (SNPD'08), Phuket, Thailand, IEEE Computer Society, Washington, DC, Aug 6–8 (2008)
7. Parsa, S., Vahidi-Asl, M., Asadi-Aghbolaghi, M.: Hierarchy-debug: a scalable statistical technique for fault localization. Softw. Qual. J. **22**(3), 427–466 (2014)
8. Parsa, S., Asadi-Aghbolaghi, M., Vahidi-Asl, M.: Statistical debugging using a hierarchical model of correlated predicates. In: Proceedings of the 3rd International Conference on Artificial Intelligence and Computational Intelligence (AICI'11), Taiyuan, China, September 24–25 (2011)
9. Seok Hong, H., Ural, H.: Dependence testing: extending data flow testing with control dependence. In: IFIP International Conference on Testing of Communicating Systems (2005)
10. Foreman, L.M., Zweben, S.H.: A study of the effectiveness of control and data flow testing strategies. J. Syst. Softw. **21**(3), 215–228 (1993)
11. Tonella, P., Antoniol, G., Fiutem, R., Merlo, E.: Variable precision reaching definitions analysis for software maintenance. In: Proceedings of the European Conference on Software Maintenance and Reengineering (CSMR), pp. 60–67 (1997)
12. Naser Masud, A., Ciccozzi, F.: Towards constructing the SSA form using reaching definitions over dominance frontiers. In: Conference: IEEE 19th International Working Conference on Source Code Analysis and Manipulation (SCAM) (2019)

Chapter 9
Coincidentally Correct Executions

9.1 Introduction

Spectrum-based statistical fault-localization techniques locate fault-relevant statements in a program by contrasting the statistics of the values of individual statements between successful and failure-causing runs. The main catch is that the statistics of associations of statements with correct and faulty executions can go wrong if there are coincidentally correct executions. An execution is coincidentally correct if the program yields correct results despite executing the faulty statement. The question is, how can it be that execution outputs correct results while executing the incorrect statement? You may find the answer in Sect. 9.2.

Correct execution paths may be identified by comparing and contrasting passing and failing paths. Clustering techniques are applied to partition execution paths based on their similarities. The similarities are measured in terms of the cross-entropies between probabilities of the n-grams in each execution path. Then by further analyzing the cross-entropy of sequences in each cluster, a series of suspicious locations are identified, and finally, using majority voting among clusters, faulty locations are reported to the programmer as faulty subpath(s).

© The Author(s), under exclusive license to Springer Nature Switzerland AG 2023
S. Parsa, *Software Testing Automation*,
https://doi.org/10.1007/978-3-031-22057-9_9

9.2 Coincidental Correctness

A program may yield correct results despite executing a faulty statement. Such executions are called coincidentally correct executions, provided that the faulty statement is covered by the backward dynamic slice of the correct result. Coincidentally correct test cases oppose the accuracy of fault suspiciousness formulas such as Ochiai and Tarantula. The catch is that executing a faulty statement/expression does not necessarily cause failure. For instance, consider the following faulty statement.

```
If( i > 20)  // should be (i> 30)
    printf(" i = %d", ++i);
else
    printf(" i = %d", --i);
```

The observation is that the fault does not reveal itself unless the value of i is greater than 30 or less than 20.

Spectrum-based fault localization (SFL) techniques seek to pinpoint faulty program elements by ranking them according to their fault suspiciousness scores. The fault suspiciousness score of each program element, i.e., statement, depends on the statistics of effective participation of that element in the erroneous or, in other words, unsuccessful runs and error-free or successful runs of the program. For instance, consider the C++ code in Table 9.1.

Table 9.1 The statement at line number 5 is faulty

Code \ Test cases	x:6 y: 4	x:7 y: 4	x:8 y: 4	x:3 y: 4	x:4 y:4	x:9 y:4
1. int main() {						
2. int y= 3;	♣	♣	♣	♣	♣	♣
3. int x;						
4. scanf("%d",&x);		♣		♣		♣
//faulty statement:						
5. x =x % 2; //should be x=x/2		♣		♣		♣
6. If(x>0)		♣		♣		♣
7. y++;		♣		♣		♣
8. return y;	♣	♣	♣	♣	♣	♣
9. }						
Coincidentally Correct / Fail	F	CC	F	CC	F	CC

Statement 5 in Table 9.1 is incorrect. It should be "x = x/2" instead of "x = x % 2". When running the code, if x is even, the fault manifests because, in this case, the backward dynamic slice of the statement at line 8, BDS(8), covers statement 5. However, despite the execution of statement 5, when x is odd, its fault does not reveal since it is not included in BDS(8). Therefore, when using a spectrum-based fault localization technique such as Tarantula or Ochiai, the suspiciousness score of the faulty statement is not calculated correctly. Such passing executions are called Coincidentally Correct Executions (CCE) [1].

When the value assigned to the variable x is even, the condition in line 6 does not meet, and as a result, the statement in line 7 does not execute. Consequently, the backward dynamic slice, calculated based on the output statement containing the variable y (line 8), only includes the statements in lines 2 and 8 and excludes the incorrect statement in line 5.

The point that I am going to make in this chapter is that even if a statement is not included in a particular execution, we are better off considering it as an active element dependent on its overall impact on the program execution results in other executions. In this way, the coverage and, consequently, the efficiency and effectiveness of the automated fault will be improved. To access the overall behavior of a statement concerning the program results in different program runs under test, I have introduced the concept of the extended backward dynamic slice or, in short, EBDS [1, 2]. The key concept underlying the use of EBDS is to detect a passing execution dependent on its similarity to failing executions. Section 9.3 introduces another approach [3], based on entropy, to determine coincidentally correct executions based on the cross-entropy.

The following section gives a motivating example illustrating the negative impact of coincidentally correct tests on calculating the suspiciousness scores. In the continuation in Sect. 9.3, it is shown how to use cross-entropy to resolve the difficulty.

9.2.1 Impact of CC Tests on SBFL (A Motivating Example)

Despite the proven efficacy of spectrum-based fault localization (SBFL) techniques, such techniques' accuracy still degrades due to the presence of coincidentally correct (CC) tests where one or more passing test cases exercise a faulty statement. Such tests result in inaccurate statistics to compute the suspiciousness score for the faulty statement. For instance, consider the code given in Table 9.2:

The function 'f' in Table 9.2 accepts three input parameters, 'a', 'b,' and 'op,' and depending on the value of 'op' being 1, 2, or 3, outputs the minimum, average, or maximum of 'a' and 'b,' respectively. The statements covered by the dynamic backward slices of the output statements are marked with a ' ✐ ' symbol. The dynamic backward slicing criterion depending on the value of 'op,' can be (min, S10), (avg, S12), or (max, S14).

As shown in Table 9.2, suspiciousness values computed for the statements covered by the backward dynamic slices of the faulty outputs do not identify the faulty

Table 9.2 Susspiciousness scores computed for backward dynamic slices

Test case:	T1	T2	T3	T4	T5	T6	T7	T8	T9	T10	T11	T12	Suspiciousness	
x (the first number):	2	5	-2	0	7	1	3	8	2	5	2	-6	Ochiai	Tarantula
y (the second number)	7	-3	4	-6	7	3	5	-6	2	-3	4	-6		
op (1:max, 2:av, 3:max)	1	1	1	1	1	2	1	2	1	3	1	3		
S1 def f(int x, int y, int op):	●	●	●	●	●	●	●	●	●	●	●	●	0.577	0.500
S2 avg=(x+y) / 2	*	*	*	*	*	●	*	●	*	*	*	*	0.000	0.000
S3 min = y	*	*	*	*	*	*	*	*	*	*	*	●	0.000	0.000
S4 max = y	*	*	*	*	*	*	*	*	*	*	*	●	0.000	0.000
S5 if (x <= y):	●	*	●	*	●	*	●	*	●	*	●	*	0.408	0.500
S6 min=y #min=x	●		●		●	*	●			●		*	0.408	0.500
S7 if (x > y):	*	*	*	*	*	*	*	*	*	●	*	*	0.000	0.000
S8 max = x		*		*						●			0.000	0.000
S9 if (op==1):	●	●	●	●	●	*	●	*	●	*	●	*	0.707	0.667
S10 print(min)	●	●	●	●	●		●		●		●		0.707	0.667
S11 if (op==2):	*	*	*	*	*	●	*	●	*	*	*	*	0.000	0.000
S12 print(avg)						●		●					0.000	0.000
S13 if (op==3):	*	*	*	*	*	*	*	*	*	●	*	●	0.000	0.000
S14 print(max)										●		●	0.000	0.000
Results: P: Pass /F =Failed	F	P	F	P	P	P	F	P	P	P	F	P		

statement. In the first step, the coincidentally correct tests should be identified before suspiciousness values can be computed appropriately.

A passing test is coincidentally correct, provided that the backward dynamic slice of the passing result covers the erroneous statement. For instance, in Table 9.2, the passing test T6 is not coincidentally correct because the faulty statement, 'S6', is not covered by the backward dynamic slice. The two tests, T5 and T9, are coincidentally correct because their backward dynamic slices cover the faulty statement, S6. Therefore, T5 and T9 should be considered failing executions when computing suspiciousness scores. In this case, as shown in Table 9.3, the suspiciousness values of the erroneous statement, S6, computed by both the Tarantula and Ochiai formulas, are higher than the other statements, and the faulty statement is detected appropriately.

The question is how to detect coincidentally correct executions before knowing the erroneous statements. The following sections present an approach to identifying coincidentally correct executions. The approach is used in Sect. 9.5 to identify the coincidentally correct tests in Fig. 9.2.

Table 9.3 Considering coincidentally correct executions as failing tests

Test case:	T1	T2	T3	T4	T5	T6	T7	T8	T9	T10	T11	T12	Suspiciousness	
x (the first number):	2	5	-2	0	7	1	3	8	2	5	2	-6	Ochiai	Tarantula
y (the second number)	7	-3	4	-6	7	3	5	-6	2	-3	4	-6		
op (1:max, 2:av, 3:max)	1	1	1	1	1	2	1	2	1	3	1	3		
S1 def f(int x, int y, int op):	✓	✓	✓	✓	✓	✓	✓	✓	✓	✓	✓	✓	0.471	0.333
S2 avg=(x+y) / 2	*	*	*	*	*	✓	*	✓	*	*	*	*	0.000	0.000
S3 min = y	*	*	*	*	*	*	*	*	*	*	*	✓	0.000	0.000
S4 max = y	*	*	*	*	*	*	*	*	*	*	*	✓	0.000	0.000
S5 if (x <= y):	✓	*	✓	*	✓	*	✓	*	✓	*	✓	*	1.000	1.000
S6	✓		✓		✓	*	✓	*	✓		✓	*	1.000	1.000
S7 if (x > y):	*	*	*	*	*	*	*	*	*	✓	*	*	0.000	0.000
S8 max = x		*		*					✓				0.000	0.000
S9 if (op==1):	✓	✓	✓	✓	✓	*	✓	*	✓	*	✓	*	0.866	0.750
S10 print(min)	✓	✓	✓	✓	✓		✓		✓		✓		0.866	0.750
S11 if (op==2):	*	*	*	*	*	✓	*	✓	*	*	*	*	0.000	0.000
S12 print(avg)						✓		✓					0.000	0.000
S13 if (op==3):	*	*	*	*	*	*	*	*	*	✓	*	✓	0.000	0.000
S14 print(max)										✓		✓	0.471	0.333
Results: P: Pass /F =Failed	F	P	F	P	F	P	F	P	F	P	F	P		

9.3 Detecting Coincidentally Correct Executions

This section uses a running example, shown in Table 9.3, to illustrate the applicability of cross-entropy evaluation to detect coincidentally correct tests [4, 5]. The key concept used here is entropy, which can be used to measure the inherent uncertainty and disorder in a set of observed predicates. This is the key concept in a field of research called information theory. This section shows how entropy can be applied to detect coincidentally correct tests based on the probability of predicates in the test traces.

Many fault localization methods assume the existence of a faulty run and a larger number of passing/correct executions and then select the passing execution that most resembles the faulty execution according to a distance criterion [6, 9]. The difference between the spectra of these runs isolates the bug's location. The bug's location can be isolated by contrasting the spectra of these runs. N-grams representing sequences of branching conditions can be used to represent the differences between a faulty execution path and its nearest neighbors.

9.3.1 Computing the Probability of N-grams

N-grams are continuous sequences of tokens where the letter 'n' indicates the number of the tokens in the sequence. Each token can be a predicate in an execution path represented as a sequence of predicates. Therefore, an execution path of length k is k-grams. Execution paths differ because of predicates or, in other words, conditional branches. 2-g can capture such differences.

Executing the faulty statement may not always cause the failure of the test case. There might be many test cases in which the faulty statement is executed without any harm to the execution results. However, often, the failure depends on the execution sequence. A specific subsequence or subpath of execution may cause the program to fail. This subsequence will be more common in the failing traces than in the passing ones. The probability of the occurrences of a predicate, pm, after a subsequence of predicates, $p_1, p_2, p_3, \ldots, p_m$, is computed as follows:

$$P(pm|p1, p2, , .., pm - 1) = \frac{P(p1, p2, , .., pm - 1, pm)}{P(p1, p2, , .., pm - 1)}$$
$$= \frac{count(p1, p2, , .., pm - 1, pm)}{count(p1, p2, , .., pm - 1)} \tag{9.1}$$

Faults propagate through the dependencies in sequences of predicates (i.e., n-grams) in the failing execution paths. Therefore, The chained probabilities are computed for the n-grams in the target failing path. Generally, the probability of a series of predicates can be calculated from the chained probabilities of its history:

$$P(p_1, p_2, \ldots, p_n) = P(p_1) * \prod_{i=2}^{n} P(p_i|p_1, p_2, \ldots, p_{i-1}) \tag{9.2}$$

Considering the above relation, the chained probability $P(p_1, p_2, p_3)$ is computed as follows:

$$P(p_1, p_2, p_3) = P(p_1) * P(p_2|p_1) * P(p_3|p_1, p_2) \tag{9.3}$$

where $P(p_2|p_1)$ indicates the probability of P_2 conditioned to P_1.

For instance, in Table 9.4, the code in Table 9.3 is instrumented. There are five branching conditions in this code. In addition, the entry to the function is also accompanied by a predicate P1. A fault is seeded in line S6 of the function f in Table 9.3. According to Table 9.4, the execution path for different test cases T1 and T2 are:

Table 9.4 Instrumenting the function f by inserting "print" statements after decisions

		Test case:	T1	T2	T3	T4	T5	T6	T7	T8	T9	T10	T11	T12
		x (the first number):	2	5	-2	0	7	1	3	8	2	5	2	-6
		y (the second number)	7	-3	4	-6	7	3	5	-6	2	-3	4	-6
		op (1:max, 2:av, 3:max)	1	1	1	1	1	2	1	2	1	3	1	3
S1		def f(int x, int y, int op):	●	●	●	●	●	●	●	●	●	●	●	●
S2		avg=(x+y) / 2	*	*	*	*	*	●	*	●	*	*	*	*
S3		min = y	*	*	*	*	*	*	*	*	*	*	*	●
S4	P1	max = y	*	*	*	*	*	*	*	*	*	*	*	●
		print("p1")												
S5		if (x <= y):	●	*	●	*	●	*	●	*	●	*	●	*
S6	P2	min=y #Error: min=x	●		●		●	*	●	*	●		●	*
		print("p2")												
S7		if (x > y):	*	*	*	*	*	*	*	*	*	●	*	*
S8	P3	max = x		*		*					●			
		print("p3")												
S9		if (op==1):	●	●	●	●	●	*	●	*	●	*	●	*
S10	P4	print(min)	●	●	●	●	●		●		●		●	
		print("p4")												
S11		if (op==2):	*	*	*	*	*	●	*	●	*	*	*	*
S12	P5	print(avg)						●		●				
		Print("p5")												
S13		if (op==3):	*	*	*	*	*	*	*	*	*	●	*	●
S14	P6	print(max)										●		●
		print("p6")												
		Results: P: Pass /F =Failed	F	P	F	P	P	P	F	P	P	P	F	P

T1 : (P1, P2, P4) failing test case,   T2 : (P1, P3, P4) passing test case.
T3 : (P1, P2, P4) failing test case,   T4 : (P1, P3, P4) passing test case.
T5 : (P1, P2, P4) Coincidentally correct, T6 : (P1, P2, P5) passing test case.
T7 : (P1, P2, P4) failing test case,   T8 : (P1, P2, P5) passing test case.
T9 : (P1, P2, P4) Coincidentally correct, T10 : (P1, P2, P6) passing test case.
T11 : (P1, P2, P4) failing test case,   T12 : (P1, P3, P6) passing test case.

It is observed that all the execution paths are tri-grams. Apparently, the fault should be sought in the failing execution paths. The search for the root cause of failure is conducted by comparing the probability of occurrence of each n-gram in failing and passing executions. To this end, each time, a failing path is selected as a target path. Then, the probability and chain probability of each uni- to 3-g of predicates in the target path is computed. The probability and chain probability of each n-gram in the target path is compared with its probability in the nearest passing test path.

Starting with the execution trace (p1, p2, p4) of T1 as the first target failing path, the probability of each predicate, as a uni-gram, in the execution trace is computed as follows:

– Target path: $T1_{path} = (P1, P2, P4)$

(1) Uni-grams: p1, p2, and p4

$$P(p1) = \frac{Count(p1)}{no.predicates} = \frac{1}{3}, P(p2) = \frac{Count(p2)}{no.predicates} = \frac{1}{3},$$
$$P(p4) = \frac{Count(p4)}{no.predicates} = \frac{1}{3}$$

(2) Bi-grams: (p1, p2), (p2, p4)

$$P(p1, p2) = P(p1|p1) = \text{probability of p2 after p1} = \frac{Count(p1, p2)}{count(p1)} = \frac{1}{1} = 1$$

The chained probability of (p1, p2) is computed as follows:

$$P(p2|p1) = \frac{P(p1, p2)}{P(p1)} \Rightarrow P(p1, p2) = P(p2|p1) * P(p1) = 1 * \frac{1}{3} = \frac{1}{3}$$

Similarly, the chained probability of (p2, p4) is computed as follows:

$$P(p2|p1) = \frac{P(p1, p2)}{P(p1)} \Rightarrow P(p1, p2) = P(p2|p1) * P(p1) = 1 * \frac{1}{3} = \frac{1}{3}$$

(3) Tri-grams: (p1, p2, p4)

$$P(p1, p2, p4) = P(p4|p1, p2) = \text{probability of p4 after (p1, p2)}$$
$$= \frac{Count(p1, p2, p4)}{count(p1, p2)} = \frac{1}{1} = 1$$

Similarly, the chained probability of (p1, p2, p4) is computed as follows:

$$P(p4|p1, p2) = \frac{P(p1, p2, p4)}{P(p1, p2)} \Rightarrow P(p1, p2, p4) = P(p4|p1, p2) * P(p1, p2)$$
$$\Rightarrow P(p1, p2, p4) = P(p4|p1, p2) * P(p2|p1) * p(p1)$$

$$\Rightarrow P(p1, p2, p4) = 1 * 1 * \frac{1}{3} = \frac{1}{3}$$

Generally, one way to calculate the joint probability of a string like P(w1, w2, w3,...,) is to use the chain rule of probability:

$$P(X_1 \ldots X_n) = P(X_1)P(X_2|X_1)P\left(X_3|X_1^2\right)\ldots P\left(X_n|X_1^{n-1}\right)$$
$$= \prod\nolimits_{k=1}^{n} P\left(X_k|X_1^{k-1}\right) \qquad (9.4)$$

In the above equation X_1^{k-1} indicates the sequence $X_1, \ldots, X_{k-1}$. After computing the probability and chained probability for each predicate in the target path, the conditional probability of each passing test case is computed. For instance, the conditional probability for uni- to tri-grams in the T2 execution path (p1, p3, p4) is as follows:

$$P(p1) = \frac{Count(p1)}{Count(uni-gram)} = \frac{1}{3} \simeq 0.33,$$

$$P(p2) = \frac{Count(p2)}{Count(uni-gram)} = \frac{1}{3} \simeq 0.33$$

$$p(p3) = \frac{Count(p3)}{Count(uni-gram)} = \frac{1}{3} \simeq 0.33$$

$$P(P1, P3) = P(P3|P1) = \frac{Count(P1, P3)}{Count(P1)} = \frac{1}{1} = 1$$

$$P(P3, P4) = P(P4|P3) = \frac{Count(P3, P4)}{Count(P3)} = \frac{1}{1} = 1$$

$$P(P1, P3, P4) = P(P4|P1, P3) = \frac{Count(P1, P3, P4)}{Count(P1, P3)} = \frac{1}{1} = 1$$

The probability of occurrence of an n-gram in failing and passing execution Cross-entropy can provide a realistic measure of the similarity between the n-grams based on their probability models. The most similar n-gram in a passing execution with the same one in a failing execution can be the most fault-suspicious region of the code.

9.3.2 Computing Cross-Entropies to Measure the Disorders

Entropy, by definition, is a measure of disorder, randomness, or lack of predictability. Faulty code is relatively more entropy (i.s. unnatural), becoming less so as bugs are fixed [4]. Considering the value of each predicate as a random variable, $X \in [0,1]$, the amount of average information inherent in the predicate, i.e., entropy(X), is the probability of the number of bits required to represent the probability of X:

$$H(X) = -\sum\nolimits_{\in[0,1]} P(X)log_2(P(X))$$

The negative sign ensures that the entropy is always positive or zero.

The entropy of an n-gram is significant in measuring uncertain information about the probability of the n-gram. Entropy is usually used to depict the degree of uncertainty of one object and is very important for measuring uncertain information.

Cross entropy is a measure of the difference between two probability distributions based on their difference in entropy. Cross entropy employs the concept of entropy. Cross-entropy can be used to measure the entropy difference between two test case execution paths, represented as sequences of uni- to trigrams. Mathematically, the cross-entropy is computed as follows:

$$H(p, q) = -\sum_{x \in X} P(x) \log_2 q(x) \tag{9.5}$$

In the above equation, x indicates an n-gram, and X is the sequence of all n-grams; $P(x)$ represents the chained probability of x in the target path, and $q(x)$ is the conditional probability of x in a passing path to be compared with the target path. When comparing the n-grams in the target path (T1) with a passing path (T2), if an n-gram, x, of the target path is not in the passing path:

$$H(p, q) = -\sum_{x \in X} P(x) \log_2(q(x))$$

x ∈ target path ∧ x ∉ correct/passing execution path

$$\Rightarrow q(x) = \frac{1}{count(n - grams) + 1} \tag{9.6}$$

Cross-entropy can depict the discrimination degree of two n-grams of predicates, and the similarity can be judged from it. Cross entropy measures the similarity of two n-grams, X and Y, of predicates as the disorder observed between the two n-grams:

$$H(X, Y) = -\sum_{W*} P_X(w_i^n) \log P_Y(w_n | w_i^{n-1}) \tag{9.7}$$

In the above equation, W^* indicates all n-grams of the test path X and $P_X(w_1^n)$ is calculated by applying the chain rule to n-grams in the target failing test path:

$$P(w_1^n) = P(w_1)P(w_2|w_1)P(w_3|w_1^2) \dots P(w_n|w_1^{n-1})$$

The above equation calculates the distance or dissimilarities between execution paths as uni-grams.

$$P_Y(w_n|w_1^{n-1}) = \frac{Count(w_1^{n-1} w_n)}{Count(w_1^{n-1})} = \frac{Count(w_1^n)}{Count(w_1^{n-1})} \tag{9.8}$$

The chained probability for the n-grams in the target execution path is as follows:

p(x): The chained probability of n-grams in T1 (target)					
['P1']	['P2']	['P4']	['P1', 'P2']	['P2', 'P4']	['P1', 'P2', 'P4']
0.33	0.33	0.33	0.33	0.33	0.33

As a result:

$$p = [0.33, \; 0.33, \; 0.33, \; 0.33, \; 0.33, \; 0.33]$$

The conditional probability for the n-grams in the target execution path is as follows:

q(x): The conditional probability of n-grams in T1 (target)					
['P1']	['P2']	['P4']	['P1', 'P3']	['P3', 'P4']	['P1', 'P3', 'P4']
0.33	0.33	0.33	0.1428	0.1428	0.1428

As a result:

$$q = [0.33, \; 0.1428, \; 0.33, \; 0.1428, \; 0.1428, \; 0.1428]$$

In the next step, the cross entropy equation is applied to the chained probability, p, of the target path, T1, and the conditional probability, q, of T2 to compute the distance between T1 and T2:

$$H(X, Y) = -\sum_{W*} P_X(w_i^n) log \; P_Y(w_n | w_i^{n-1})$$
$$= 0.33 * log(0.33) + 0.33 * log(0.33) + 0.33 * log(0.33)$$
$$+ 0.33 * \frac{1}{1+6} + 0.33 * 0.33 * \frac{1}{1+6} + 0.33 * \frac{1}{1+6}$$
$$= 4.799782$$

Similarly, the cross entropy between the chained probability of the target path T1 and all the conditional probability of all the passing and failing test cases is calculated as shown above.

9.3.3 Detecting Coincidentally Correct Executions

Coincidentally correct execution is very similar to a failing execution. If the distance between a passing execution path, Tp, and a failing pass, Tf, is less than a given threshold, e.g., 0.001, then Tf is a coincidentally correct execution and should be considered as an incorrect/failing execution. The cross-entropy of the target path with itself, H, is reduced from the cross-entropy between the chained probability of the target path and the conditional probability of the other paths:

$$H = H \ ((\text{chained\_probability}, \text{T1}), \text{conditional} \ (\text{probability} \ (\text{T1})) = 1.584963$$

$$\text{Distance} \ (\text{T1}, \text{T1}) = 1.584963 - 1.584963 = 0,$$

$$\text{Distance} \ (\text{T1}, \text{T2}) = H \ ((\text{chained\_probability}, \text{T1}), \text{conditional} \ (\text{probability} \ (\text{T2})) - H$$
$$= 4.799782 - 1.584963 = 3.214819,$$

$$\text{Distance} \ (\text{T1}, \text{T3}) = H \ ((\text{chained\_probability}, \text{T1}), \text{conditional} \ (\text{probability} \ (\text{T3})) - H$$
$$= 1.584963 - 1.584963 = 0,$$

$$\text{Distance} \ (\text{T1}, \text{T4}) = H \ ((\text{chained\_probability}, \text{T1}), \text{conditional} \ (\text{probability} \ (\text{T4})) - H$$
$$= 4.799782 - 1.584963 = 3.214819,$$

$$\text{Distance} \ (\text{T1}, \text{T5}) = H \ ((\text{chained\_probability}, \text{T1}), \text{conditional} \ (\text{probability} \ (\text{T5})) - H$$
$$= H(\text{T1}, \text{T5}) = 1.584963 - 1.584963 = 0,$$

$$\text{Distance} \ (\text{T1}, \text{T6}) = H \ ((\text{chained\_probability}, \text{T1}), \text{conditional} \ (\text{probability} \ (\text{T6})) - H$$
$$= H(\text{T1}, \text{T6}) = 3.863997 - 1.584963 = 2.279034,$$

$$\text{Distance} \ (\text{T1}, \text{T7}) = H \ ((\text{chained\_probability}, \text{T1}), \text{conditional} \ (\text{probability} \ (\text{T7})) - H$$
$$= H(\text{T1}, \text{T7}) = 1.584963 - 1.584963 = 0,$$

$$\text{Distance} \ (\text{T1}, \text{T8}) = H \ ((\text{chained\_probability}, \text{T1}), \text{conditional} \ (\text{probability} \ (\text{T8})) - H$$
$$= H(\text{T1}, \text{T8}) = 3.863997 - 1.584963 = 2.279034,$$

$$\text{Distance} \ (\text{T1}, \text{T9}) = H \ ((\text{chained\_probability}, \text{T1}), \text{conditional} \ (\text{probability} \ (\text{T9})) - H$$
$$= H(\text{T1}, \text{T9}) = 1.584963 - 1.584963 = 0,$$

$$\text{Distance} \ (\text{T1}, \text{T10}) = H \ ((\text{chained\_probability}, \text{T1}),$$
$$\text{conditional} \ (\text{probability} \ (\text{T10})) - H$$
$$= H \ (\text{T1}, \ \text{T10}) = 5.207246 - 1.584963 = 3.622283,$$

$$\text{Distance} \ (\text{T1}, \text{T11}) = H \ ((\text{chained\_probability}, \text{T1}),$$
$$\text{conditional} \ (\text{probability} \ (\text{T11})) - H$$
$$= H \ (\text{T1}, \ \text{T11}) = 1.584963 - -1.584963 = 0,$$

$$\text{Distance} \ (\text{T1}, \text{T12}) = H \ ((\text{chained\_probability}, \text{T1}),$$

$$\text{conditional (probability (T12))} - H$$
$$= H(T1, T12) = 3.863997 - 1.584963 = 2.279034,$$

The coincidentally correct execution paths are those passing/correct paths that are very close to the failing target path. Therefore, in the next step, the test cases are clustered based on their distances:

Clustre1 = [T1, T3, T5, T7, T9, T11]
Clustre2 = [T2, T4]
Clustre3 = [T6, T8, T12]
Clustre4 = [T10]

All the test cases residing in the same cluster as the target failing execution path are coincidentally correct and should be considered as failing execution paths:
Detected coincidentally correct executions = [T5, T9].

9.3.4 Locating Suspicious N-grams

The suspicious n-grams are those sequences of n-grams that occur relatively more in failing executions than the passing ones. The suspiciousness score of each n-gram is calculated as the average distance between the n-gram in target failing paths and the other paths as follows::

$$\forall t \in target_{paths} \forall n : n\_gram \in t \wedge n\_gram \text{ not repeated}$$
$$Dist_{failed}^t(n) = \sum_{f \in faileds} Chained_p(n \in t) \times \log_2 Conditional(n \in f)$$

For example, the distance of "p1" in T1 from all the failing executions, T1, T3, T5, T7, T9, and T11, is calculated as follows:

$$Dist_{failed}^{T1}(\text{p1})$$
$$= \left(-\frac{count(''P1'', T1)}{count(unigram, T1)} \times \log_2\left(\frac{count(''P1'', T1)}{count(unigram, T1)} \right) \right)$$
$$+ \left(-\frac{count(''P1'', T1)}{count(unigram, T1)} \times \log_2\left(\frac{count(''P1'', T3)}{count(unigram, T3)} \right) \right)$$
$$+ \left(-\frac{count(''P1'', T1)}{count(unigram, T1)} \times \log_2\left(\frac{count(''P1'', T5)}{count(unigram, T5)} \right) \right)$$
$$+ \left(-\frac{count(''P1'', T1)}{count(unigram, T1)} \times \log_2\left(\frac{count(''P1'', T7)}{count(unigram, T7)} \right) \right)$$

$$+ \left(-\frac{count(''P1'', T1)}{count(unigram, T1)} \times \log_2 \left(\frac{count(''P1'', T9)}{count(unigram, T9)} \right) \right)$$

$$+ \left(-\frac{count(''P1'', T1)}{count(unigram, T1)} \times \log_2 \left(\frac{count(''P1'', T11)}{count(unigram, T11)} \right) \right)$$

$$= \left(\left(-\frac{1}{3} \right) \times \log_2 \left(\frac{1}{3} \right) \right) + \left(\left(-\frac{1}{3} \right) \times \log_2 \left(\frac{1}{3} \right) \right) + \left(\left(-\frac{1}{3} \right) \times \log_2 \left(\frac{1}{3} \right) \right)$$

$$+ \left(\left(-\frac{1}{3} \right) * \log_2 \left(\frac{1}{3} \right) \right) + \left(\left(-\frac{1}{3} \right) * \log_2 \left(\frac{1}{3} \right) \right) + \left(\left(-\frac{1}{3} \right) * \log_2 \left(\frac{1}{3} \right) \right)$$

$$= (-0.33 \times -1.585) \times 6 \simeq 3.1383$$

Similarly, the overall distance of the "p1, p2" sub-path in T1 from all the failing execution paths is calculated as follows:

$$Dist_{failed}^{T1}(p1, \ p2)$$

$$= \left(-\frac{count(''P1'', T1)}{count(unigram, T1)} \times \frac{count(''P1, P2'', T1)}{count\ (''P1'', T1)} \right.$$

$$\left. \times \log \left(\frac{count(''P1, P2'', T1)}{count\ (''P1'', T1)} \right) \right)$$

$$+ \left(-\frac{count(''P1'', T1)}{count(unigram, T1)} \times \frac{count(''P1, P2'', T1)}{count\ (''P1'', T1)} \right.$$

$$\left. \times \log \left(\frac{count(''P1, P2'', T3)}{count\ (''P1'', T3)} \right) \right)$$

$$+ \left(-\frac{count(''P1'', T1)}{count(unigram, T1)} \times \frac{count(''P1, P2'', T1)}{count\ (''P1'', T1)} \right.$$

$$\left. \times \log \left(\frac{count(''P1, P2'', T5)}{count\ (''P1'', T5)} \right) \right)$$

$$+ \left(-\frac{count(''P1'', T1)}{count(unigram, T1)} \times \frac{count(''P1, P2'', T1)}{count\ (''P1'', T1)} \right.$$

$$\left. \times \log \left(\frac{count(''P1, P2'', T7)}{count\ (''P1'', T7)} \right) \right)$$

$$+ \left(-\frac{count(''P1'', T1)}{count(unigram, T1)} \times \frac{count(''P1, P2'', T1)}{count\ (''P1'', T1)} \right.$$

$$\left. \times \log \left(\frac{count(''P4, P2'', T9)}{count\ (''P4'', T9)} \right) \right)$$

$$+ \left(-\frac{count\left("P1", T1\right)}{count(unigram, T1)} \times \frac{count\left("P1, P2", T1\right)}{count\left("P1", T1\right)} \right.$$

$$\left. \times \log\left(\frac{count\left("P4, P2", T11\right)}{count\left("P4", T11\right)}\right) \right)$$

$$= \left(\left(-\frac{1}{3}\right) \times \log_2(1)\right) + \left(\left(-\frac{1}{3}\right) \times \log_2(1)\right)$$

$$+ \left(\left(-\frac{1}{3}\right) \times \log_2(1)\right) + \left(\left(-\frac{1}{3}\right) * \log_2(1)\right)$$

$$+ \left(\left(-\frac{1}{3}\right) * \log_2(1)\right) + \left(\left(-\frac{1}{3}\right) * \log_2(1)\right)$$

$$= (-0.33 \times 0) \times 6 = 0$$

Table 9.5 represents the cross entropies and the sum of the cross entropies, $Dist_{failed}^{T1}(n)$, for all the n-grams, n, in the target path T1 in Table 9.4.

The following equation computes the distance of each n-gram in the target path from the passing execution paths:

$$\forall t \in target_{paths} \forall n : n\_gram \in t \wedge n\_gram \ not \ repeated$$

Table 9.5 The distance of all n-grams in the target path, T1, from failing paths

| | Failing/Incorrect executions cross entropies | | | | | | Sum |
	T1	T3	T5	T7	T9	T11	$Dist_{failed}^{T1}$
P1	0.5283	0.5283	0.5283	0.5283	0.5283	0.5283	3.1699
P2	0.5283	0.5283	0.5283	0.5283	0.5283	0.5283	3.1699
P4	0.5283	0.5283	0.5283	0.5283	0.5283	0.5283	3.1699
P1P2	0.0000	0.0000	0.0000	0.0000	0.0000	0.0000	0.0000
P2P4	0.0000	0.0000	0.0000	0.0000	0.0000	0.0000	0.0000
P1P2P4	0.0000	0.0000	0.0000	0.0000	0.0000	0.0000	0.0000

Table 9.6 The distance of all n-grams in the target path, T1, from passing/correct paths

| | Passing/Correct executions cross entropies | | | | | | Sum |
	T2	T4	T6	T8	T10	T12	$Dist_{passed}^{T1}$
P1	0.5283	0.5283	0.5283	0.5283	0.5283	0.5283	3.1699
P2	0.9358	0.9358	0.5283	0.5283	0.9358	0.5283	4.3923
P4	0.5283	0.5283	0.9358	0.9358	0.9358	0.9358	4.7998
P1P2	0.9358	0.9358	0.0000	0.0000	0.9358	0.0000	2.8074
P2P4	0.9358	0.9358	0.9358	0.9358	0.9358	0.9358	5.6147
P1P2P4	0.9358	0.9358	0.9358	0.9358	0.9358	0.9358	5.6147

$$Dist_{passed}^{t}(n) = \sum_{p \in passeds} Chained_p(n \in t) \times \log_2 Conditional(n \in p)$$

For example, the distance of "p1" in T1 from all the passing/correct executions, T2, T4, T6, T8, T10, and T12, is calculated as follows:

$$Dist_{passsed}^{T1}(p1)$$

$$= \left(-\frac{count("P1", T1)}{count(unigram, T1)} \times \log\left(\frac{count("P1", T2)}{count(unigram, T2)}\right)\right)$$

$$+ \left(-\frac{count("P1", T1)}{count(unigram, T1)} \times \log\left(\frac{count("P1", T4)}{count(unigram, T4)}\right)\right)$$

$$+ \left(-\frac{count("P1", T1)}{count(unigram, T1)} \times \log\left(\frac{count("P1", T6)}{count(unigram, T6)}\right)\right)$$

$$+ \left(-\frac{count("P1", T1)}{count(unigram, T1)} \times \log\left(\frac{count("P1", T8)}{count(unigram, T8)}\right)\right)$$

$$+ \left(-\frac{count("P1", T1)}{count(unigram, T1)} \times \log\left(\frac{count("P1", T10)}{count(unigram, T10)}\right)\right)$$

$$+ \left(-\frac{count("P1", T1)}{count(unigram, T1)} \times \log\left(\frac{count("P1", T12)}{count(unigram, T12)}\right)\right)$$

$$= \left(\left(-\frac{1}{3}\right) \times \log_2\left(\frac{1}{3}\right)\right) + \left(\left(-\frac{1}{3}\right) \times \log_2\left(\frac{1}{3}\right)\right)$$

$$+ \left(\left(-\frac{1}{3}\right) \times \log_2\left(\frac{1}{3}\right)\right) + \left(\left(-\frac{1}{3}\right) * \log_2\left(\frac{1}{3}\right)\right)$$

$$+ \left(\left(-\frac{1}{3}\right) * \log_2\left(\frac{1}{3}\right)\right) + \left(\left(-\frac{1}{3}\right) * \log_2\left(\frac{1}{3}\right)\right)$$

$$= (-0.33 \times -1.585) \times 6 \simeq 3.1383$$

Similarly, the distance of the "p1, p2" bi-gram in T1 from all the passing/correct execution paths is calculated as follows:

$$Dist_{passsed}^{T1}(p1, p2) =$$

$(-(count("P1", T1))/(count(unigram, T1)) \times (count("P1, P2", T1))$
$/(count("P1", T1)) * log((count("P1, P2", T2))/(count("P1", T2))))$
$+ (-(count("P1", T1))/(count(unigram, T1)) \times (count("P1, P2", T1))$
$/(count("P1", T1)) * log((count("P1, P2", T4))/(count("P1", T4))))$
$+ (-(count("P1", T1))/(count(unigram, T1)) \times (count("P1, P2", T1))$
$/(count("P1", T1)) * log((count("P1, P2", T6))/(count("P1", T6))))$

Table 9.7 The suspiciousness score of all the n-grams in the target failing path

Subpath	suspiciousness	Df	Dp
P1	1	3.169925001	3.169925001
P2	1.385621875	3.169925001	4.392317423
P4	1.514162499	3.169925001	4.799781563
P1P2	2.807354922	1	2.807354922
P2P4	5.614709844	1	5.614709844
P1P2P4	5.614709844	1	5.614709844

$$
\begin{aligned}
&+ (-(\text{count}(''\text{P1}'', \text{T1}))/(\text{count}(\text{unigram}, \text{T1})) \times (\text{count}(''\text{P1}, \text{P2}'', \text{T1})) \\
&/(\text{count}(''\text{P1}'', \text{T1})) * \log((\text{count}(''\text{P1}, \text{P2}'', \text{T8}))/(\text{count}(''\text{P1}'', \text{T8})))) \\
&+ (-(\text{count}(''\text{P1}'', \text{T1}))/(\text{count}(\text{unigram}, \text{T1})) \times (\text{count}(''\text{P1}, \text{P2}'', \text{T1})) \\
&/(\text{count}(''\text{P1}'', \text{T1})) * \log((\text{count}(''\text{P1}, \text{P2}'', \text{T10}))/(\text{count}(''\text{P1}'', \text{T10})))) \\
&+ (-(\text{count}(''\text{P1}'', \text{T1})))/(\text{count}(\text{unigram}, \text{T1})))(\text{count}(''\text{P1}, \text{P2}'', \text{T1})) \\
&/(\text{count}(''\text{P1}'', \text{T1})) * \log((\text{count}(''\text{P1}, \text{P2}'', \text{T12}))/(\text{count}(''\text{P1}'', \text{T12})))) \\
&= 2.8074
\end{aligned}
$$

Table 9.6 represents the cross-entropies and the sum of the cross-entropies, $Dist^{T1}_{passed}(n)$, for all the n-grams, n, in the target path.

Once the distances from failing and passing executions for each n-gram, n, as a subpath of the target failing path, are computed, the suspiciousness score of n is calculated as follows:

$$
Score(n) = \begin{cases} Dist^{T1}_{passed}(n), & Dist^{T1}_{failed}(n) = 0 \\ \dfrac{Dist^{T1}_{passed}(n)}{Dist^{T1}_{failed}(n)} & Dist^{T1}_{passed}(n) \neq 0 \end{cases} \tag{9.9}
$$

The suspiciousness score for all the n-grams is shown in Table 9.7. It is observed that the fault propagates through the predicate p1 to p2 and then p4.

As shown in Table 9.4, due to the coincidentally correct executions, the suspiciousness value calculated by the SBFL equations Ochiai and Tarantula formulas was wrong, and the faulty statement was not localized. As shown in Sect. 9.3.3, T5 and T9 are coincidentally correct executions. Therefore, Table 9.8 shows that these test cases are considered failing executions, and the suspiciousness values are recalculated.

Table 9.9 shows that the faulty statement is localized correctly this time. As shown in the table, the suspiciousness values computed for s5 and s6 are the highest compared to the other statements.

Table 9.8 Despite correct output results, T5 and T9, are considered as failing/incorrect tests

		Test case:	T1	T2	T3	T4	T5	T6	T7	T8	T9	T10	T11	T12
		x (the first number):	2	5	-2	0	7	1	3	8	2	5	2	-6
		y (the second number)	7	-3	4	-6	7	3	5	-6	2	-3	4	-6
		op (1:max, 2:av, 3:max)	1	1	1	1	1	2	1	2	1	3	1	3
S1		def f(int x, int y, int op):	✓	✓	✓	✓	✓	✓	✓	✓	✓	✓	✓	✓
S2		avg=(x+y) / 2	*	*	*	*	*	✓	*	✓	*	*	*	*
S3		min = y	*	*	*	*	*	*	*	*	*	*	*	✓
S4	P1	max = y	*	*	*	*	*	*	*	*	*	*	*	✓
		print("p1")												
S5		if (x <= y):	✓	*	✓	*	✓	*	✓	*	✓	*	✓	*
S6	P2	min=y #Error: min=x	✓		✓		✓	*	✓		✓		✓	*
		print("p2")												
S7		if (x > y):	*	*	*	*	*	*	*	*	*	✓	*	*
S8	P3	max = x		*		*						✓		
		print("p3")												
S9		if (op==1):	✓	✓	✓	✓	✓	*	✓	*	✓	*	✓	*
S10	P4	print(min)	✓	✓	✓	✓	✓		✓		✓		✓	
		print("p4")												
S11		if (op==2):	*	*	*	*	*	✓	*	✓	*	*	*	*
S12	P5	print(avg)						✓		✓				
		Print("p5")												
S13		if (op==3):	*	*	*	*	*	*	*	*	*	✓	*	✓
S14	P6	print(max)										✓		✓
		print("p6")												
		Results: P: Pass /F =Failed	F	P	F	P	_F_	P	F	P	_F_	P	F	P

Table 9.9   Statements S5 and S9 have the highest suspiciousness values

#stat	def Rishedaraje2():	passed	failed	Suspiciousness (Ochiai)	Suspiciousness (Tarantula)
	Total Passed = 6			**Total Failed = 6**	
S1	x = int(input(" x: ")) y = int(input("y: ")) op = int(input("z: "))	8	4	0.471	0.333
S2	avg=(x+y) / 2	2	0	0.000	0.000
S3	min = x	2	0	0.000	0.000
S4	max = y	1	0	0.000	0.000
S5	if (x < y):	0	6	1.000	1.000
S6	min = y #correct:	0	6	1.000	1.000
S7	if (x > y):	1	0	0.000	0.000
S8	max = x	1	0	0.000	0.000
S9	if (op==1):	2	6	0.866	0.750
S10	print(min)	2	6	0.866	0.750
S11	if (op==2):	2	0	0.000	0.000
S12	print(avg)	2	0	0.000	0.000
S13	if (op==3):	2	0	0.000	0.000
S14	print(max)	2	0	0.471	0.333

9.4   Writing a Program to Locate Faults[1]

In this section, another example is used to describe the main components of a Python program to detect coincidentally correct executions and locate faults in C++ functions. Based on this heuristic that a coincidentally correct (CC) execution path is very similar to failing execution paths, passing execution paths are compared with each failing path selected as a target path. The comparison is made by clustering the execution paths based on their similarities. Each execution path is represented as a sequence of predicates, indicating which decision options are made at a branching point. N-gram models are built for each execution trace. The models are further analyzed using cross-entropy to compute the similarities among their corresponding execution paths [8]. By counting elements as unigrams in each execution path, the most likelihood, MLE, probabilities to create N-gram models, known as Markov models, can be computed. Figure 9.1 represents the main function to compute cross entropies.

---

[1] Download the source code from: https://doi.org/10.6084/m9.figshare.21321708

```
def main():
    # 1- Instrument the software under test (SUT) and run the instrumented code
    tests = instrument_sut()
    # 2- Compute unigrams, bigrams, and trigrams for each test path
    tests.compute_n_grams()
    # 3- Select a failing test case as the target and compute n-gram probabilities
    for k in range(tests.number_of_test_cases):
        if not tests.test_runs_status[k]:
            target_test_case = k
            tests.compute_probability(target_test_case)
            # 4- Compute the chained probability of the n-grams in target paths
            tests.compute_target_probability(target_test_case)
            # 5- Equate the length of test case paths with the target path
            tests.equate_paths_with_target(target_test_case)
            # 6- Compute cross-entropies
            tests.compute_corss_entropy(target_test_case)
            # 7- Compute the distance of each test case from the target failing test case
            tests.compute_d_distance(target_test_case)
            # 8- Cluster test cases based on their distances from the target test case
            tests.Clustring()
            # 9- identify coincidentally correct tests
            tests.identify_coincidentally_correct()
            # 10- Evaluate Distp and Distf to identify suspicious regions
            tests.calculate_subpaths_entropy(target_test_case)
            # 11- evaluate Dp and Df for distinct uni_ to tri-grams
            tests.Suspicious()
            # 12- Display suspicious branches
            tests.print_candidate_suspiciousness()

if __name__ == "__main__":
    main()
```

Fig. 9.1 The main body of the program to compute CC tests and locate faults

9.4.1 Source Code Instrumentation

This section presents a running example to demonstrate the use of the proposed algorithm to detect coincidentally correct executions. Consider the function f, presented in Fig. 9.2, with two faulty statements, s9 and s10.

The function, f(), in Fig. 9.2, takes two matrices, $\times 41$ and $\times 2$, and two flags, flag1, and flag2. The function checks whether the two matrices, $\times 1$ and $\times 2$, are invertible, provided that the value of flag1 is greater than zero. If both the matrices are invertible, the program checks whether one of them is the inverse of the other. The function also checks whether the matrix, $\times 3 = \times 1 * \times 2$, is invertible if flag2 is greater than zero. As shown in Fig. 9.2, the two statements, S9 and S10, are faulty.

	void f(int x1[2][2], x2[2][2], x3[2][2], flag1, flag2){
S1	det1=x1[0][0]*x1[1][1]-x1[0][1]*x1[1][0];
S2	if (det1!=0){
S3	for(i=0;i<2;i++)
S4	for(j=0;j<2;j++){
S5	x3[i][j]=0;
S6	for(k=0;k<2;k++)
S7	x3[i][j]+= x1[i][k]* x2[k][j] }
S8	if (flag1>0)
S9	if (x2[0][0]* x2[1][1]!= x1[0][1]* x2[1][0])
	// it should be x2[0][1] instead of x1[0][1]
S10	if (x1[0][0]!=1 || x3[1][1]!=1 || (x3[0][1]!=0 || x3[1][0]!=0))
	// it should be x3[0][0] instead of x1[0][0]
S11	print ('Both are invertible, but x1 is not the inverse of x2');
S12	else print('x1 is the inverse of x2');
S13	else print('x2 is not invertible');
S14	if (flag2>0)
S15	if (x3[0][0]* x3[1][1]!= x3[0][1]* x3[1][0])
S16	print ('the multiplication result of x1 and x2 is invertible');
S17	else print('the multiplication result of x1 and x2 is not invertible'); }
S18	else print('x1 is not invertible'); }

Fig. 9.2 Statements S10 and S10 are incorrect

The first faulty statement, S9, should be $\times 2[0][1]$ instead of $\times 1[0][1]$. The second fault is in S19, where $\times 1[0][0]! = 1$ should be replaced with $\times 3[0][0]! = 1$.

The traces of executing the function "f," in Fig. 9.2, with the test data set, represented in Table 9.10, is given in Table 9.11. Statistical approaches determine the fault suspiciousness of individual program statements based on the statistics of the statement's cooperation in failing and passing executions. Intuitively statements executed more frequently by failing test cases than the passing ones are more likely to be faulty. The likelihood of a statement, St_i, being faulty is computed by contrasting its involvement in the execution traces of failing and passing executions.

Below, D-Star(D*) [9], Ochiai [10], and Op2 [11] fault localization methods are used to compute the suspiciousness score of the statements. A recent study [3] indicates the following are the most effective spectrum-based fault localization methods for locating real and seeded faults.

$$D - Star \ susp(s) = N_{CF}(s)^2/(N_{UF}(s) + N_{CS}(s)) \qquad (9.10)$$

$$Ochiai \ susp(s) = N_{CF}(s)/\sqrt{N_F \times (N_{CS}(s) + N_{CF}(s))} \qquad (9.11)$$

$$Op^2 \ susp(s) = N_{CF}(s) - N_{CS}(s)/(N_S + 1) \qquad (9.12)$$

Table 9.10 Test data sets

TC #1	TC #2	TC #3	TC #4	TC #5
X1 = $\begin{bmatrix} 1 & 2 \\ 1 & -1 \end{bmatrix}$	X1 = $\begin{bmatrix} 1 & 1 \\ -1 & 0 \end{bmatrix}$	X1 = $\begin{bmatrix} 1 & 0 \\ 1 & 1 \end{bmatrix}$	X1 = $\begin{bmatrix} 1 & 0 \\ 1 & 1 \end{bmatrix}$	X1 = $\begin{bmatrix} -1 & 1 \\ 0 & -1 \end{bmatrix}$
X2 = $\begin{bmatrix} 1 & 0 \\ 1 & -1 \end{bmatrix}$	X2 = $\begin{bmatrix} 1 & 2 \\ 1 & 1 \end{bmatrix}$	X2 = $\begin{bmatrix} 1 & 0 \\ 2 & 0 \end{bmatrix}$	X2 = $\begin{bmatrix} 1 & -1 \\ 1 & -1 \end{bmatrix}$	X2 = $\begin{bmatrix} -1 & 1 \\ 0 & -1 \end{bmatrix}$
flag1 > 0	flag 1 = 0	flag 1 > 0	flag 1 > 0	flag1 > 0
flag 2 > 0	flag 2 > 0	flag 2 > 0	flag 2 > 0	flag2 > 0
TC #6	TC #7	TC #8	TC #9	TC #10
X1 = $\begin{bmatrix} 1 & 1 \\ 1 & 2 \end{bmatrix}$	X1 = $\begin{bmatrix} 1 & -1 \\ 0 & 1 \end{bmatrix}$	X1 = $\begin{bmatrix} 1 & 1 \\ -1 & 0 \end{bmatrix}$	X1 = $\begin{bmatrix} 1 & 1 \\ 1 & 1 \end{bmatrix}$	X1 = $\begin{bmatrix} 1 & 0 \\ 1 & 1 \end{bmatrix}$
X2 = $\begin{bmatrix} 1 & 0 \\ 1 & 1 \end{bmatrix}$	X2 = $\begin{bmatrix} -1 & 1 \\ 0 & 1 \end{bmatrix}$	X2 = $\begin{bmatrix} 1 & 1 \\ 0 & 0 \end{bmatrix}$	X2 = $\begin{bmatrix} 1 & 0 \\ 0 & 1 \end{bmatrix}$	X2 = $\begin{bmatrix} -1 & 0 \\ 1 & 0 \end{bmatrix}$
flag1 > 0	flag1 > 0	fag1 > 0	flag1 > 0	flag1 = 0
flag2 > 0	flag2 > 0	flag2 = 0	flag2 = 0	flag2 > 0

In the above equations, s stands for a statement; $N_{CF}(s)$ is the number of failed tests that execute s; $NUF(s)$ is the number of failed tests that do not execute p; $NCS(s)$ is the number of passed tests that execute p, and $NUS(s)$ is the number of passed tests that do not executes. As shown in Table 9.11, there are ten tests, T1 to T10, from which T1, T3, and T5 are coincidentally correct (CC).

It is observed that despite executing the faulty statements S9 and S10, the results of running the function "f" with T4, 1nd T5, and T6 are correct. In fact, these tests are coincidentally correct. In the following sections, it is shown how to detect coincidentally correct executions.

Test runs can be tracked by instrumenting the code under test. Automated instrumentation tools probe the program under test with instructions to keep a log of the test runs using the test cases. For instance, the C++ function, shown in Fig. 9.2, is instrumented in Table 9.12.

The instrumented code is executed with the test cases presented in Table 9.4 to keep a log of the executions. The execution paths are represented in Table 9.6. As shown in Table 9.13, some predicates are repeated in the execution path due to the for loops.

A test trace is generated during the execution of a test case. In fact, a test trace is the log of a test case execution together with the result of that test case execution as a failing (incorrect result) or passing (correct result) execution.

Table 9.11 Statements and predicates covered by the test data given in Table 9.2

St. No.	Predicate #	Slice coverage										Suspiciousness score		
		T1	T2	T3	T4	T5	T6	T7	T8	T9	T10	D-Star	Ochiai	Op2
S1		●	●	●	●	●	●	●	*	●	●	1.286	0.548	2.125
S2	P1	●	●	●	●	●	●	●	*	●	●	1.286	0.548	2.125
S3	P2	●	●	●	●	●	●	●	*		●	1.500	0.577	2.250
S4	P3	●	●	●	●	●	●	●	*		●	1.500	0.577	2.250
S5		●	●	●	●	●	●	●	*		●	1.500	0.577	2.250
S6	P4	●	●	●	●	●	●	●	*	●		1.500	0.577	2.250
S7		●	●	●	●	●	●	●	*	●		1.500	0.577	2.250
S8	P5	●	*	●	●	●	●	●	●		*	1.800	0.612	2.375
S9	P6	●		●	●	●	●	●	●			2.250	0.655	2.500
S10	P7	●			●	●		●				1.333	0.577	1.750
S11		●			●	●						1.333	0.577	1.750
S12	P8							●				0.000	#DIV/0!	0.000
S13	P9			●			●	●				0.250	0.333	0.750
S14	P10	●	●	●	●	●	●	*	*		●	0.667	0.436	1.375
S15	P11	●	●	●	●	●	●				●	0.667	0.436	1.375
S16		●	●			●	●					0.000	#DIV/0!	0.000
S17	P12			●	●						●	0.667	0.436	1.375
S18	P13									●		0.000	0.000	-0.125
		CC	P	CC	F	CC	F	F	P	P	P	CC : Coincidetally		
		P : Pass & F: Fail										Correct		

9.4.2   Computing N-grams

In the next step, Coincidentally correct execution paths are detected by selecting a failing execution path as the target and comparing all the passing executions with the failing one. The similarity between the two paths is computed regarding their

Table 9.12 Instrumented version of the source code in Table 9.3

St. No.	Predicate #	Instrumented source code
		void f(int x1[2][2], x2[2][2], x3[2][2], flag1, flag2){
S1		det1=x1[0][0]*x1[1][1]-x1[0][1]*x1[1][0];
S2	P1	if (det1!=0){cout << "P1";
S3	P2	for(i=0;i<2;i++) { cout << "P2";
S4	P3	for(j=0;j<2;j++){ cout << "P3";
S5		x3[i][j]=0;
S6	P4	for(k=0;k<2;k++) { cout << "P4";
S7		x3[i][j]+= x1[i][k]* x2[k][j]; } }}
S8	P5	if (flag1>0) { cout << "P5";
S9	P6	if (x2[0][0]* x2[1][1]!= x1[0][1]* x2[1][0]){ cout << "P6";
		// it should be x2[0][1] instead of x1[0][1]
S10	P7	if (x1[0][0]!=1 || x3[1][1]!=1 || (x3[0][1]!=0 || x3[1][0]!=0))
		{cout << "P7";// it should be x3[0][0] instead of x1[0][0]
S11		print ('Both are invertible but x1 != x2⁻¹ ');}
S12	P8	else { print('x1 is the inverse of x2'); cout << "P8";}
S13	P9	else { print('x2 is not invertible'); cout << "P9";}}
S14	P10	if (flag2>0) { cout << "P10";
S15	P11	if (x3[0][0]* x3[1][1]!= x3[0][1]* x3[1][0]){ cout << "P11";
S16		print (' x3 = x1 * x2 is invertible');}
S17	P12	else { print('('x3=x1* x2 is not invertible'); cout << "P12";}
S18	P13	} else {print('x1 is not invertible'); cout << "P13" }}

Table 9.13 Test traces

Test	Test trace : [Execution path, test result (F = Fail, P =Pass)]	
T1	[P1,P2,P3,P4,P4,P3, P4,P4,P2,P3,P4,P4,P3,P4,P4,P5,P6,P7,P11,P12]	P
T2	[P1,P2,P3,P4,P4,P3,P4,P4,P2,P3,P4,P4,P3,P4,P4,P10,P11,P12]	P
T3	[P1,P2,P3,P4,P4,P3,P4,P4,P2,P3,P4,P4,P3,P4,P4,P5,P9,P11,P13]	P
T4	[P1,P2,P3,P4,P4,P3,P4,P4,P2,P3,P4,P4,P3,P4,P4,P5,P6,P7,P11,P13]	F
T5	[P1,P2,P3,P4,P4,P3,P4,P4,P2,P3,P4,P4,P3,P4,P4,P5,P6,P7,P11,P12]	P
T6	[P1,P2,P3,P4,P4,P3,P4,P4,P2,P3,P4,P4,P3,P4,P4,P5,P9,P11,P12]	F
T7	[P1,P2,P3,P4,P4,P3,P4,P4,P2,P3,P4,P4,P3,P4,P4,P5,P6,P8,P14]	F
T8	[P1,P2,P3,P4,P4,P3,P4,P4,P2,P3,P4,P4,P3,P4,P4,P5,P9,P14]	P
T9	[P15]	P
T10	[P1,P2,P3,P4,P4,P3,P4,P4,P2,P3,P4,P4,P3,P4,P4,P10,P11,P13]	P

crossing entropy [12]. If a passing execution path is very similar to a failing one, It can be a coincidentally correct path.

Traces of test runs of the program are collected as sequences of the predicates covered by the test cases. Once all the execution traces are available, for each trace, n-grams are computed. Figure 9.3 represents a function to compute n-grams.

The function, generate_ngrams, accepts an execution trace, exec_trace, and an integer k as input and generates an array of k_grams. The function is invoked to generate n_grams for different values of n. The result is saved in an array, n_grams. All the n_grams generated for the test traces kept when executing the function f with different test cases are saved in an array,n_grams_set, as follows:

Fig. 9.3 A function to
compute n-grams

```
def generate_ngrams(exec_trace, k):
    prediates = exec_trace.split(",")
    k_grams = []
    for i in range(len(predicates)- k+1):
        k_grams.append(predicates[i:i+k])
    return k_grams
```

```
n_grams_set = []
for i in range(no_exec_traces):
    n_grams = []
    for j in range(1,4):#generate 1_grams to 4_grams
        n_grams = n_grams +
            generate_ngrams(exec_traces[i], j)
    n_grams_set.append(n_grams)
```

For instance, Table 9.14 represents all the n_grams_set generated for the execution trace of the target failing path, T4.

In effect, a tri-gram [P4, P5, P6] shows that P6 appears after P4 and P5. The question is, what is the chance or, more precisely, the probability for the appearance of P6 after P4 and P5?

9.4.3 Compute N-gram Probabiliies

Each n-gram is a division of an execution path into sequences of n predicates. The n-grams can be used to keep some crucial relationships between the predicates. Upcoming predicates in a sequence of n-grams are affected by the previous ones.

The unsmoothed maximum likelihood estimate (MLE) of the unigram probability of the predicate P_i is its count C_i normalized by the total number of predicates, N:

$$P^{unigram}(P_i) = \frac{Count(P_i)}{\text{total number of predicates}} = \frac{C_i}{N} \tag{9.13}$$

For instance, the probabilities of the unigrams in Table 9.14 is given in Table 9.15.

The Python code for computing the MLE probability of n-grams is shown in Fig. 9.4.

To compute the MLE of the bigram probability of the predicate P_n given a previous predicate P_{n-1}, we will compute the count of the bigram (P_{n-1}, P_n) normalized by the sum of all the bigrams (P_{n-1}, P_k) that share the same first predicate P_{n-1}:

Table 9.14 The set of 1-g to n-grams generated for the test trace of the test case T4

Unigrams (1_grams)
[['P1'], ['P2'], ['P3'], ['P4'], ['P4'], ['P3'], ['P4'], ['P4'], ['P2'], ['P3'], ['P4'], ['P4'], ['P3'], ['P4'], ['P4'], ['P5'], ['P6'], ['P7'], ['P11'], ['P13']]#20

bigrams(2_grams)
[['P1', 'P2'], ['P2', 'P3'], ['P3', 'P4'], ['P4', 'P4'], ['P4', 'P3'], ['P3', 'P4'], ['P4', 'P4'], ['P4', 'P2'], ['P2', 'P3'], ['P3', 'P4'], ['P4', 'P4'], ['P4', 'P3'], ['P3', 'P4'], ['P4', 'P4'], ['P4', 'P5'], ['P5', 'P6'], ['P6', 'P7'], ['P7', 'P11'], ['P11', 'P13']]#19

trigrams(3_grams)
[['P1', 'P2', 'P3'], ['P2', 'P3', 'P4'], ['P3', 'P4', 'P4'], ['P4', 'P4', 'P3'], ['P4', 'P3', 'P4'], ['P3', 'P4', 'P4'], ['P4', 'P4', 'P2'], ['P4', 'P2', 'P3'], ['P2', 'P3', 'P4'], ['P3', 'P4', 'P4'], ['P4', 'P4', 'P3'], ['P4', 'P3', 'P4'], ['P3', 'P4', 'P4'], ['P4', 'P4', 'P5'], ['P4', 'P5', 'P6'], ['P5', 'P6', 'P7'], ['P6', 'P7', 'P11'], ['P7', 'P11', 'P13']]#18

Table 9.15 MLE probabilities computed for predicates in execution trace of T4

		Probabilities for Unigrams				
['P1']	['P2']	['P3']	['P4']	['P4']	['P3']	['P4']
$\frac{1}{20} = 0.05$	$\frac{2}{20} = 0.1$	$\frac{4}{20} = 0.2$	$\frac{8}{20} = 0.4$	$\frac{8}{20} = 0.4$	$\frac{4}{20} = 0.2$	$\frac{8}{20} = 0.4$
['P4']	['P2']	['P3']	['P13']	['P4']	['P3']	['P4']
$\frac{8}{20} = 0.4$	$\frac{2}{20} = 0.1$	$\frac{4}{20} = 0.2$	$\frac{1}{20} = 0.05$	$\frac{8}{20} = 0.4$	$\frac{4}{20} = 0.2$	$\frac{8}{20} = 0.4$
['P11']	['P4']	['P5']	['P6']	['P7']	['P11']	
$\frac{1}{20} = 0.05$	$\frac{8}{20} = 0.4$	$\frac{1}{20} = 0.05$	$\frac{1}{20} = 0.05$	$\frac{1}{20} = 0.05$	$\frac{1}{20} = 0.05$	

$$P^{\text{bigram}}(P_n|P_{n-1}) = \frac{\text{Count}(P_{n-1}, P_n)}{\sum_k P_{n-1}, P_k} \tag{9.14}$$

This equation can be simplified since the sum of all bigram counts that start with a given predicate P_{n-1} must be equal to the unigram count for that predicate P_{n-1} (It is worth the reader's time to contemplate this):

$$P^{\text{bigram}}(P_n|P_{n-1}) = \frac{\text{Count}(P_{n-1}, P_n)}{\text{Count}(P_{n-1})} \tag{9.15}$$

For instance, the MLE of the bigrams in Table 9.14 is given in Table 9.16.

Similarly, the MLE of the trigram probability of the predicate P_n given the previous sequence of predicate P_{n-2}, P_{n-1} is as follows:

$$P^{trigram}(P_n|P_{n-2}, P_{n-1}) = \frac{Count(P_{n-2}, P_{n-1}, P_n)}{Count(P_{n-2}, P_{n-1})} \tag{9.16}$$

```
#calculate the probability of UniGrams, BiGrams, and TriGrams
def compute_ngrams_probability(self,number_of_test_cases,
                              test_paths, n_grams_set, base_test_case):
    n_grams_probability_set = []
    #1- calculate unigrams Probability
    for i in range(number_of_test_cases):
        probability_set=[]
        n_gram_index = 0
        for j in range(len(test_paths[i].split(","))):
            number_of_unigram = 0
            for z in range(len(test_paths[i].split(","))):
                if n_grams_set[i][j] == n_grams_set[i][z]:
                    number_of_unigram = number_of_unigram + 1
            probability = number_of_unigram/len(test_paths[i].split(","))
            probability_set.insert(n_gram_index, probability)
            n_gram_index += 1
        target_path_length = len(test_paths[target_test_case].split(","))
        test_path_length = len(test_paths[i].split(","))
        if(target_path_length > test_path_length) :
            # fill with -1 the remaining cells in probability_set
            for p in range(test_path_length, target_path_length):
                probability_set.insert(n_gram_index, 1.0/(target_path_length +1.0))
                n_gram_index += 1
        else:
            for p in range(test_path_length, target_path_length):
                probability_set.pop()
    #2- Calculate bigrams probability (similar to unigrams)
        ...
    #3- Calculate trigrams probability (similar to unigrams)
        ...
```

Fig. 9.4   A function to compute MLE probability for n-grams in a given test path

Table 9.16   MLE probabilities computed for the bigrams in Table 9.14

[P1, P2]	[P2, P3]	[P3, P4]	[P4, P4]	[P4, P3]	[P3, P4]	[P4, P4]	
1	1	1	0.5	0.25	1	0.5	No.
0.05	0.1	0.2	0.2	0.1	0.2	0.2	Ch.
[P4, P2]	[P2, P3]	[P3, P4]	[P4, P4]	[P4, P3]	[P3, P4]	[P4, P4]	
0.125	1	1	0.5	0.25	1	0.5	No.
0.05	0.1	0.2	0.2	0.1	0.2	0.2	Ch.
[P4, P5]	[P5, P6]	[P6, P7]	[P7, P11]	[P11,P13]	Ch.:	No.:	
0.125	1	1	1	1	Chained	Normal	
0.05	0.05	0.05	0.05	0.05			

Table 9.17 MLE probabilities computed for the tigrams in Table 9.14

P1,P2, P3	P2,P3, P4	P3,P4,P4	P4, P4, P3	P4, P3, P4	P3,P4 P4	P3,P4 P2	
1	1	1	0.5	1	1	0.25	No.
0.05	0.1	0.2	0.1	0.1	0.2	0.05	Ch.
P4,P2,P3	P2,P3,P4	P3,P4 P4	P4,P4,P3	P4,P3,P4	P3,P4,P4	P4,P4,P5	
1	1	1	0.5	1	1	0.25	No.
0.05	0.1	0.2	0.1	0.1	0.2	0.05	Ch.
P7,P11,P13	P4,P5,P6	P5,P6,P7	P6,P7,P11	P7,P11,P13	Ch.:	No.:	
1	1	1	1	1	Chained	Normal	
0.05	0.05	0.05	0.05	0.05			

Accordingly, the MLE probability of k-grams is computed as follows:

$$P^{k\_gram}(P_n|P_{n-k+1}, \ldots, P_{n-1}) = \frac{Count(P_{n-k+1}, \ldots, P_{n-1}, P_n)}{Count(P_{n-k+1}, \ldots, P_{n-1})} \qquad (9.17)$$

Table 9.17 represents the MLE probabilities of the trigrams in Table 9.14 (Table 9.17).

The chained probability, shown in the above table, is calculated to determine the joint probability of the predicates in n-grams.

9.4.4 Computing Chained Probabilities

The chained probabilities are computed for the n-grams in the target failing path. Generally, the probability of a series of predicates can be calculated from the chained probabilities of its history:

$$P(p_1, p_2, \ldots, p_n) = P(p_1) * \prod_{i=2}^{n} P(p_i|p_1, p_2, \ldots, p_{i-1}) \qquad (9.18)$$

Considering the above relation, the chained probability $P(p_1, p_2, p_3)$ is computed as follows:

$$P(p_1, p_2, p_3) = P(p_1) * P(p_2|p_1) * P(p_3|p_1, p_2) \qquad (9.19)$$

where $P(p_2|p_1)$ indicates the probability of P_2 conditioned to P_1.

The chained probability for unigrams, bigrams, and trigrams computed for the execution path of the test case T4, selected as the target failing test case, is as follows:

{0.05, 0.1, 0.2, 0.4, 0.4, 0.2, 0.4, 0.4, 0.1, 0.2, 0.4, 0.4, 0.2, 0.4, 0.4, 0.05, 0.05, 0.05, 0.05, 0.05, 0.05, 0.1, 0.2, 0.2, 0.1, 0.2, 0.2, 0.05, 0.1, 0.2, 0.2, 0.1, 0.2, 0.2, 0.05, 0.05,

```
# Computes the chained probability for the uni- to tri-grams in the target path
def compute_target_probability(self,target_test_case):
    target = []
    uni_gram_last_index = (len(self.test_paths[target_test_case]))
    for i in range(uni_gram_last_index):
        target.append(self.n_grams_probability_set[target_test_case][i])

    bi_gram_last_index = 2*uni_gram_last_index - 1
    for i in range(uni_gram_last_index, bi_gram_last_index):
        target.append([self.n_grams_probability_set[target_test_case][i][0],
            self.n_grams_probability_set[target_test_case][i][1]*
        target[i- uni_gram_last_index][1]])
    tri_gram_end_index = 3*one_gram_end_index - 3
    for i in range(two_geram_end_index,three_gram_end_index):
        target.append([self.n_grams_probability_set[target_test_case][i][0],self.
            n_grams_probability_set[target_test_case][i][1]
        *target[i-len(self.test_paths[target_test_case])+1][1]])
    text = "Chained Target Probablity of T{}.txt".format(target_test_case+1)
    f                       =                       open(text,              "w")
    f.write(str(target))
    path_parent             =                os.path.dirname(os.getcwd())
    os.chdir(path_parent)
    self.target = target
```

Fig. 9.5   A function to compute the chained probability for uni- to tri-grams

0.05, 0.05, 0.05, 0.05, 0.1, 0.2, 0.1, 0.1, 0.2, 0.05, 0.05, 0.1, 0.2, 0.1, 0.1, 0.2, 0.05, 0.05, 0.05, 0.05, 0.05}

The function comute_target_probability, in Fig. 9.5, takes the target path as input and computes the chained probability for the uni- to tri-grams in the path.

9.4.5   Compute Cross-Entropy

Entropy is interpreted as the degree of disorder or uncertainty. High entropy means a highly disordered set of variables. Conversely, low entropy means an ordered set of variables. Entropy is computed as the average uncertainty of a single self-information random variable [8].

$$H(p) = H(X) = - \sum_{x \in X} P(x) \log_2 P(x) = \sum_{x \in X} P(x) \log_2 \frac{1}{P(x)} \quad (9.20)$$

$P(x)$ is the probability mass function of a random variable X over an alphabet discrete set of symbols X. When computing entropy, $\log(0)$ should be considered zero.

Information entropy measures the information a given event contains, given all possible outcomes. Entropy is often measured in bits. For instance, the amount of information in a random variable representing the result of a 16-sided dice is as follows:

$$H(X) = -\sum_{i=1}^{16} P(i) \log_2 P(i) = -\sum_{i=1}^{16} \frac{1}{16} \log_2 \frac{1}{16} = -\log_2 \frac{1}{16} = \log_2 16 = 4$$

Therefore, the most efficient way to transmit the result of rolling an eight-sided dice is to encode it as a four-digit binary message.

Cross entropy measures the difference between two probability distributions based on their difference in entropy. Cross entropy employs the concept of entropy. Cross-entropy can be used to measure the entropy difference between two test case execution paths, represented as sequences of uni- to trigrams. Mathematically, the cross-entropy is computed as follows:

$$H(p, q) = -\sum_{x \in X} P(x) \log_2 q(x) \tag{9.21}$$

In [51], the main idea of entropy is used for n-gram models, and as a result, the following equation is obtained by combining Eqs. (9.20) and (9.21):

$$H(X, Y) = -\sum_{W^*} P_X\left(w_1^n\right) \log_2 P_Y(w_n|w_1^{n-1}) \tag{9.22}$$

where W^* indicates all n-grams of the test path X and $P_X\left(w_1^n\right)$ will be obtained by applying a chain law to n-grams in the target failing test path:

$$P\left(w_1^n\right) = P(w_1)P(w_2|w_1)P\left(w_3|w_1^2\right) \ldots P\left(w_n|w_1^{n-1}\right) \tag{9.23}$$

and:

$$P_Y(w_n|w_1^{n-1}) = \frac{Count\left(w_1^{n-1}w_n\right)}{Count\left(w_1^{n-1}\right)} = \frac{Count\left(w_1^n\right)}{Count\left(w_1^{n-1}\right)} \tag{9.24}$$

Table 9.18 Cross entropy as the similarity measure

H(T4, T1)	6.89590	D (T$_4$,T$_1$)	0.00000
H(T4, T2)	8.32830	D (T$_4$,T$_2$)	1.43241
H(T4, T3)	8.51246	D (T$_4$,T$_3$)	1.61656
H(T4, T4)	6.89590	D (T$_4$,T$_4$)	0.00000
H(T4, T5)	6.89590	D (T$_4$,T$_5$)	0.00000
H(T4, T6)	8.51246	D (T$_4$,T$_6$)	1.61656
H(T4, T7)	7.35087	D (T$_4$,T$_7$)	0.45497
H(T4, T8)	8.03418	D (T$_4$,T$_8$)	1.13829
H(T4, T9)	35.12283	D (T$_4$,T$_9$)	28.22694
H(T4,T10)	8.32830	D(T$_4$,T$_{10}$)	1.43241

Below is the Python code for computing the cross-entropy:

```
#Compute cross_entropy
cross_entropy=[]
for i in range(tests.number_of_test_cases):
    tests.compute_corss_entropy( tests.target_chained_probability,
                            tests.n_grams_probability_set[i])
    cross_entropy.append(tests.cross_entropy)

#a method to evaluate cross_entropy

def compute_corss_entropy(self,p, q):
    res = -sum([p[i]*log(q[i]) for i in range(len(p))])
    self.cross_entropy = res
```

Afterward, the cross-entropy of the target failing test case, T4, is reduced from all the computed cross-entropies to measure the distance between each test case execution path and the target path. Table 9.18 represents the cross-entropy measures between the execution path of each test case, T1 to T10, with the execution path of the target failing test case, T4.

The function compute_cross_entropy in Fig. 9.6, computes the cross entropy between the target path and all other test paths.

9.4.6 Cluster Test Cases

After selecting each failing test case as the target and obtaining the similarities of other test cases, a percentage of the most similar test cases are selected and maintained as a cluster. All the test cases, "Ti," are clustered based on their cross-entropy distance, D(Ti, Tj), from the target failing test cases, "Tj."

```
#function to calculate entropy two array list
def compute_corss_entropy(self,target_test_case):
    def corss_entropy(p,q):
        return -sum([p[i]*log2(q[i]) for i in range(len(p))])
    entropy=[]
    p = [float(tar[1]) for tar in self.target]
    for i in range(self.number_of_test_cases):
        q = [float(prob[1]) for prob in self.n_grams_probability_set[i]]
        entropy.append(corss_entropy(p,q))
    with open("7- Entropy Result.txt", "w") as file:
        for i in range(self.number_of_test_cases):
            file.write("H(T{},T{}) = {}\n".format(target_test_case + 1, i + 1,
                                                  format(entropy[i], '.6f')))
    self.entropy = entropy
```

Fig. 9.6 Computes the cross entropy between the target path and all other test paths

Based on the distances from the target failing path, shown in Table 9.18, the test cases are clustered as follows:

Cluster 1 = [T1, T4, T5]   =>T1 and T5 are CC test cases.

Clustre 2 = [T2, T10]

Clustre 3 = [T3, T6]   => T3 is a CC test case

Clustre 4 = [T7]

Clustre 5 = [T8]

Clustre 6 = [T9]

T1 and T5 are two passing test cases residing in the same cluster as the failing test case T4. Therefore, T1 and T5 could be two coincidentally correct executions. Also, T3 could be a coincidentally correct test case because it is clustered with T6, a failing test case. Therefore, T1, T3, and T5 are detected as coincidentally correct test cases. Therefore, T1, T3, and T5 should be considered failing test cases when computing the suspiciousness scores for the statements.

9.4.7   Locating Suspicious N-grams

Each different uni- to n-gram in an execution path is considered a sub-path in the last step. As an example, Table 9.19 presents the distinct subpath of the target failing path T4.

In this step, considering each different uni- to trigram in the target failing path as a subpath, s, the similarity of the subpath with the same subpaths in failing and passing executions is calculated as $D_f(s)$ and $D_p(s)$, respectively. In this way, the

Table 9.19 different uni- to trigram models of the target failing test case, T4

['P1'] ,['P2'] ,['P3'] ,['P4'] ,['P5'] ,['P6'] ,['P7'] ,['P10'] ,['P12'] ,
['P1', 'P2'] ,['P2', 'P3'] ,['P3', 'P4'] ,['P3', 'P4'] ,['P4', 'P4'] ,['P4', 'P3'] ,['P4', 'P2'] ,['P4', 'P5'] ,['P5', 'P6'] ,['P6', 'P7'] ,['P7', 'P10'],['P10', 'P12'] ,
['P1', 'P2', 'P3'] ,['P2', 'P3', 'P4'] ,['P3', 'P4', 'P4'] ,['P4', 'P4', 'P3'] ,['P4', 'P3', 'P4'] ,['P4', 'P4', 'P2'] ,['P4', 'P2', 'P3'] ,['P4', 'P4', 'P5'] ,['P4', 'P5', 'P6'] ,['P5', 'P6', 'P7'] ,['P6', 'P7', 'P10'] ,['P7', 'P10', 'P12']
#32

suspiciousness of each subpath is computed as:

$$\text{Suspiciousness (s)} = \frac{D_f(s)}{D_p(s)} \tag{9.25}$$

$$D_f(s) = \sum_{i=1}^{\text{no.passing tests}} H(s, q_i) \tag{9.26}$$

where q_i addresses an occurrence of s in the ith failing test path

$$D_p(s) = \sum_{i=1}^{\text{no.failing tests}} H(s, r_i) \tag{9.27}$$

where r_i addresses an occurrence of s in the ith passing test path

For instance, as shown in Table 9.19, the predicate 'p1' is a subpath of the target path, T4. The distance of 'p1' in T4 from the failing execution paths, T1, T3, T4, T5, T6, and T7 are calculated as follows:

$$
\begin{aligned}
D_f(p1) = &\left(-\frac{count\left("P1", T4\right)}{count(unigram, T4)} \times \log\left(\frac{count\left("P1", T1\right)}{count(unigram, T1)}\right)\right) \\
&+ \left(-\frac{count\left("P1", T4\right)}{count(unigram, T4)} \times \log\left(\frac{count\left("P1", T3\right)}{count(unigram, T3)}\right)\right) \\
&+ \left(-\frac{count\left("P1", T4\right)}{count(unigram, T4)} \times \log\left(\frac{count\left("P1", T4\right)}{count(unigram, T4)}\right)\right) \\
&+ \left(-\frac{count\left("P1", T4\right)}{count(unigram, T4)} \times \log\left(\frac{count\left("P1", T5\right)}{count(unigram, T5)}\right)\right) \\
&+ \left(-\frac{count\left("P1", T4\right)}{count(unigram, T4)} \times \log\left(\frac{count\left("P1", T6\right)}{count(unigram, T6)}\right)\right) \\
&+ \left(-\frac{count\left("P1", T4\right)}{count(unigram, T4)} \times \log\left(\frac{count\left("P1", T7\right)}{count(unigram, T7)}\right)\right)
\end{aligned}
$$

$$= \left(\left(-\frac{1}{20}\right) * \log_2\left(\frac{1}{20}\right)\right) + \left(\left(-\frac{1}{20}\right) * \log_2\left(\frac{1}{19}\right)\right)$$

$$+ \left(\left(-\frac{1}{20}\right) * \log_2\left(\frac{1}{20}\right)\right) + \left(\left(-\frac{1}{20}\right) * \log_2\left(\frac{1}{20}\right)\right)$$

$$+ \left(\left(-\frac{1}{20}\right) * \log_2\left(\frac{1}{19}\right)\right) + \left(\left(-\frac{1}{20}\right) * \log_2\left(\frac{1}{18}\right)\right)$$

$$= (-0.05 \times -4.322) + (-0.05 \times -4.248)$$

$$+ (-0.05 \times -4.322) + (0.05 \times -4.322)$$

$$+ (0.05 \times -4.248) + (0.05 \times -4.17) = 1.2816$$

Similarly, the sum of cross-entropies of the unigram 'P1' in the target path with each presence of 'P1' in the passing test cases is computed to measure its distance from the passing test cases:

$Dp(P1)$

$$= \left(-\frac{count("P1", T4)}{count(unigram, T4)} * \log\left(\frac{count("P1", T2)}{count(unigram, T2)}\right)\right)$$

$$+ \left(-\frac{count("P1", T4)}{count(unigram, T4)} * \log\left(\frac{count("P1", T8)}{count(unigram, T8)}\right)\right)$$

$$+ \left(-\frac{count("P1", T4)}{count(unigram, T4)} * \log\left(\frac{count("P1", T9)}{count(unigram, T9)}\right)\right)$$

$$+ \left(-\frac{count("P1", T4)}{count(unigram, T4)} * \log\left(\frac{count("P1", T10)}{count(unigram, T10)}\right)\right) = 0.90601$$

The suspiciousness score of 'P1' is computed as:

$$\text{Suspiciousness}(p1) = \frac{D_f(s)}{D_p(s)} = \frac{1.2816}{0.90601} = \frac{1.2816}{0.90601} = 1.4145$$

The distance of the "P4, P2" subpath in the target failing path from the other failing test paths is computed as follows:

$$H(p, q) = - \sum_{p4p2} chained(P4, P2)log_2 q(P4, P2)$$

$$= P("P4")P("P2"|"P4")log_2 q("P4, P2")$$

$$= -\frac{count("P4", T4)}{count(unigram, T4)} \times \frac{count("P4, P2", T4)}{count("P4", T4)}$$

$$\times \; log_2\left(\frac{count\left("P4,\,P2",\,T6\right)}{count\left("P4",\,T6\right)}\right)$$

$$\mathrm{Df} = \left(-\frac{count\left("P4",\,T4\right)}{count\left(unigram,\,T4\right)} \times \frac{count\left("P4,\,P2",\,T4\right)}{count\left("P4",\,T4\right)}\right.$$

$$\times \; log\left(\frac{count\left("P4,\,P2",\,T4\right)}{count\left("P4",\,T4\right)}\right)\bigg)$$

$$+ \left(-\frac{count\left("P4",\,T4\right)}{count\left(unigram,\,T4\right)} \times \frac{count\left("P4,\,P2",\,T4\right)}{count\left("P4",\,T4\right)}\right.$$

$$\times \; log\left(\frac{count\left("P4,\,P2",\,T1\right)}{count\left("P4",\,T1\right)}\right)\bigg)$$

$$+ \left(-\frac{count\left("P4",\,T4\right)}{count\left(unigram,\,T4\right)} \times \frac{count\left("P4,\,P2",\,T4\right)}{count\left("P4",\,T4\right)}\right.$$

$$\times \; log\left(\frac{count\left("P4,\,P2",\,T3\right)}{count\left("P4",\,T3\right)}\right)\bigg)$$

$$+ \left(-\frac{count\left("P4",\,T4\right)}{count\left(unigram,\,T4\right)} \times \frac{count\left("P4,\,P2",\,T4\right)}{count\left("P4",\,T4\right)}\right.$$

$$\times \; log\left(\frac{count\left("P4,\,P2",\,T4\right)}{count\left("P4",\,T4\right)}\right)\bigg)$$

$$+ \left(-\frac{count\left("P4",\,T4\right)}{count\left(unigram,\,T4\right)} \times \frac{count\left("P4,\,P2",\,T4\right)}{count\left("P4",\,T4\right)}\right.$$

$$\times \; log\left(\frac{count\left("P4,\,P2",\,T6\right)}{count\left("P4",\,T6\right)}\right)\bigg)$$

$$+ \left(-\frac{count\left("P4",\,T4\right)}{count\left(unigram,\,T4\right)} \times \frac{count\left("P4,\,P2",\,T4\right)}{count\left("P4",\,T4\right)}\right.$$

$$\left.\times \; log\left(\frac{count\left("P4,\,P2",\,T7\right)}{count\left("P4",\,T7\right)}\right)\right) = 0.9$$

The distance of the "P4, P5, P6" subpath in the target failing path from the other failing test paths is computed as follows:

$$H(p,q) = -\sum_{p4\,p5\,p6} p(\text{P4, P5, P6}) log_2 q\,(\text{P4, P5, P6})$$

$$= P(\text{P4})\,P(\text{P5}|\text{P4})\,P(\text{P6}|\text{P4, P5}) \, \log q\,(P4,\,\text{P5, P6})$$

$$= -\frac{count(P4, T4)}{count(unigram, T4)} \times \frac{count(P4, P5, T4)}{count(P4, T4)}$$

$$\times \frac{count(P4, P5, P6, T4)}{count(P4, P5, T4)} \times \log\left(\frac{count(P4, P5, P6, T6)}{count(P4, P5, T6)}\right)$$

$$Df = \left(-\frac{count(P4, T4)}{count(unigram, T4)} \times \frac{count(P4, P5, T4)}{count(P4, T4)} \times \frac{count(P4, P5, P6, T4)}{count(P4, T4)}\right.$$

$$\left. \times \log\left(\frac{count(P4, P5, P6, T1)}{count(P4, p5, T1)}\right)\right) +$$

$$\left(-\frac{count(P4, T4)}{count(unigram, T4)} \times \frac{count(P4, P5, T4)}{count(P4, T4)} \times \frac{count(P4, P5, p6, T4)}{count(P4, p5, T4)}\right.$$

$$\left. \times \log\left(\frac{count(P4, P5, P6, T3)}{count(P4, p5, T3)}\right)\right) +$$

$$\left(-\frac{count(P4, T4)}{count(unigram, T4)} \times \frac{count(P4, P5, T4)}{count(P4, T4)} \times \frac{count(P4, P5, P6, T4)}{count(P4, p5, T4)}\right.$$

$$\left. \times \log\left(\frac{count("P4, P5, P6", T4)}{count(P4, p5, T4)}\right)\right) +$$

$$\left(-\frac{count(P4, T4)}{count(unigram, T4)} \times \frac{count(P4, P5, T4)}{count(P4, T4)} \times \frac{count(P4, P5, P6, T4)}{count(P4, p5, T4)}\right.$$

$$\left. \times \log\left(\frac{count(P4, P5, P6, T5)}{count(P, p54, T5)}\right)\right) +$$

$$\left(-\frac{count(P4, T4)}{count(unigram, T4)} \times \frac{count(P4, P5, T4)}{count(P4, T4)} \times \frac{count(P4, P5, p6, T4)}{count(P4, p5, T4)}\right.$$

$$\left. \times \log\left(\frac{count(P4, P5, P6, T6)}{count(P4, p5, T6)}\right)\right) +$$

$$\left(-\frac{count(P4, T4)}{count(unigram, T4)} \times \frac{count(P4, P5, T4)}{count(P4, T4)} \times \frac{count(P4, P5, P6, T4)}{count(p4, p5, T4)}\right.$$

$$\left. \times \log\left(\frac{count(P4, P5, P6, T7)}{count(P4, p5, T7)}\right)\right) = 0.5858$$

The function is_suspicious in Fig. 9.6 computes the distance of the chained probability of the n-grams in the target path from their probability in all the failing and passing execution paths and saves the results in the lists "df" and "df," respectively.

It then computes the suspiciousness score for each n-gram in the target path using the Eq. (9.25).

```
def is_suspicious(self):
    ##evaluate Dp and Df
    dp = []
    df = []
    for i in range(len(self.unique_n_grams_in_target_test)):
        sum_entropy_true_test = 0
        sum_entropy_false_test = 0
        for j in range(self.number_of_test_cases):
            if(self.test_runs_status_copy[j]):
                sum_entropy_true_test += self.subpaths_entropy[i][j]
            else:
                sum_entropy_false_test += self.subpaths_entropy[i][j]
        dp.append(sum_entropy_true_test)
        if sum_entropy_false_test == 0:
            df.append(1)
        else:
            df.append(sum_entropy_false_test)
    Suspiciousness = []
    for i,j in zip(dp,df):
        Suspiciousness.append(i/j)
    self.Suspiciousness = Suspiciousness
```

9.4.8   Computing Suspiciousness Scores

After computing D_p and D_f for all the subpaths, the suspiciousness score, Susp, of each subpath, s, is computed as:

$$\text{Susp}(s) = \frac{D_p(s)}{D_f(s)} \tag{9.28}$$

The suspiciousness values computed for all the subpaths in the target execution path are given in Table 9.20.

It is observed that the suspiciousness value computed for the following subpaths are relatively higher than the rest of the subpaths:

[P5, P6], [P6, P7], [P10, P11], [P4, P5, P6], [P5, P6, P7]

The suspiciousness score computed for all the above subpaths is 2. However, P5, P6, and P7 appear more frequently than the other predicates in the above subpaths. Therefore, the subpath [P5, P6, P7] can be the faulty path. As a result, statements S9 and S10 can be the root cause of the failure.

The suspiciousness for each statement is computed using Op2, D*(where * equals 2), and Ochiai, shown in Table 9.21. It is observed that the suspiciousness score, computed by the Ochiai formula for statement S9, is the highest value (0.65). Thus,

Table 9.20 The suspiciousness score of all the predicates of the target faulty path, T4

Subpath	D_F	D_P	$\dfrac{D_p}{D_f}$
P1	1.2816	0.90601	0.706937
P2	1.96314	1.51205	0.77022
P3	2.72634	2.42407	0.88913
P4	3.05262	3.64816	1.195091
P5	1.2816	1.08307	0.845092
P6	1.4426	1.1716	0.812145
P7	1.4426	1.1716	0.812145
P10	1.527	1.1716	0.767256
P12	1.527	1.1716	0.767256
P1P2	1	0.2929	0.2929
P2P3	1	0.5858	0.5858
P3P4	1	1.1716	1.1716
P4P4	1.2	1.7716	1.476333
P4P3	1.2	1.1858	0.988167
P4P2	0.9	0.7429	0.825444
P4P5	0.9	0.8787	0.976333
P5P6	0.5858	1.1716	2
P6P7	0.5858	1.1716	2
P7P10	0.8787	1.1716	1.333333
P10P12	0.2929	0.5858	2
P1P2P3	1	0.2929	0.2929
P2P3P4	1	0.5858	0.5858
P3P4P4	1	1.1716	1.1716
P4P4P3	0.6	0.8858	1.476333
P4P3P4	1	0.5858	0.5858
P4P4P2	0.6	0.5929	0.988167
P4P2P3	1	0.2929	0.2929
P4P4P5	0.6	0.9787	1.631167
P4P5P6	0.5858	1.1716	2
P5P6P7	0.5858	1.1716	2
P6P7P10	0.8787	1.1716	1.333333
P7P10P12	0.8787	1.1716	1.333333

the highest position in the ranking is assigned to statement S9, which is one of two faulty statements in our example. However, the Ochiai score of the statements S3 to S8 and S12 is the same as the second faulty statement, S10 (Table 9.21).

Table 9.22 shows the suspiciousness scores for the statements covered by the backward dynamic slices.

9.5   Summary

Faulty statements are detected based on the statistics of their association with correct and incorrect executions. However, sometimes despite executing the faulty statement, the execution result is correct. Such executions are called coincidentally correct.

Coincidentally correct execution paths can be detected by contrasting the probability of the presence of their subpaths in failing and passing executions. The distance

Table 9.21 Computation of suspiciousness using OP2, D2, and Ochiai

	Statement coverage										Suspiciousness score		
	T1	T2	T3	T4	T5	T6	T7	T8	T9	T10	Op2	D2	Ochiai
S1	*	*	*	*	*	*	*	*	*	*	2.13	1.29 (2)	0.55
S2	*	*	*	*	*	*	*	*	*	*	2.13	1.29 (2)	0.55
S3	*	*	*	*	*	*	*	*		*	2.25 (2)	1.29 (2)	0.58 (2)
S4	*	*	*	*	*	*	*	*		*	2.25 (2)	1.29 (2)	0.58 (2)
S5	*	*	*	*	*	*	*	*		*	2.25 (2)	1.29 (2)	0.58 (2)
S6	*	*	*	*	*	*	*	*		*	2.25 (2)	1.29 (2)	0.58 (2)
S7	*	*	*	*	*	*	*	*		*	2.25 (2)	1.29 (2)	0.58 (2)
S8	*	*	*	*	*	*	*	*		*	2.25 (2)	1.29 (2)	0.58 (2)
S9	*		*	*	*	*	*	*			2.50 (1)	1.50 (1)	0.65 (1)
S10	*			*	*		*				1.75	0.57	0.58 (2)
S11	*			*	*						0.75	0.14	0.33
S12						*					1.00	0.20	0.58 (2)
S13			*		*		*				0.75	0.14	0.33
S14	*	*	*	*	*	*	*	*		*	2.25 (2)	0.82	0.58 (2)
S15	*	*	*	*	*	*				*	1.38	0.57	0.44
S16	*	*			*	*					0.63	0.14	0.29
S17			*	*						*	0.75	0.14	0.33
S18									*		-0.13	0.00	0.00
	CC	P	CC	F	CC	F	F	CC	P	P	CC : Coincidetally		
	P : Pass & F: Fail										Correct		

between the two probabilities can be calculated as their cross-entropy. The term
'cross-entropy' is derived from the field of information theory and generally represents the difference between two probability distributions. Spectrum-based techniques provide more accurate suspiciousness scores after detecting coincidentally

Table 9.22 Backward dynamic slice coverage

	Slice coverage										Suspiciousness score		
	T1	T2	T3	T4	T5	T6	T7	T8	T9	T10	Op2	D-Star	Ochiai
S1	☑	☑	☑	☑	☑	☑	☑	*	☑	☑	1.29 (1)	0.58	2.25
S2	☑	☑	☑	☑	☑	☑	☑	*	☑	☑	1.29 (1)	0.58	2.25
S3	☑	☑	☑	☑	☑	☑	☑	*		☑	1.29 (1)	0.61	2.38
S4	☑	☑	☑	☑	☑	☑	☑	*		☑	1.29 (1)	0.61	2.38
S5	☑	☑	☑	☑	☑	☑	☑	*		☑	1.29 (1)	0.61	2.38
S6	☑	☑	☑	☑	☑	☑	☑	*		☑	1.29 (1)	0.61	2.38
S7	☑	☑	☑	☑	☑	☑	☑	*		☑	1.29 (1)	0.61	2.38
S8	☑	*	☑	☑	☑	☑	☑	☑		*	1.29 (1)	0.65 (1)	2.50 (1)
S9	☑		☑	☑	☑	☑	☑	☑			1.29 (1)	0.65 (2)	2.50 (2)
S10	☑			☑	☑		☑				0.57 (2)	0.58	1.75
S11	☑			☑	☑						0.14	0.33	0.75
S12							☑				0.14	0.58	1.00
S13			☑			☑		☑			0.14	0.33	0.75
S14	☑	☑	☑	☑	☑	☑	*	*		☑	0.57 (2)	0.44	1.38
S15	☑	☑	☑	☑	☑	☑				☑	0.57 (2)	0.44	1.38
S16	☑	☑			☑	☑					0.14	0.29	0.63
S17			☑	☑						☑	0.14	0.33	0.75
S18								☑			0.00	0.00	-0.13
	CC	P	CC	F	CC	F	F	CC	P	P	CC : Coincidetally		
	P : Pass　& F: Fail										Correct		

correct test cases and considering them as failing ones. Each sub-path in an execution path is an n-gram.

The technique generates N-grams, subsequences of predicates, from program spectra. Each N-gram's conditional probability relating it to faulty and correct executions is computed. A list of the most suspicious predicates is obtained from the most suspicious N-grams.

9.6 Exercises

(1) The probability of occurrences of the predicates covered by the test cases T1 to T5 are as follows:

	P1	P2	P3	P4
T1	0.4	0.1	0.25	0.25
T2	0.25	0.25	0.25	0.25
T3	0.4	0.1	0.1	0.4

To compute entropy for the test cases T1 to Tn, write to Python functions, Entropy. Write another function, H, to compute cross-entropy. Call H(T1, T2) to compute the cross entropy for the test cases T1 and T2.

(2). Write a function to compute n-grams for the following execution paths and then compute the cross entropies to compute the distance of all test cases from T1.

Test case	Execution path as a sequence of predicates
T1	P1 P2 P3 P4 P5
T2	P1 P2 P3 P4
T3	P1 P2 P3 P4 P6
T4	P1 P2 P3 P4 P5 P6
T5	P1 P2 P1 P2 P3 P1 P2 P4
T6	P1 P2 P3 P4 P6 P7 P8 P9

(3) Use the Cross entropy approach described in this chapter to locate faulty the statement in the following function:

```
Mid() {

  int x,y,z,m;

  read("Enter 3 numbers:",x,y,z);

  m = z;

  if (y<z){

    if (x<y)

      m = y;

    else

      if (x<z)

        m = y; //*** bug ***

  }else

    if (x>y)

      m = y;

    else

      if (x>z)

        m = x;

  print("Middle number is:",m);

}
```

Test	(x,y,z)	Result
t_1	(3,3,5)	Pass
t_2	(1,2,3)	Pass
t_3	(3,2,1)	Pass
t_4	(5,5,5)	Pass
t_5	(5,3,4)	Pass
t_6	(2,1,3)	Fail

(4). Below is an instrumented code fragment. Follow the procedure described in this chapter to locate the faulty statement.

```
Arr = [int(x) for x in input("Enter multiple value: ").split()]

for I in range(len(arr)):

    minimum = i

    print('P1')

    for j in range(I + 1 , len(arr)):

        print('P2')

        if arr[j] < arr[min]-1: #correct:if arr[j] < arr[min]:

            print('P3')

            minimum = j

        else:

            print('P4')

    arr[minimum], arr[i] = arr[i], arr[minimum]

print('P5')

print(arr)
```

The test suite and test results are as follows:

Test case	Input	Output	Result	Execution path
T1	[9, 0]	[0, 9]	Fail	P1 , P2 , P3 , P1 , P5
T2	[3, 1, 2]	[1, 3, 2]	Fail	P1,P2,P3, P2,P4,P1, P2, P4, P1,P5
T3	[2, 3, 1]	[2, 1, 3]	Fail	P1, P2,P4, P2,P4,P1, P2,P3,P1, P5
T4	[2, 8]	[2, 8]	Pass	P1,P2,P4, P1,P5
T5	[7, 7]	[7, 7]	Pass	P1, P2 , P4, P1, P5
T6	[1, 2, 3]	[1, 2, 3]	Pass	P1,P2,P4, P2,P4,P1, P2, P4,P1,P5
T7	[3, 2, 1]	[1, 2, 3]	Pass	P1,P2,P4, P2,P3,P1, P2,P4,P1,P5
T8	[7, 7, 7]	[7, 7, 7]	Pass	P1,P2,P4, P2,P4,P1, P2,P4,P1,P5

References

1. Parsa, S.: A program slicing-based method for effective detection of coincidentally correct test cases. Computing. Archives for Informatics and Numerical Computation; Wien **100**(9) (2018)
2. Parsa, S.: Kernel-based detection of coincidentally correct test cases to improve fault localization effectiveness. Int. J. Appl.Pattern Recogn. **5**(2), 119–136 (2018)
3. Hajibaba, M., Parsa, S.: Software fault localization using cross-entropy and N-gram models. Soft Comput. J. **2**(1), Number 3 (2013)
4. Shannon, C.E.: A mathematical theory of communication. Bell Syst. Tech. J. **27**(3), 379–423 (1948). https://doi.org/10.1002/j.1538-7305.1948
5. Shannon, C.E.: A mathematical theory of communication. Bell Syst. Tech. J. **27**(4), 623–656 (1948). https://doi.org/10.1002/j.1538-7305.1948
6. Mousavian, Z., Vahidi-Asl, M., Parsa, S.: Finding software fault relevant subgraphs a new graph mining approach for software debugging. 24th Canadian Conference on Electrical and Computer Engineering (CCECE) (2011)
7. Mousavian, Z., Vahidi-Asl, M., Parsa, S.: Scalable graph analyzing approach for software fault-localization. Proceedings of the 6th International Workshop on Automation of Software Test (2011)
8. Wang, Y., Yang, H., Qin, K., Wang, Y.: The consistency between cross-entropy and distance measures in fuzzy sets. Symmetry **11**(3), 386 (2019)
9. Wong, W.E., Debroy, V., Gao, R., Li, Y.: The DStar method for effective software fault localization. IEEE Trans. Reliab. **63**(1), 290–308 (2014)
10. Abreu, R., Zoeteweij, P., Golsteijn, R., & Van Gemund, A.J.: A practical evaluation of spectrum-based fault localization. J. Syst. Softw. **82**(11), 1780–1792 (2009)
11. Naish, L., Lee, H.J., Ramamo hanarao, K.: A model for spectra-based software diagnosis. ACM Trans. Softw. Eng. Methodol. (TOSEM) **20**(3), 11 (2011)
12. Esposti, M.D., Altmann, E.G., Pachet, F.: Creativity and universality in language. Springer (2016)

Chapter 10
Graph and Frequency-Based Fault Localization

10.1 Introduction

Spectrum-based fault localization techniques do not consider the number of iterations of statements in each failing or passing execution as a factor when computing
suspiciousness scores of the statements. However, a rule of thumb suggests that
the more frequent a statement cooperates in the computation of a faulty result, the
more suspicious the statement will be. There have not been many considerations for
statement frequencies in the fault localization literature.

Regions of code suspicious to include the root cause of the failure could be those
appearing in a faulty execution path but not in any passing paths very similar to it.
In other words, we should look for suspicious statements regarding discrepancies
between failing executions with the passing ones. This chapter presents a graph-
based approach to fault localization. Enough hints are given to help the readers to
develop their own graph-based fault localization tools.

© The Author(s), under exclusive license to Springer Nature Switzerland AG 2023   409
S. Parsa, *Software Testing Automation*,
https://doi.org/10.1007/978-3-031-22057-9_10

```
S0.    Def Invert():
S1.    arr = [int(x) for x in input("Enter multiple value: ").split()]
S2.    for i in range(len(arr)):
S3.        minimum = i
S4.        print(' P1 ')
S5.        for j in range(i + 1 , len(arr)):
S6.            print(' P2 ')
S7.            if arr[j] < arr[minimum]-1: #correct:if arr[j] < arr[minimum]:
S8.                print(' P3 ')
S9.                minimum = j
S10.        else:
S11.            print(' P4 ')
S12.            arr[minimum], arr[i] = arr[i], arr[minimum]
S13.        print('P5')
S14.        print(arr)
```

Fig. 10.1 A fault is seeded in line 7 of the code fragment

10.2 Graph-Based Fault Localization[1]

Representing program execution paths as subgraphs of the program control flow graph (CFG), the edges of the graph can be weighted based on their similarity to failing and passing execution subgraphs of their nearest neighbors. Each execution subgraph is represented as a vector to identify the nearest neighbors or, in other words, the closest execution graphs. The similarities are measured by contrasting the nearest neighbor failing and passing execution graphs. The Jaccard and Cosine similarity measures are combined to identify the nearest graphs [1, 2]. The scoring formula takes advantage of null hypothesis testing for ranking weighted transitions. The main advantage of the proposed technique is its scalability which makes it work on large and complex programs.

10.2.1 A Running Example

As usual, a running example is used to describe the algorithm. For instance, consider the code fragment given in Fig. 10.1. A fault is seeded in statement S7.

The aim is to locate the fault in S7. The function is executed with each one of the test cases, T1 to T8. The execution paths are shown in Table 10.1. When running the function with a test case, if the result is as expected, the test case is passing;

[1] Download the source code from: https://doi.org/10.6084/m9.figshare.21321735

Table 10.1 Test paths of the failing and passing test cases

Test case	Input	Expected output	Test paths representing the execution of test cases	Exec. Result
T1	9,0	0,9	[P1,P2,P3,P1,P5]	Failing
T2	3,1,2	2,1,3	[P1,P2,P3,P2,P4,P1,P2,P4,P1,P5]	Failing
T3	7, 8	8,7	[P1,P2,P4,P2,P4,P1,P2,P3,P1,P5]	Failing
T4	2,8	8,2	[P1,P2,P4,P1,P5]	Passing
T5	7,7	7,7	[P1,P2,P4,P1,P5]	Passing
T6	1,2,3	3,2,1	[P1,P2,P4,P2,P4,P1,P2,P4,P1,P5]	Passing
T7	3,2,1	1,2,3	[P1,P2,P4,P2,P3,P1,P2,P4,P1,P5]	Passing
T8	7,7,7	7,7,7	[P1,P2,P4,P2,P4,P1,P2,P4,P1,P5]	Passing

otherwise, it is a failing test case. The execution results are shown in the rightmost column of Table 10.1.

10.3 Execution Graphs

As shown in Table 10.1, the execution paths could be lengthy, specifically when there are long loops. The execution paths are shortened by removing control flow graph branches that do not appear in any of the execution paths. Each execution path is represented as an n*n matrix, where n is the number of predicates. Each matrix element (Pi, Pj) represents a control flow graph branch connecting the predicate Pi to Pj. The (Pi, Pj) branch is labeled with the number of times the branch is observed in the execution path. For example, Fig. 10.2 shows the matrix and graph representation of the test path of T6. The test path representing the execution of T6 is shown in Table 10.1, as well.

$$T_6 = [P1, P2, P4, P2, P4, P1, P2, P4, P1, P5]$$
(a) Execution path as a sequence of predicates

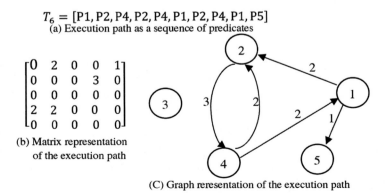

(b) Matrix representation of the execution path

(C) Graph reresentation of the execution path

Fig. 10.2 Matrix and graph representation of the execution path corresponding to T6

Table 10.2 Matrix representation of branches between consecutive predicates

T 1: Passing	T2: Passing	T3: Passing	T4: Failing
[0, 1, 0, 0, 1]	[0, 2, 0, 0, 1]	[0, 2, 0, 0, 1]	[0, 1, 0, 0, 1]
[0, 0, 1, 0, 0]	[0, 0, 1, 2, 0]	[0, 0, 1, 2, 0]	[0, 0, 0, 1, 0]
[1, 0, 0, 0, 0]	[0, 1, 0, 0, 0]	[1, 0, 0, 0, 0]	[0, 0, 0, 0, 0]
[0, 0, 0, 0, 0]	[2, 0, 0, 0, 0]	[1, 1, 0, 0, 0]	[1, 0, 0, 0, 0]
[0, 0, 0, 0, 0]	[0, 0, 0, 0, 0]	[0, 0, 0, 0, 0]	[0, 0, 0, 0, 0]
T5: Failing	T 6: Failing	T 7: Failing	T 8: Failing
[0, 1, 0, 0, 1]	[0, 2, 0, 0, 1]	[0, 2, 0, 0, 1]	[0, 2, 0, 0, 1]
[0, 0, 0, 1, 0]	[0, 0, 0, 3, 0]	[0, 0, 1, 2, 0]	[0, 0, 0, 3, 0]
[0, 0, 0, 0, 0]	[0, 0, 0, 0, 0]	[1, 0, 0, 0, 0]	[0, 0, 0, 0, 0]
[1, 0, 0, 0, 0]	[2, 2, 0, 0, 0]	[1, 1, 0, 0, 0]	[2, 1, 0, 0, 0]
[0, 0, 0, 0, 0]	[0, 0, 0, 0, 0]	[0, 0, 0, 0, 0]	[0, 0, 0, 0, 0]

All the matrices generated for the T1 to T8 test paths are shown in Table 10.2. As described above, the ith row and the ith column of the matrix represent all the branches beginning and ending with Pi, respectively (Table 10.2).

All the red elements, $P_{i,j}$, of the T1 matrix are also zero in all the other matrices. This means no branches connect Pi to Pj in the test paths. Therefore, as shown in Table 10.3, all these matrices are firstly linearized. Afterward, as shown in Table 10.4, all the elements whose values are zero in all the matrices are removed, and the matrices are simplified.

After removing the zero elements, each matrix is represented as a vector shown in Table 10.4. Note that the first row represents the remaining branches.

The question is how these vectors are constructed. As shown in the first row there are 8 branches left. For, instance consider the T6 execution graph in Fig. 10.2. There is an edge labeled 2 from P1 to P2 and another edge labeled 1 to P5. Therefore, the first two branches [P1→P2, P5], in the T6 execution path, are represented as [2, 1]. Similarly considering the graph representation of the T6 execution graph the remaining branches are as follows:

[P2→P3, P4] : [0, 3]
[P3→P1, P2] : [0, 0]

Table 10.3 All the branches in red does not appear in any of the test paths

T1$_p$: [0 1 0 0 1] [0 0 1 0 0] [1 0 0 0 0] [0 0 0 0 0] [0 0 0 0 0]
T2$_p$: [0 2 0 0 1] [0 0 1 2 0] [0 1 0 0 0] [2 0 0 0 0] [0 0 0 0 0]
T3$_p$: [0 2 0 0 1] [0 0 1 2 0] [1 0 0 0 0] [1 1 0 0 0] [0 0 0 0 0]
T4$_p$: [0 1 0 0 1] [0 0 0 1 0] [0 0 0 0 0] [1 0 0 0 0] [0 0 0 0 0]
T5$_p$: [0 1 0 0 1] [0 0 0 1 0] [0 0 0 0 0] [1 0 0 0 0] [0 0 0 0 0]
T6$_p$: [0 2 0 0 1] [0 0 0 3 0] [0 0 0 0 0] [2 1 0 0 0] [0 0 0 0 0]
T7$_p$: [0 2 0 0 1] [0 0 1 2 0] [1 0 0 0 0] [1 1 0 0 0] [0 0 0 0 0]
T8 $_p$: [0 2 0 0 1] [0 0 0 3 0] [0 0 0 0 0] [2 1 0 0 0] [0 0 0 0 0]

Table 10.4 Branches that are covered by at least one test path
[P1→P2, P5] [P2→P3, P4] [P3→P1, P2] [P4→P1, P2]

$T1_{vec}$: [1, 1] [1, 0] [1, 0] [0, 0] = [1, 1, 1, 0, 1, 0, 0, 0]: Passing test path
$T2_{vec}$: [2, 1] [1, 2] [0, 1] [2, 0] = [2, 1, 1, 2, 0, 1, 2, 0]: Passing test path
$T3_{vec}$: [2, 1] [1, 2] [1, 0] [1, 1] = [2, 1, 1, 2, 1, 0, 1, 1]: Passing test path
$T4_{vec}$: [1, 1] [0, 1] [0, 0] [1, 0] = [1, 1, 0, 1, 0, 0. 1, 0]: Failing test path
$T5_{vec}$: [1, 1] [0, 1] [0, 0] [1, 0] = [1, 1, 0, 1, 0, 0, 1, 0]: Failing test path
$T6_{vec}$: [2, 1] [0, 3] [0, 0] [2, 1] = [2, 1, 0, 3, 0, 0, 2, 1]: Failing test path
$T7_{vec}$: [2, 1] [1, 2] [1, 0] [1, 1] = [2, 1, 1, 2, 1, 0, 1, 1]: Failing test path
$T8_{vec}$: [2, 1] [0, 3] [0, 0] [2, 1] = [2, 1, 0, 3, 0, 0, 2, 1]: Failing test path

[P4→P1, P2] : [2, 2]

Putting all these branches next to each other the T6 execution vector will be as
follows:

$T6_{vec}$: [2, 1] [0, 3] [0, 0] [2, 1] = [2, 1, 0, 3, 0, 0, 2, 1] : Failing test path

As shown in Fig. 10.3, each test path is simplified to an array of integer values,
where each value represents a branch shown in the first row of the table. For instance,
the $T8_{vec}$ vector can be represented as a graph shown in Fig. 10.3.

Listed in Table 10.4 are eight vectors representing test paths as sequences of predicates. Each vector represents an execution graph corresponding to a test execution
path. The following section describes how to compute the vector's similarities to
identify passing test paths similar to each failing execution path.

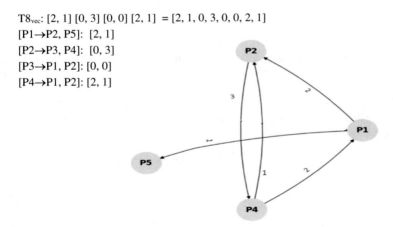

$T8_{vec}$: [2, 1] [0, 3] [0, 0] [2, 1] = [2, 1, 0, 3, 0, 0, 2, 1]
[P1→P2, P5]: [2, 1]
[P2→P3, P4]: [0, 3]
[P3→P1, P2]: [0, 0]
[P4→P1, P2]: [2, 1]

Fig. 10.3 Graph representation of the T8 test path

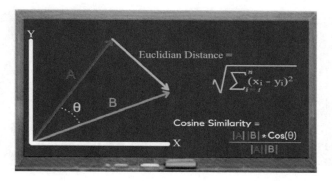

10.4   Vector Similarity Measurement

Presumably, faults should be sought in dissimilarities between failing and passing executions. The lesser the dissimilarities, the higher the probability of finding the root cause of the failure in the discrepancies. It is shown in the following sub-sections that Cosine similarity, together with Extended Jaccard similarity, provides appropriate measures for comparing the vector representation of the execution paths represented as vectors of predicates.

10.4.1   Euclidian and Cosine Distances

After simplifying the execution paths as vectors shown in Table 10.4, the vectors corresponding to each failing test case are compared with those of the passing ones. The branches in the failing vector, which are different from the same ones in the passing vectors, are fault-suspicious branches of the code. The most similar passing test path to a failing path may be detected by computing the Euclidian and Jaccard distances between the failing and passing vectors. Euclidian distance between two vectors V1 and V2 is computed as follows:

$$\text{dist}(V1, \ V2) = \sqrt{\sum_{i=1}^{n}(q_i - p_i)^2} \tag{10.1}$$

The shorter the Euclidean distance between two vectors, the more similar vectors. This measure is perfect for comparing vectors when the dimensionality is low but not when the number of dimensions is high. For instance, Table 10.5 represents the dimensions of four different vectors in a ten-dimensional space. Each vector represents a test path as a sequence of predicates.

Table 10.5 Test paths represented as vectors of predicates

Vector/Dlm	P1	P2	P3	P4	P5	P6	P7	P8	P9	P10
Test_Path_1	3	0	0	0	0	0	0	0	0	0
Test_Path_2	0	0	0	0	0	0	0	0	0	4
Test_Path_3	3	2	4	0	1	2	3	1	2	0
Test_Path_4	0	2	4	0	1	2	3	1	2	4

Test_Path_3 and Test_Path_4 have more predicates (dimensions) in common than Test_Path_1 and Test_Path_2. Therefore, the distance between Test_Path_3 and Test_Path_4 is expected to be less than Test_Path_1 and Test_Path_2. However, contrary to expectation, their Euclidean distance is the same:

$$d(\text{Test\_path\_1}, \text{Test\_path\_2}) = \sqrt{\sum_{i=1}^{n}(q_i - p_i)^2}$$

$$= \sqrt{(3-0)^2 + (0-0)^2 + + (0-0)^2 + (0-0)^2 + (0-0)^2 +}$$
$$(0-0)^2 + (0-0)^2 + (0-0)^2 + (0-0)^2 + (0-4)^2} = \sqrt{25} = 5.00$$

$$d(\text{Test\_path\_3}, \text{Test\_path\_4}) = \sqrt{\sum_{i=1}^{n}(q_i - p_i)^2}$$

$$= \sqrt{(3-0)^2 + (2-2)^2 + (4-4)^2 + (0-0)^2 + (1-1)^2 +}$$
$$(2-2)^2 + (3-3)^2 + (1-1)^2 + (2-2)^2 + (0-4)^2} = \sqrt{25} = 5.00$$

Euclidean distance is a proper metric for measuring the distance between two vectors. But, as shown above, it is not appropriate for measuring the similarity between two relatively long vectors.

Cosine similarity is another metric used to measure the distance between vectors as the cosine of the angle between the vectors. Cosine similarity looks at the angle between two vectors, while Euclidean similarity is the actual distance between two points. Compared to the unbounded Euclidean distance, cosine distance is often preferred since it is bounded to the range $[-1, 1]$. The cosine similarity of two vectors v and w is computed as follows:

$$\text{CSim}(V, W) = \cos(\theta) = \frac{V \cdot W}{||V|| \cdot ||W||} = \frac{\sum_{i=1}^{n} V_i \times W_i}{\sqrt{\sum_{i=1}^{n} V_i^2} \times \sqrt{\sum_{i=1}^{n} W_i^2}} \tag{10.2}$$

The cosine similarity of vectors Test_Path_1 and Test_Path_2 in Table 10.5 is zero. It follows that these two vectors are the least similar to one another. The similarity of the two vectors Test_Path_3 and Test_Path_4 is equal to 0.759. As expected, the cosine similarity of Test_Path_1 and Test_Path_2 is less than Test_Path_3 and Test_Path_4, while their Euclidean distance is the same.

Despite getting better similarity results from cosine similarity still, there are certain situations does not provide appropriate similarity measures.

10.4.2   Jaccard and Cosine Distances

As described above, the Euclidean distance does not provide an appropriate measure of similarity in certain circumstances. Similarly, Cosine similarity cannot be used in some cases to compute vector similarity. Therefore, my suggestion is to use a combination of Jaccard and Cosine similarities. As an example, consider the following three vectors:

$$V1 = [0,\ 1,\ 0,\ 1,\ 1,\ 0,\ 0,\ 0,\ 1,\ 1,\ 0,\ 0,\ 1,\ 1]$$
$$V2 = [1,\ 1,\ 1,\ 1,\ 0,\ 0,\ 1,\ 1,\ 0,\ 2,\ 0,\ 0,\ 2,\ 2]$$
$$V3 = [0,\ 1,\ 0,\ 1,\ 0,\ 1,\ 0,\ 0,\ 1,\ 0,\ 0,\ 1,\ 1,\ 1]$$

Below, the cosine similarity between V1 and V2 and then V1 and V3) is computed. It is observed that despite having more elements in common, the Cosine similarity of V1 and V3, Sim(V1, V3) is less than V1 and V2, Sim(V1, V2).

$$CSim(V1,\ V2) = \frac{\sum_{i=1}^{14} V1_i \times V2_i}{\sqrt{\sum_{i=1}^{14} V1_i^2} \times \sqrt{\sum_{i=1}^{14} V2_i^2}}$$

$$= \frac{\begin{array}{c} 0*1 + 1*1 + 0*1 + 1*1 + 1*0 + 0*0 \\ +0*1 + 0*1 + 1*0 + 1*2 + 0*0 + 0*0 + 1*2 + 1*2 \end{array}}{\begin{array}{c} \sqrt{0^2 + 1^2 + 0^2 + 1^2 + 1^2 + 0^2 + 0^2} \\ \sqrt{+0^2 + 1^2 + 1^2 + 0^2 + 0^2 + 1^2 + 1^2} \times \sqrt{18} \end{array}}$$

$$= \frac{8}{\sqrt{7} \times \sqrt{14}} = \frac{8}{\sqrt{98}} = \frac{8}{9.89949493661} = 0.808$$

$$CSim(V1,\ V3) = \frac{\sum_{i=1}^{14} V1_i \times V3_i}{\sqrt{\sum_{i=1}^{14} V1_i^2} \times \sqrt{\sum_{i=1}^{14} V3_i^2}} = \frac{5}{\sqrt{7} \times \sqrt{7}} = \frac{5}{7} = 0.714$$

Such a deficiency can be resolved by considering Jaccard distance in addition to the cosine similarity. Two vectors are similar only if the Jaccard distance and cosine similarity measures indicate their similarity. The Jaccard distance measures the similarity of the two vectors as the intersection of their elements divided by the union of the vector elements:

$$Jaccard(V,\ W) = \frac{V \cap W}{V \cup W} \tag{10.3}$$

The Jaccard index applies only to the binary vectors. Therefore, its modified version, the Tanimoto index, should be used to measure the similarity between the vectors, representing test paths as sequences of predicates. The Tanimoto Distance is a synonym for extended Jaccard Distance (1- Jaccard_similarity). The Tanimoto similarity between two vectors, V and W, is defined as follows:

$$\text{Tsim}(V, \ W) \frac{V \cdot W}{||V||^2 + ||V||^2 - V \cdot W} = \frac{\sum_{i=1}^{n} V_i \times W_i}{\sum_{i=1}^{n} V_i^2 + W_i^2 - V_i \times W_i} \qquad (10.4)$$

The Tanimoto similarity of the vector V1 with V2 and V3, for which the cosine similarity is given above, is as follows:

$$\text{Tsim}(V1, V2) \frac{8}{7 + 14 - 8} = \frac{8}{21 - 8} = \frac{8}{13} = 0.615$$

$$\text{Tsim}(V1, \ V3) = \frac{5}{7 + 7 - 5} = \frac{5}{14 - 5} = \frac{5}{9} = 0.555$$

10.5 Selecting the Most Similar Failing and Passing Path

The main idea behind identifying similarities between passing and failing execution graphs is to identify the cause of failure in the discrepancies between graphs. Tables 10.6 represent the similarities between the test paths in Table 10.4.

Fig. 10.4 lists Python functions to compute Cosine and Extended Jaccard similarities and the Euclidean distance between the two vectors V and W.

Table 10.6 shows the Cosine similarity of the two execution paths vectors $T1_{vec}$ and $T4_{vec}$ is 0.500, while their Extended Jaccard similarity is 0.333. Considering the execution paths for T1 and T4, in Fig. 10.5, the Cosine similarity metric is more accurate and realistic than the Extended Jaccard (Tanimoto). The Cosine and Jaccard similarities of T2 and T4 are 0.904 and 0.583, respectively. The Cosine and Extended

Table 10.6 Cosine & Tanimoto similarities between failing and passing test paths

Cosine & Extended Jaccard (Tanimoto) Similarities						
P\F	$T4_{vec}$	$T5_{vec}$	$T6_{vec}$	$T7_{vec}$	$T8_{vec}$	
$T1_{vec}$	0.500	0.500	0.344	0.693	0.344	Cosine Similarity
$T2_{vec}$	0.904	0.904	0.889	0.859	0.889	
$T3_{vec}$	0.832	0.832	0.891	1.000	0.891	
$T1_{vec}$	0.333	0.333	0.15	0.417	0.15	Tamino Simlarity
$T2_{vec}$	0.583	0.583	0.789	0.75	0.789	
$T3_{vec}$	0.545	0.545	0.778	1	0.778	

```
import numpy as np
def cosine_similarity(v, w):
    return np.dot(v, w) / (np.sqrt(np.dot(v, v)) * np.sqrt(np.dot(w, w)))
def extended_jacard_similarity(v, w):
    return np.dot(v, w) / (np.dot(v, v) + np.dot(w, w) - np.dot(v, w))
def euclidean_distance(x, y):
    return np.sqrt(np.sum((v - w) ** 2))
def convert_euclidean_to_cosine_sim(dist, sim)
    retuen (1 – (dist ** 2 ) / 2.0)
def convert_euclidean_to_cosine_sim(dist, sim)
        retuen (2 – (sim * 2 )) ** 0.5
```

Fig. 10.4 Python functions to compute similarities between the two vectors v and w

Jaccard similarities of T2 and T4 are relatively high, and the execution graphs are also very similar. However, despite the relatively high value of Cosine similarity between T2 and T4, their Extended Jaccard similarity is relatively low. Using the Cosine similarity to compare different execution graphs seems reasonable, provided that it falls within the Extended Jaccard similarity range (Fig. 10.5).

The root cause of the failure should be sought in the discrepancies between the failing test paths with their most similar passing execution paths. In Table 10.7, the failing execution paths are ordered based on their cosine similarities with passing execution paths. The Jaccard similarity is also taken into account to ensure cosine similarity accuracy. The cosine similarity between two execution paths will not be reliable if its difference with the Extended Jaccard similarity is high.

The function select_closest_failing_passing_path in Fig. 10.6 selects the pairs of most similar failing and passing test paths. The selection considers the Cosine and Tanimoto similarities between each failing and passing test path. The function operates on an array, self.selected_test_paths, shown in Table 10.7. The function begins by looking for the pairs of failing and passing test paths with a cosine similarity value greater than 0.9 and Extended Jaccard similarity less than 0.1. If it does not

Fig. 10.5 Execution graphs of T1 to T4 test cases in Table 10.4

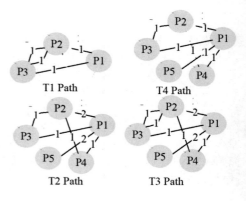

Table 10.7 Cosine and Tanimoto similarities between failing and passing test paths

Executions		Cosine	Tanimoto	Cosine - Tanimoto
Failing	Passing	Sim	Sim	
$T7_{vec}$	$T3_{vec}$	1	1	0
$T4_{vec}$	$T2_{vec}$	0.904	0.583	0.321
$T5_{vec}$	$T2_{vec}$	0.904	0.583	0.321
$T6_{vec}$	$T3_{vec}$	0.891	0.778	0.113
$T8_{vec}$	$T3_{vec}$	0.891	0.778	0.113
$T6_{vec}$	$T2_{vec}$	0.889	0.789	0.100
$T8_{vec}$	$T2_{vec}$	0.889	0.789	0.100
$T7_{vec}$	$T2_{vec}$	0.839	0.750	0.089
$T4_{vec}$	$T3_{vec}$	0.832	0.545	0.287
$T5_{vec}$	$T3_{vec}$	0.832	0.545	0.287
$T7_{vec}$	$T1_{vec}$	0.693	0.417	0.276
$T4_{vec}$	$T1_{vec}$	0.500	0.333	0.167
$T5_{vec}$	$T1_{vec}$	0.500	0.333	0.167
$T6_{vec}$	$T1_{vec}$	0.344	0.150	0.194
$T8_{vec}$	$T1_{vec}$	0.344	0.150	0.194

```
# Select a failing path that is the most similar to a passing one
Def select_closest_failing_passing_path()
        selected_test_paths = []
        alfa = 1.0
        while(len(selected_test_paths) == 0 and alfa!=0.1):
          alfa -= 0.1
          # self.selected_test_paths = [failing, passing, Cosine, Jaccard]
          for i in range(len(self.selected_test_paths)):
            if((self.selected_test_paths[i][2] >= alfa)
                and (abs(self.selected_test_paths[i][2] –
                  self. selected_test_paths [i][3]) < 0.2)):
                selected_test_paths.append(self. selected_test_paths [i])
        return selected_test_paths
```

Fig. 10.6 Selecting the most similar failing and passing execution path

succeed, the accepted Cosine similarity value is reduced by 0.1, and the accepted distance from the Jaccard value is increased by 0.1. This process repeats as far as there are no more test cases to be compared or a pair of the failing and passing test paths satisfying the minimum values for the Cosine similarity and the maximum acceptable values for the Extended Jaccard.

As a result of applying the function 'select_closest_failing_passing_path()' to the Similarity values given in Table 10.7, the following test paths are selected:

Test paths: [Falling, Passing]	Similarities: [cosine, Jaccard]
[T3$_{vec}$, T7$_{vec}$]	[1.0, 1.0]

10.6   Selecting the Most Suspicious Transitions (Branches)

The suspicious transitions in a failing run are sought in the selected failing and passing test paths, e.g., T3$_{vec}$. A suspicious transition, 'edge,' occurs in relatively more erroneous executions than the correct ones. If the frequency or, in other words, the number of iterations of an edge to the total number of occurrences of the edge in failing runs is less than its participation in the passing runs, then the pass is suspicious to be faulty:

$$\frac{frequency_{failing\_runs}(edge)}{total\_number\_of_{failing\_runs}} \geq \frac{frequency_{passinng_{runs}}(edge)}{total\_number\_of_{passing\_runs}}$$

If the above relation holds, then the suspiciousness score of the edge, Susp(edge), is the frequency of the edge in the failing runs divided by the total number of failing runs:

$$\text{Susp(edge)} = \frac{frequency_{failing\_runs}(edge)}{total\_number\_of_{failing\_runs}} \qquad (10.4)$$

Table 10.8 represents the suspiciousness scores calculated by Eq. 10.4. In all the execution graphs, only the three transitions, 3, 5, and 6, are suspicious of faults. For instance, for transition 3, P2 → P3, suspiciousness is calculated as follows:

$$\text{Susp(3)} = \frac{frequency_{failing\_runs}(edge)}{total\_number\_of_{failing\_runs}} = \frac{3}{3} > \frac{frequency_{passing_{runs}}(edge)}{total\_number\_of_{passing\_runs}} = \frac{1}{5}$$

In Sect. 10.6, it is shown that T3 is the faulty path with the highest similarity with the passing path, T7. As shown in Table 10.8, the transitions P2 → P3, P3 → P1, and P3 → P2, are suspicious of faults in T7.

The detected suspicious transitions are ranked based on their relevance to the observed fault. The ranking is based on how closely the transitions correlate with failing versus the passing run. The transition is scored based on the weight in the selected execution graphs, T3 and T7. The weight of a transition is the number of its occurrences in the corresponding test run. If a transition is not in an execution graph, its weight will be zero.

Table 10.8 Finding suspicious edges

Transitions in execution graph = Edges of CFG = Branches of code								
	P1 → P2	P1 → P5	P2 → P3	P2 → P4	P3 → P1	P3 → P2	P4 → P1	P4 → P2
Edge	1	2	3	4	5	6	7	8
$T1_{vec}$	1	1	1	0	1	0	0	0
$T2_{vec}$	2	1	1	2	0	1	2	0
$T3_{vec}$	2	1	1	2	1	0	1	1
Total failed	3	3	3	2	2	1	2	1
$T4_{vec}$	1	1	0	1	0	0	1	0
$T5_{vec}$	1	1	0	1	0	0	1	0
$T6_{vec}$	2	1	0	3	0	0	2	1
$T7_{vec}$	2	1	1	2	1	0	1	1
$T8_{vec}$	2	1	0	3	0	0	2	1
Total passed	5	5	1	5	1	0	5	3
Susp(fail)	$\frac{3}{3}=1$	$\frac{3}{3}=1$	$\frac{3}{3}=1$	$\frac{2}{3}=0.66$	$\frac{2}{3}=0.66$	$\frac{1}{3}=0.33$	$\frac{3}{3}=1$	$\frac{1}{3}=0.33$
Ratio(pass)	$\frac{5}{5}=1$	$\frac{5}{5}=1$	$\frac{1}{5}=0.2$	$\frac{5}{5}=1$	$\frac{1}{5}=0.2$	0	$\frac{5}{5}=1$	$\frac{3}{5}=0.6$
Suspicious			✓		✓	✓		

A zero point is given to a suspicious transition with identical weights in the selected failing execution graph and its most similar passing one. Otherwise, if the weight of the transition in the failing execution graph is more than the passing one, it will be scored one provided that the following relation holds:

$$\frac{Weight(transition)_{failing\_graph} - Weight(transition)_{passing\_graph}}{Weight(transition)_{failing\_graph}} > 0.25$$

The weight of the selected suspicious transition, as shown in Fig. 10.7 is equal to one. Therefore, the score of the selected suspicious transitions is zero, and no judgment can be made at this stage. Hence, null hypothesis testing is used to select the most suspicious transition.

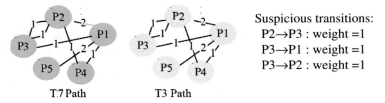

T.7 Path   T3 Path

Suspicious transitions:
P2→P3 : weight =1
P3→P1 : weight =1
P3→P2 : weight =1

Fig. 10.7 Suspicious failing execution graph T7 and its similar passing path T3,

10.7   Summary

Test paths can be represented as labeled subgraphs of the program control flow graph where the label of each edge indicates the number of its occurrences in the test run. Depending on whether a test case is failing or passing, its corresponding subgraph is labeled afailing or passing, respectively. Given two sets of software failing and passing subgraphs, the aim is to find some edges or, in other words, partial subgraphs which are highly indicative of bugs. Intuitively, a partial subgraph is related to a bug with a high probability if it frequently appears in faulty subgraphs but rarely in the correct ones. The root cause of the failure can be sought in the discrepancies between the failing test paths with their most similar passing execution paths. The cosine similarity between two execution paths will not be reliable if its difference with the Extended Jaccard similarity is high. Therefore, a combination of Cosine and extended Jaccard is used. Once the most similar passing execution graph to a passing subgraph is detected, the transitions in the failing sub-graph, which are different from the failing, are given higher priority to look for the root cause of the failure.

10.8   Exercises

1. Use the guidelines and code presented in this chapter to write a program to take as input a faulty program to be debugged and outputs the suspicious basic block.
2. Discriminate subgraphs in failing and passing executions are used to detect suspicious edges. However, it is observed [3] that a failure context does not necessarily appear in a discriminative subgraph. When contrasting a failing graph with its nearest neighbor passing graph, each edge, E, in the failing graph is scored as follows:

    If (E ∈ failing_graph ∧ E ∉ passing_graph) ⟹ score(E)++;

    Else if (E ∉ failing_graph ∧ E ∈ passing_graph) ⟹ score(E) += 0;

    Else if (E ∉ failing_graph ∧ E ∈ passing_graph) ⟹

    $$\text{Score(E)} += \frac{Weight(E)_{failing\_graph} - Weight(E)_{passing\_graph}}{Weight(E)_{failing\_graph}} > 0.25$$

    $$\text{Score(E)} += (Weight(E)_{failing\_graph} - Weight(E)_{passing\_graph}) * 100;$$

    The most discriminative subgraph is the one with the highest average edge score. Find the most discriminative subgraph for the running example in this chapter.
3. Use the discriminative power formula [5] given below to determine the most discriminative execution subgraph for the working example in Fig. 10.1. The

point that I have made in [5] is that frequent sub-graphs do not provide a reasonable basis for differentiating failing from passing execution graphs. Statistical significance should be considered instead of the frequency of graph patterns.

Suppose for each execution graph, ex_g, the set of execution graphs, Cex_g, contains it as a subgraph, and Čex_g indicates those not containing ex_g:

$$\text{Cex\_g} = \{\text{ex\_g} \in \text{execution\_graphs} \mid \exists \text{ex\_g}' \supseteq \text{ex\_g}:$$

$$\forall [\text{edge, weight}] \in \text{ex\_g}, [\text{edge, weight}'] \in \text{ex\_g}'.\text{weight} \leq \text{weigh}' \}$$

$$\text{no.Container} = \mid \text{Cex\_g} \mid$$

In other words, no.Container(ex_g) is the number of execution graphs in which ex_g is a subgraph. Let ex_g denote the Jth execution graph then:

$$f^i_{ex\_g} = \begin{cases} 1 & \exists \text{ ex}'_g. \supseteq \text{ex\_g}: \forall [\text{edge, weight}] \in \text{ex\_g}, [\text{edge, weight}'] \\ & \in \text{ ex\_g}'.\text{weight} \leq \text{weigh}'\} \\ 0 & else \end{cases}$$

Suppose m is the number of execution graphs. The mean of subgraphs containing ex_g is computed as follows:

$$\mu_{ex\_g} = \frac{1}{m} |Cex\_G|, \quad m = \text{no.passed} + \text{no.failed execution graphs}$$

$$\mu^{failed}_{ex\_g} = \frac{1}{no.failed} |Cex\_G|, \quad \text{ex\_g is a failed execution}$$

$$\mu^{passed}_{ex\_g} = \frac{1}{no.passed} |Cex\_G|, \quad \text{ex\_g is a passing execution.}$$

The variance of the failing and passing executions is as follows:

$$\left(\sigma^{failed}_{ex\_g}\right)^2 = \frac{1}{no.failed} \sum_{i=1}^{no.failed} (f^i_{ex\_g} - \mu^{failed}_{ex\_g})^2$$

$$\left(\sigma^{passed}_{ex\_g}\right)^2 = \frac{1}{no.passed} \sum_{i=1}^{no.passed} (f^i_{ex\_g} - \mu^{passed}_{ex\_g})^2$$

The discriminative power of an execution graph, ex_g, is computed as follows:

$$\text{discriminative\_power(ex\_g)} =$$

$$\frac{no.failed \times \left(\mu^{failed}_{ex\_g} - \mu_{ex\_g}\right)^2 - no.passed \times \left(\mu^{passed}_{ex\_g} - \mu_{ex\_g}\right)^2}{no.failed \times \left(\sigma^{failed}_{ex\_g}\right)^2 + no.passed \times \left(\sigma^{passed}_{ex\_g}\right)^2}$$

4. Information gain is the reduction in entropy, and the amount of information covered by an event is entropy. According to information gain, as the frequency difference between the edges of a subgraph in the faulty executions and the correct executions increases, the subgraph becomes more discriminative. Use information gain to determine the most discriminative subgraph [2].

References

1. Mousavian, Z., Vahidi-Asl, M., Parsa, S.: Scalable graph analyzing approach for software fault-localization. In: Proceedings of the 6th International Workshop on Automation of Software Test (2011)
2. Parsa, S., Arabi, S., Ebrahimi, N., Vahidi-Asl, M.: Finding discriminative weighted sub-graphs to identify software bugs. CCIS 70, pp. 305–309. Springer-Verlag Berlin Heidelberg (2010)
3. Parsa, S., Mousavian, Z., Vahidi-Asl, M.: Analyzing dynamic program graphs for software fault localization. In: 5th International Symposium on Telecommunications. Publisher, IEEE (2010)
4. Cheng, H., Lo, D., Zhou, Y.: Identifying bug signatures using discriminative graph mining. ISSTA'09, July 19–23 (2009)
5. Parsa, S., Arabi, S., Ebrahimi, N.: Software fault localization via mining execution graphs. In: ICCSA 2011. Part II, LNCS 6783, pp. 610–623 (2011)

Part III
Automatic Test Data Generation

Behaviors are not instantaneous but continuous for a while. They are triggered by specific events but do not change instantaneously. Therefore, precise analysis of unexpected behaviors, especially in cyber-physical systems, requires the domain of inputs triggering the behavior. Unexpected behaviors in a cyber-physical system are sometimes due to unknowns where mathematical models cannot offer judgments. Let us try to cross the unknown by supplying cyber-physical systems with software testing tools to receive test data from the physical system in action and predict unexpected behaviors while extracting the underlying governing physical rules.

This part of the book focuses on developing software tools for test data generation. The reader will get insights into how to develop their own software tools to generate test data for a program under test. Domain coverage adequacy criteria introduced in this part form the basis for evaluating the coverage obtained by the test data.

Chapter 11
Search-Based Testing

11.1 Introduction

Meta-heuristic search algorithms have proposed a suitable alternative for developing test data generators. Genetic algorithms are the most popular heuristic search technique applied to test data generation problems. Search-based testing involves looking in a **program** input space for test data, satisfying a test adequacy criterion encoded as a fitness function. A well-defined test adequacy criterion should lead the search toward generating a minimum set of inputs while maximizing the test achievements. The objective of test data generation is implemented with a fitness function. This chapter describes the details of a white box test data generator implementing genetic algorithms. The term "white box" describes testing based on source code inspection. This chapter shows how to write your own test data generators.

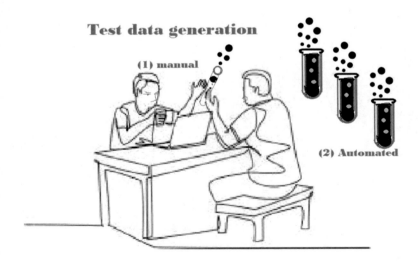

Test data generation
(1) manual
(2) Automated

© The Author(s), under exclusive license to Springer Nature Switzerland AG 2023
S. Parsa, *Software Testing Automation*,
https://doi.org/10.1007/978-3-031-22057-9_11

11.2   Encoding

Test data generation involves searching a program input space to find test data satisfying a test objective. The question is, what is a program input space?

Consider a function, f, with two inputs, x and y, where x and y are both in the range [3, 10]. Now, considering x and y as the axes of Cartesian space, the input space will be a rectangle. In this case, test data satisfying a search objective, such as path coverage, is a point inside the rectangle. A genetic search over the rectangular input space could be applied to find points as test data sets satisfying the search objective. Therefore, test data generation is, in fact, a search problem.

Search-based test data generation techniques use generic algorithms to search for appropriate test data targeted at exercising desired program regions under test. For instance, consider the code snippet and its corresponding control flow graph, shown in 0.

There are six execution paths in test_function():

Path 1 2 - 6 - 7 – 9
Path 2 - 6 - 7 – 8 – 9
Path 3 2 - 3 - 5 – 7
Path 4 2 - 3 - 5 - 7 – 8 – 9
Path 5 2 - 3 - 4 – 7 – 9
Path 6 2 - 3 - 4 - 7 – 8 – 9

A test data for the function, test_function in Fig. 11.1, is a triple (x, y, z) covering one of the above paths. For instance, the test data set $\{(1, 2, 1), (10, 2, 5), (3, 16, 4)\}$ covers the execution paths: {path-5, path-6, path-5}. The path coverage for the data set is $2/6 = 33\%$. The total number of branches covered by the data set is $4 + 2 = 6$. Hence, the branch coverage is $6/10 = 60\%$.

Suppose all three input parameters, x, y, and z are in the interval (0, 10]. In this case, the input space of this function is a cube shown in 0. If the parameters x, y,

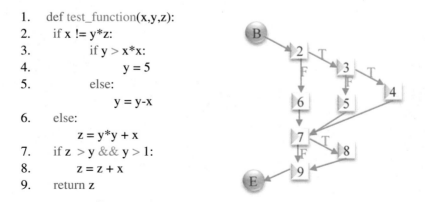

```
1.   def test_function(x,y,z):
2.     if x != y*z:
3.         if y > x*x:
4.             y = 5
5.         else:
               y = y-x
6.     else:
           z = y*y + x
7.     if z > y && y > 1:
8.         z = z + x
9.     return z
```

Fig. 11.1 A sample function and its control flow graph

and z are integers, the size of the input space will be 10*10*10 = 1000. We have to
search in this space to provide a suitable test data set with a certain coverage.

The space of all feasible solutions is called search space. In the case of test
data generation, the search objective is to provide test data covering certain regions
of code, and input space is the most appropriate search space. Every point in the
input space represents one possible solution. Therefore each possible solution can be
labeled with its fitness value, depending on the search objectives. Genetic algorithms
are used to search for the fittest solutions.

Each individual or chromosome in a genetic population is an array of inputs to
the component under test. For instance, each individual for test_func() in Fig. 11.2
is an array of length 3:

To restrict the search space, we define lower and upper bounds for each input
variable as an axis of the input space. The input space for the test data covering a
path consists of input data satisfying the path constraint as a conjunction of branching
conditions along the path. For instance, consider the following path constraint:

$$(10 * x \leq y^2) \text{ and } (25 * \sin(30x) \leq y) \text{ and} (15 * \cos(40x) \leq y)$$

The sub-space of inputs satisfying the path constraint is shown in Fig. 11.3.

Every point (x, y) in shaded regions in Fig. 11.3 satisfies the path constraint. A
path constraint is a predicate over inputs such that if the path is feasible and the
predicate is satisfied, the path will be executed.

Fig. 11.2 The input space of
test_function

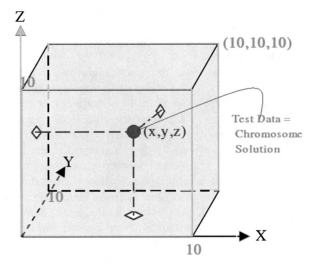

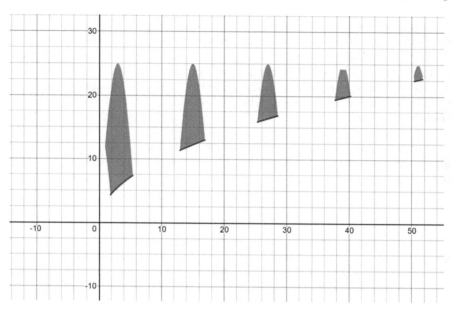

Fig. 11.3 The input space consists of shaded closed regions

11.3   Test Data Generation Using Genetic Algorithms

Genetic algorithms are search algorithms based on how species evolve through the natural selection of the fittest individuals. A population of chromosomes represents the possible solutions to the search problem. A chromosome is a sequence of genes where each gene represents a specific state of a characteristic to be optimized. Figure 11.4 presents the main body of a Python program to generate test data for Python functions.

In the main function of the genetic algorithm, in Fig. 11.4, the population size is 40. The question is, what is the most appropriate size for the population? Population size has a considerable impact on the performance of genetic algorithms. In general, a value between 50 and 100 seems to be suggested by most practitioners [1]. Table 11.1 shows the genetic algorithm parameters, population size (ps), crossover probability (cp), and mutation probability (mp) suggested in the literature.

On the one hand, a greater population size avoids a fast and undesired convergence, which contributes to exploring the solutions' space. However, large populations need relatively more time to exploit the most promising solutions. On the other hand, smaller populations lead to faster exploitation and convergence. If this optimal value is substantial, the algorithm likely encounters difficulties exploring the solutions space and preserving the population diversity. The algorithm is, therefore, unlikely if this is the case. The optimal population size depends on the execution time and the algorithm facilities to maintain a diverse population.

```
# the main function
# 1- Instrument the input source, testfunc.
instrument_input_file(testfunc)
# 2- Generate test data
generate_testdata(population_size=40, no_generations=15000,
                  min_val_test_data=-100, max_val_test_data=100)
x = []
y = []
print("best of all generations:", best_of_generations)
for i in best_of_generations:
    x.append(i[0])
    y.append(i[1])
plt.plot(x, y, 'bo')
plt.show()d
```

Fig. 11.4 The main body of a test data generator

Table 11.1 Parameter used in previous literature

Ref	ps	cp	mp	Ref	ps	cp	mp
[2]	30	0.95	0.01	[10]	40	0.6	0.1
[2]	80	0.45	0.01	[11]	100	0.8	0.3
[3]	100	0.9	1.00	[12]	100	0.6	0.02
[4]	100	0.8	0.005	[13]	30	0.6	0.05
[5]	50	0.9	0.03	[14]	76	0.8	0.05
[5]	50	1.0	0.03	[15]	30	0.6	0.001
[6]	50	0.8	0.01	[16]	50	0.8	0.2
[7]	15	0.7	0.05	[17]	50	0.9	0.1
[8]	100	1.0	0.003	[18]	30	0.9	0.1
[9]	30	0.8	0.07	[13]	20	0.8	0.02

11.3.1   Source Code Instrumentation

Source code instrumentation is a commonly used technique to track the execution of program elements such as predicates, lines, and functions. In the rest of this Section, at first, the use of a Python library function to automatically track the execution of lines of Python programs is described. Then the code for writing a software tool to automatically instrument predicates is described.

1. **Line instrumentation**

   The Python library function, sys.settrace() [19], takes a function called during several events. One is the "line" event. Before the Phyton interpreter executes a line, it calls the trace function with the current frame and the string "line." The

sys.setrace(traceit) call automatically invokes the "traceit" function whenever a
new line of the function under test is executed.

```
# Invoked whenever a newline of code
# is to be executed.
def traceit(frame, event, arg):
    count=0;
    if event == "line":
        lineno = frame.f_lineno;
        linenos.append(lineno);
    for i in exec_path:
        count=count+1;
        if count in linenos:
            exec_path[count-1] += 1
    return traceit
```

The "f_lineno" attribute of the frame object is the line number about to run.
Whenever the "line" event is raised, the line number f_lineno is added to the
execution path. The "line" event is raised whenever a new line from the code
under test is about to run. Whenever "traceit" is invoked, it adds the number
of the line to be executed in a vector, "execPathVector." This vector is a global
variable that is initially filled with zeros. The size of "execPathVector" equals the
number of lines in the function under test. If a line is executed, its corresponding
cell in the "execPathVector" is set to '1'; otherwise, it is '0'.

2. **Predicate instrumentation**

Generally, the instrumented function is invoked by the fitness function to compute
the fitness of a solution. Thus, depending on the objective of the test, different
instrumentation points and techniques are used. Here, the objective is to estimate
how close a test data is to evaluating a branching condition to true or false. The
function instrumen_input_file, in Fig. 11.5, takes a function, func, as input and
instruments the function by adding probes to compute the branch distance for
each condition within a predicate.

In Fig. 11.6, BranchTransformer is a subclass of a Python class,
ast.NodeTransformer [19]. NodeTransformer walks through an abstract syntax
tree and calls a visitor function to visit each node it meets. When visiting a node,
a function is invoked to transform the node depending on the node type. For
example, as shown in Fig. 11.6, the visi_compare method is invoked whenever
the visit() method visits a node, including a comparison operator.

The ast.NodeTransformer [13] class uses the visitor pattern where there is one
visit_* function for each node type in the abstract syntax tree. The NodeTransformer class allows programmers to override the visit methods and modify the
method's return value. The node corresponding to a return statement is removed
from the tree when the return value is None. The return value will be the original
node if no replacement is done to the return value. For instance, the visit_compare
method of the BranchTransformer class in Fig. 11.6 returns the object returned
by ast.Call(). The ast.Call() method transforms each condition in a predicate

```python
def instrument_input_file(func):
    # converts the source code into a stream
    source = inspect.getsource(func)
    # Build a parse tree for the input source
    tree = ast.parse(source)
    # While traversing the parse tree instrument, the AST
    BranchTransformer().visit(tree)
    # Removes extra blank lines and spaces
    tree = ast.fix_missing_locations(tree)
    # Instrument the source code based on the AST instrumentation
    save_as_instrumented_python(astor.to_source(tree),
                        "nameeeeeeeeeeeeeeeeeeeeeee")
    # We are in the main
    current_module = sys.modules[__name__]
    code = compile(node, filename="<ast>", mode="exec")
    exec(code, current_module.__dict__)
```

Fig. 11.5   Instruments a given function, func

```python
class BranchTransformer(ast.NodeTransformer):
    branch_num = 0
    # this method is automatically invoked when a function definition is met
    def visit_FunctionDef(self, node):
        # change the function name
        node.name = node.name + "_instrumented"
        return self.generic_visit(node)
    # Invoked when the tree walker meets a comparison operator
    def visit_Compare(self, node):
        # if it is a set comparison operator, then return
        if node.ops[0] in [ast.Is, ast.IsNot, ast.In, ast.NotIn]:
            return node

        self.branch_num += 1
        # ast.call adds a function call to AST at the current node
        # ast.load loads the current node attributes
        # ast.Num adds branch_num to the parameters of evaluate_condition
        # ast.Str(node.op[0] takes the operator from the condition
        # node.comparator[0] = right operand
        # three optional parameters could be sent to a Python function
        return ast.Call(func=ast.Name("evaluate_condition", ast.Load()),
                args=[ast.Num(self.branch_num),
                ast.Str(node.ops[0].__class__.__name__), node.left,
                node.comparators[0]], keywords=[], starargs=None, kwargs=None)
```

Fig. 11.6   The BranchTransformer class instruments all the conditions in each predicate

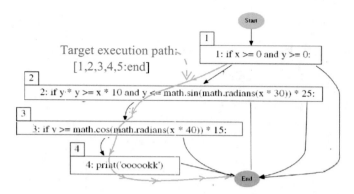

Fig. 11.7   A sample function and its corresponding instrumented function

into a call to a function called evaluate_condition. For instance, in Fig. 11.6 the condition:

$$y*y >= x*10$$

is transformed to:

$$evaluate\_condition(3, \textbf{'GtE'}, y*y, x*10)$$

Figure 11.7 represents a function testfuc() and its corresponding instrumented source code:

(a)   Original code

```
1.   def testfunc(x, y):
2.       if x >= 0 and y >= 0:
3.           if y*y >= x*10  and y <= math.sin(math.radians(x*30))*25:
4.               if y >= math.cos(math.radians(x*40))*15:
5.                   print('oooookk')
```

(b)   Instrumented code

```
1. def testfunc_instrumented(x, y):
2.   if evaluate_condition(1, 'GtE', x, 0) and evaluate_condition(2, 'GtE', y, 0):
3.     if evaluate_condition(3, 'GtE', y * y, x * 10) and evaluate_condition(
4.       'LtE', y, math.sin(math.radians(x * 30)) * 25):
4.       if evaluate_condition(5, 'GtE', y, math.cos(math.radians(x * 40 )) * 15):
5.         print('oooookk')
```

(c)   Control flow graph
A predicate consists of one or more clauses/conditions as follows:

predicate: clause I predicate logical_operator clause;
clause: expression relational_operator expression I expression;

For instance, the following predicate contains two clauses:

$$(x \geq 0) \text{ and } (y \geq 0)$$

The predicate is instrumented as follows:

evaluate_condition(1, **'GtE'**, x, 0) and evaluate_condition(2, **'GtE'**, y, 0)

In this way, when executing the instrumented function, test-func_instrumented(), the function evaluate_condition() is invoked to compute the branch distance of the condition/clause passed to it as its input parameters.

11.3.2 Branch Distance

The fitness function, derived from our search objective, minimizes the distance from branches in a particular path defined by the user [20]. The evaluate_condition function implements Table 11.2 to compute the extent to which an individual (test data) should be altered to cover a branch condition, "lhs op rhs," where "lhs" and "rhs" are the operands of the relational operator "op."

In practice, a condition may evaluate to as either false or true. Hence, each condition has two distance estimates, indicating how close it is to true and how close it is to false. If the condition is true, the actual distance is 0; if the condition is false, then the false distance is 0. Table 11.2 shows how to calculate the distance for different types of comparison.

In Table 11.2, K is an optional constant that is often set to zero. When executing an instrumented code with new test data as a solution, all the branching conditions are executed by calling the evaluate_condition function, shown in Fig. 11.8. The

Table 11.2 Branch distance for different conditions

Predicate	Branch distance	
	If the true branch is covered	If the false branch is covered
Boolean	K	K
$a = b$	K	$abs(a - b)$
$a \neq b$	$abs(a - b)$	K
$a < b$	$(b - a)$	$(a - b) + K$
$a \leq b$	$(b - a) + K$	$(a - b)$
$a > b$	$(a - b)$	$(b - a) + K$
$a \geq b$	$(a - b) + K$	$(b - a)$
$a \vee b$	$min(d(a), d(b))$	
$a \wedge b$	$d(a) + d(b)$	
$\neg a$	*Negation propagated over a*	

function takes as input a unique id, num, that identifies a given condition, op that is a relational operator, and the two operands lhs and rhs.

The function calculates two distances for a given condition: The distance to the condition evaluating to true, distance_to_true, and the distance to the condition evaluating to false, distance_to_false. It is always true that one of the two outcomes is true, with a distance of 0. Depending on which distance is 0, the function returns true or false, as it replaces the original comparison. That means the example expression "if a >= f(b)*5" would be replaced by "if evaluate_condition(0, "GtE", a, 2 * f(b))" such that the arguments are evaluated once, and thus side-effects are handled properly.

The function computes the difference between the values of the left and right operands of the conditional operator to measure the distance to the condition evaluating true and false. For instance, if the condition is "a >= b" and "a" and "b" are respectively 10 and 90, then according to Table 11.2, the distance is "(b − a) = 80".

The evaluate_condition() function does not yet store the distances observed. The values should be stored somewhere to be accessed from the fitness function. The function store_distances, in Fig. 11.9, stores the distances to the conditions of interest in two global dictionaries, distances_true and distances_false.

Each condition has a unique identifier, condition_num. The true and false distances are stored in the corresponding dictionaries for the first time a particular condition is executed. However, a test may execute a condition multiple times in a loop. In such cases, the minimum distance is kept in the distances_true and distances_false dictionaries.

11.3.3 Main Function[1]

The function generate_test_data_set, implementing a genetic algorithm for generating test data, is shown in Fig. 10.11. Code written in Python is understandable even for novice programmers. That is why I prefer to start with the actual executable code rather than the pseudocode (Fig. 11.10).

The main body of the genetic algorithm to generate test data, shown in Fig. 10.11, comprises the following steps:

1 Create initial population
2 Evaluate the population fitnesses
3 While (Termination condition not satisfied)

    3.1 Generate a new population of individuals.
        3.1.1 Use the tournament operator to select two candidate individuals.
        3.1.2 Crossover the selected individuals (chromosomes) with the crossover probability to generate two offspring.
        3.1.3 Mutate the generated offspring with the mutation probability.

---

[1] Download the source code from: https://doi.org/10.6084/m9.figshare.21323295

```
# evaluates the condition (lhs op rhs), e.g. (2*sin(x) <= 3*cos(x*y)
def evaluate_condition(num, op, lhs, rhs):
    # Two global variables are set to zero
    distance_to_true = 0
    distance_to_false = 0

    if isinstance(lhs, str):
        lhs = ord(lhs)
    if isinstance(rhs, str):
        rhs = ord(rhs)

    if op == "Eq":
        if lhs == rhs:
            distance_to_false = 1
        else:
            distance_to_true = abs(lhs - rhs)

    elif op == "NotEq":
        if lhs != rhs:
            distance_to_false = abs(lhs - rhs)
        else:
            distance_to_true = 1

    elif op == "Lt":
        if lhs < rhs:
            distance_to_false = rhs - lhs
        else:
            distance_to_true = lhs - rhs + 1
    elif op == "LtE":
        if lhs <= rhs:
            distance_to_false = rhs - lhs + 1
        else:
            distance_to_true = lhs - rhs

    elif op == "Gt":
        if lhs > rhs:
            distance_to_false = lhs - rhs
        else:
            distance_to_true = rhs - lhs + 1

    elif op == "GtE":
        if lhs >= rhs:
            distance_to_false = lhs - rhs + 1
        else:
            distance_to_true = rhs - lhs
```

Fig. 11.8 A helper function to compute distances for a given condition

```
      elif op == "In":
          minimum = sys.maxsize
              for elem in rhs.keys():
                  distance = abs(lhs - ord(elem))
                  if distance < minimum:
                      minimum = distance

              distance_true = minimum
              if distance_to_true == 0:
                  distance_to_false = 1

          store_distances(num, distance_to_true, distance_to_false)

          if distance_to_true == 0:
              return True
          else:
              return False
```

Fig. 11.8 (continued)

Fig. 11.9 A function to save
the distances to true and false
for the desired conditions

```
def store_distances(condition_num, to_true, to_false):
    global distances_true, distances_false

    if condition_num in distances_true.keys():
        distances_true[condition_num] = min(
            distances_true[condition_num], to_true)
    else:
        distances_true[condition_num] = to_true

    if condition_num in distances_false.keys():
        distances_false[condition_num] = min(
            distances_false[condition_num], to_false)
    else:
        distances_false[condition_num] = to_false
```

3.2   Replace population with new population
3.3   Evaluate the population
3.4   Apply the elitism strategy to reduce genetic drift

Step 1 generates the initial population of randomly selected test data. The initial population is the subject of Sect. 11.3.4. In step 2, the function evaluate_population() is invoked to compute the fitness of each chromosome in the genetic population generated randomly. The fitness function is described in Sect. 11.3.7. In step 3, the evolutionary process repeats as far as the termination condition is satisfied. The termination condition is the subject of Sect. 11.3.5. Step 3.1 creates a new generation by

```python
def generate_test_data_set(npop, ngen, minn, maxx):
    generation = 0

    # 1. Initialize population
    generation = 0
    print('Initialize population...')
    # population = list of chromosomes where each chromosome represents a test
    # data
    population = create_population(npop, minn, maxx)
    print("Initialized domain_generated successfully. \n")
    #  chromosome[0] = test data, chromosome[1] = fitness, population[i] = the ith
    # chromosome

    # 2. Evaluate population
    fitness = evaluate_population(population)
    best = min(fitness, key=lambda item: item[1])
    best_individual = best[0]
    best_fitness = best[1]
    print("Best fitness of initial population: {a} - {b} ".format(a=best_individual,
                                                                   b=best_fitness))

    # 3. While (Termination conditions not satisfied )
    while generation < ngen:
        new_population = []
        # 3.1 Generate new population
        while len(new_population) < len(population):
            # Selection
            offspring1 = selection(fitness, 10)
            offspring2 = selection(fitness, 10)
            # Crossover
            offspring1 = offspring1[0]
            offspring2 = offspring2[0]
            if random.random() < 0.7:
                (offspring1, offspring2) = \
                    crossover(offspring1, offspring2)
            # Mutation
            offspring1 = mutate(offspring1, minn, maxx)
            offspring2 = mutate(offspring2, minn, maxx)

            new_population.append(offspring1)
            new_population.append(offspring2)
        # 3.2 Replace population with new population
        generation += 1
        population = new_population
```

Fig. 11.10   The main body of the genetic algorithm

```
# 3.3 Evaluate the population
fitness = evaluate_population(population)
print(fitness)
for i in fitness:
    if i[1] == 0:
        best_of_generations.append(i[0])

best = min(fitness, key=lambda item: item[1])
best_individual = best[0]
best_fitness = best[1]
# best_of_generations.append(best[0])

print("Best fitness at generation {a}: {b} - {c}".format(a=generation,
                            b=best_individual, c=best_fitness))
print("Best individual: {a}, fitness {b}".format(a=best_individual,
                            b=best_fitness))
```

Fig. 11.10   (continued)

applying the genetic operator's selection, crossover, and mutation. Genetic operators are described in Sect. 11.3.6.

11.3.4   Initial Population

Genetic algorithms begin by generating a random population or initial generation of possible solutions to the problem. Population members (individuals) are known as chromosomes (phenotypes) and are represented as solutions to the problem at hand.

When generating a genetic population, care should be taken to maintain the diversity of the population; otherwise, it may lead to premature convergence [21]. An important observation is that the diversity is most probably maintained if the initial population is generated randomly rather than by heuristics. The following function generates the initial population of test data, each considered as a solution at random:

```
def create_population(size, minn, maxx):
    return [[random.uniform(minn, maxx), random.uniform(minn, maxx)]
            for i in range(size)]
```

For instance, consider the function F(int x, int y), where the input parameters x and y are in the range [1.0.100]. The input space contains $100*100 = 10,000$ chromosomes, where each chromosome is a tuple (x,y). The above function creates a population of input variables x and y in the range [minn = 1, maxx = 100].

11.3.5   Termination Condition

The termination condition of a genetic algorithm determines when the algorithm's run ends. Generally, genetic algorithms progress rapidly in the initial stages, with better solutions emerging every few iterations, but this slows down in the later stages, where improvements are marginal. Usually, a termination condition ending the run with optimal solutions is desirable [22].

In most cases, the evolution stops when:

- the fitness of the best individual in the population reaches some limit,
- there are no improvements in the population for some iterations,
- a specified number of generations have been created,
- a maximum execution time limit is reached, or
- the fitness of all the individuals in the genetic population is almost the same value.

For instance, a counter may record the generations where there has been no progress in the population in a genetic algorithm. Initially, this counter is set to zero. The counter is increased by the number of offspring whose fitness is less than other individuals. It is reset to zero if an offspring's fitness is greater than at least one individual in the population.

Like other genetic parameters, the termination condition is also highly problem-specific, and the G.A. designer should try different options to find the most appropriate one.

11.3.6   Genetic Operators

Mutation, crossover, and selection are the three leading genetic operators that work in conjunction to succeed in searching the input space for a test data set with acceptable coverage.

The two cornerstones of problem-solving by the search are exploration and exploitation. Evolutionary algorithms explore the search space for new solutions in new regions by using (genetic) crossover and mutation operators, while exploitation is carried out by selecting appropriate material in the current set. An evolutionary search algorithm should explore the problem space through crossover and mutation, but it should do so by preferencing solutions near other reasonable ones. Finding the right balance is always a challenge. Excessive exploitation often ends up with getting trapped in local maxima and going too far into exploration, wasting time on solutions that are less likely suitable and ignoring the information already gathered.

– **Selection operator**

The natural selection principle of the evolutionary theory suggests the survival of the fittest and the elimination of less fit [23]. Accordingly, the selection operator

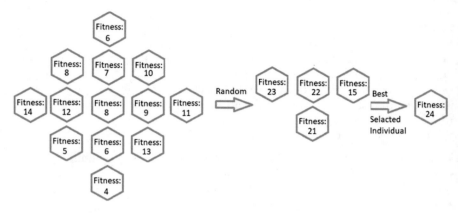

Fig. 11.11 Shows the procedure for tournament selection

selects and matches the fittest individuals (chromosomes) to breed offspring for an upcoming generation.

There are several genetic selection operators, amongst which an operator called Tournomount is handy [24]. Due to its efficiency and simple implementation, Tournament selection is doubtlessly one of the most popular selection operators [25]. The Tournament initially selects n individuals randomly from the genetic population and compares their fitness. The individual with the highest fitness wins and is selected. The size of a Tournament refers to the number of participants Fig. 11.11 illustrates the mechanism of tournament selection.

The time complexity is low susceptibility to take over by dominant individuals and no requirement for fitness scaling or sorting [25]. In general, individuals with high fitness scores win the match. The following code implements the Tournament operator.

```
def selection(self, evaluated_population, tournament_size):
    competition = random.sample(evaluated_population, tournament_size)
    winner = min(competition, key=lambda item: item[1])
    return winner[:]
```

There are different techniques to implement selection in Genetic Algorithms [7]. They are [26]:

1   Roulette wheel selection
2   Tournament selection
3   Rank selection
4   Proportionate selection
5   Steady-State Selection
6   Boltzmann Selection
7   Stochastic Universal Sampling
8   Linear Rank Selection

9   Exponential Rank Selection
10   Truncation Selection.

– **Crossover operator.**

The selection operator selects the fittest individuals for mating [27]. New offspring is generated by applying the crossover operator to pairs of selected individuals. The operator combines the characteristics of two selected individuals (parents) to generate a new solution (offspring).

The simplest way to do crossover is to randomly choose a crossover point, swap everything before the chosen point between the two parents, and then swap everything after the crossover point. The following function selects a crossover point, pos, at random and swaps the tails of the two parents, parnt1 and parnt2, to get new offspring.

```
def crossover(parent1, parent2):
    pos = random.randint(1, len(parent1))
    offspring1 = parent1[:pos] + parent2[pos:]
    offspring2 = parent2[:pos] + parent1[pos:]
    return offspring1, offspring2
```

Genes contain instructions for building and maintaining cells and passing genetic traits to offspring. The process of mating between two individual parents to swap genes is called crossover [28]. This operator is inspired by how the genetic code of one individual is inherited by its descendants in nature.

– **Mutation operator.**

Natural selection is known as the leading cause of evolution. Mutation speeds up evolution by promoting diversity in the population. Mutation operators are typically regarded as random changes in individual chromosomes, so they can escape from being trapped in local optima. Over time, these mutations become a source of lasting diversity. Below, the mutation operator is applied to newly generated individuals with a probability of mutation p_mutation.

```
def mutate(chromosome, minn, maxx):
    mutated = chromosome[:]
    p_mutation = 1.0 / len(mutated)
    for pos in range(len(mutated)):
        if random.random() < p_mutation:
            mutated[pos] = random.uniform(minn, maxx)
    return mutated
```

In fact, the goal of the mutation operator is to provide more diversity and, consequently, more exploration and exploitation. The most crucial issue regarding evolutionary algorithms is establishing a trade-off between exploration and exploitation. Exploration or reliability helps the algorithm explore the problem space to find promising regions, while exploitations enhance finding the optimum in that region.

11.3.7   Fitness Functions

Search-based software testing uses metaheuristic search techniques to look for test data sets, satisfying a test adequacy criterion. A test adequacy criterion differentiates a good test set from a bad one. Test adequacy criteria are transformed into fitness functions for the metaheuristic search techniques. Fitness functions use test adequacy criteria as test requirements to evaluate the fitness of solutions generated during a metaheuristic search. Coverage criteria are adequacy measures often used to determine if a test objective is met when a test case is executed on the software under test. Test data sets are evaluated on their capability to detect faults.

When using a genetic algorithm to search for test data, the objective of the search could be to provide test data covering certain regions of the code under test. The objective is defined and implemented as a fitness function. The fitness function guides the search in the input space. For instance, in the example given in Fig. 11.12, the objective of test data generation is to cover a particular execution path, including branch numbers 1 to 4. The extent to which a given test data cover the path is the sum of the branch distances for these branches.

However, a potential problem with adding up branch distances is that if one condition relies on very large values and another relies on small values, then an improvement of the large values would likely improve fitness, thus biasing the search. The problem is resolved by applying the following normalization to the branches before the sum-up:

```
def normalize(a_branch_distance):
    return a_branch_distance / (1.0 + a_branch_distance)
```

The get_fitness function, in Fig. 11.13, takes a solution, i.e., test data, as input and returns a value indicating how far the solution is from covering the desired execution path. The function runs the instrumented function under test with the input data,

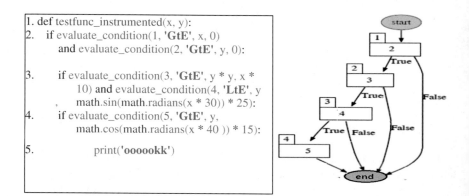

Fig. 11.12 An instrumented code and its control flow graph

x and y, it receives as input. It then adds up the normalized values of the relevant branches. If the test data do not cover the desired condition, its normalized fitness is set to $+1$.

The selection operator prefers the fitter individuals. Therefore, the fitness value for each individual in the genetic population is computed. The evaluate_population() function computes the fitness for each solution. The function returns a list of *tuples*, where each tuple consists of a solution and its fitness value.

```
def evaluate_population(population):
    fitness = [get_fitness(x, y) for x, y in population]
    return list(zip(population, fitness))
```

```
def get_fitness(x, y):
    # Reset any distance values from previous executions
    global distances_true, distances_false
    distances_true = {}
    distances_false = {}
    # Run the function under the test
    try:
        testfunc_instrumented(x, y)
    except BaseException:
        pass
    # Sum up branch distances
    fitness = 0.0
    for branch in [1, 2, 3, 4, 5]:
        if branch in distances_true:
            # normalize(x) = x / (1.0 + x)
            fitness += normalize(distances_true[branch])
        else:
            fitness += 1.0

    for branch in []:
        if branch in distances_false:
            fitness += normalize(distances_false[branch])
        else:
            fitness += 1.0

    return fitness
```

Fig. 11.13 The fitness function executes the instrumented code with test data (x,y)

11.3.8 Execution Results

The genetic code described in this Section is, in fact, generates the test data covering the domain of inputs for a given Python function. Figure 11.14 shows the regions of input space covering the execution path [1–5] of the function test_func in Fig. 11.7a. The execution path is shown in Fig. 11.7c.

A part of the input space, including the region satisfying the desired path constraint, is shown in Fig. 11.14. All the shaded regions' points (x,y) satisfy the path constraint. In other words, when running test_func with any test data, represented as a point in the shaded region, the desired path, [1–5], shown in Fig. 11.7, will be examined.

Figure 11.14 shows that the path constraint's input domain is discontinuous, and the algorithm is not trapped in local optima.

11.4 MCMC Search Method

An efficient approach to look for suitable test data satisfying a particular coverage criterion is to use a Markov chain Monte Carlo (MCMC) sampling method. MCMC iteratively finds test data as samples with stable frequencies derived from a fixed proposal distribution by generating random walks in the input space.

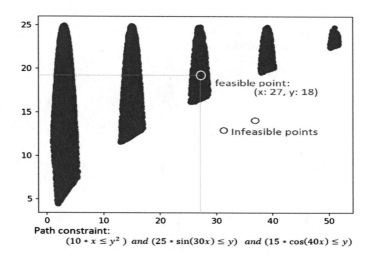

Fig. 11.14 The feasible region of inputs satisfying the path constraint

11.4.1 MCMC Sampling Method

In a Markov chain, each variable depends solely on the previous one, i.e., $P(\theta_k|\theta_1,\ldots,\theta_{k-1}) = P(\theta_k|\theta_{k-1})$ [29, 30]. Considering, as θ_{k-1} a point randomly selected in the input space, then the normalized distance of θ_{k-1} is the probability, $P(\theta_{k-1})$, of reaching the desired execution path from θ_{k-1}. As described above, the following normalization is applied to the distances:

$$\text{normalized}_d\text{istance}(\text{aPoint}) = \frac{\text{aPoint}}{1 + \text{aPoint}}$$

Therefore, the next point, θ_k, should be selected so that compared with θ_{k-1}, it is closer to the target execution path. If the distance of θ_k from the target is less than the θ_{k-1} distance, its normalized distance will be greater. In other words, $P(\theta_k) > P(\theta_{k-1})$. Hence, given the current location θ_{t-1} in the input space, a probability distribution, $J(\theta_t|\theta_{t-1})$ may define the probability distribution of the next location, θ_t, in the next time step.

The distribution $J(\theta^*|\theta)$ proposes the conditional probability of θ^* given the point θ and the acceptance distribution $A(\theta^*|\theta)$ is the probability of accepting the proposed point θ^* in the input space. The transition probability is written as:

$$P(\theta^*|\theta) = J(\theta^*|\theta)^* A(\theta^*|\theta)$$

Given that $P(\theta^*|\theta) * P(\theta) = P(\theta|\theta^*) * P(\theta^*)$ we have:

$$\frac{A(\theta^*, \theta)}{A(\theta, \theta^*)} = \frac{P(\theta^*|\theta) * J(\theta|\theta^*)}{J(\theta^*|\theta) * P(\theta|\theta^*)} = \frac{P(\theta^*|\theta) * J(\theta|\theta^*)}{P(\theta|\theta^*) * J(\theta^*|\theta)} = \frac{P(\theta^*) * J(\theta|\theta^*)}{P(\theta) * J(\theta^*|\theta)}$$

According to the Metropolis-hasting algorithm, an acceptance ratio that fulfills the condition above is:

$$A(\theta^*, \theta) = \min\left(1, \frac{P(\theta^*) * J(\theta|\theta^*)}{P(\theta) * J(\theta^*|\theta)}\right)$$

One of the most common and perhaps the simplest choices for the jumping/proposal distribution, $J(\theta^*|\theta)$ is:

$$J(\theta^*|\theta) = N(\theta, \sigma^2)$$

Data near the mean are more likely to occur than data far from the mean in a normal distribution, also known as the Gaussian distribution. Using a normal distribution $J(\theta^*|\theta)$ will be equal to $J(\theta|\theta^*)$ as a result of which:

$$A(\theta^*, \theta) = \min\left(1, \frac{P(\theta^*)}{P(\theta)}\right)$$

The ratio $\frac{P(\theta^*)}{P(\theta)}$ represents how much more likely the proposal θ^* is under the target distribution than the current sample θ.

$$\frac{p(\theta^*)}{p(\theta)} = \frac{normalised\_distance(\theta^*)}{normalised\_distance(\theta)}$$

11.4.2   Metropolis Hasting Algorithm[2]

Metropolis is a relatively simple yet widely used method to implement the Markov chain Monte Carlo (MCMC) method to obtain a sequence of random samples from a probability distribution from which direct sampling is difficult [31, 32]. An implementation of the Metropolis algorithm for generating test data is presented in Fig. 11.15. The main steps of the algorithm are as follows:

1   Initial configuration:
    Use a uniform distribution to select a point, $\theta 0$, randomly.
    Set time step $t = 0$.
    Set the normal distribution variance $\sigma = $ max_step_size_of_random_walk.
2   Iterate 4000 times:

$$t = t + 1$$

Use the normal distribution $N(\mu = \theta_{t-1}, \sigma)$ to select θ^*.
Calculate the acceptance probability $\alpha = A(\theta^*, \theta)$.
Use the uniform distribution to select $u \in [0,1]$ randomly.
Set $\theta_t = (\alpha >= u)? \theta^* : \theta_{t-1}$

The algorithm uses a random walk incremental strategy to find its way through the input space. In a random walk, each time increment in position is determined by the current position, but it is unrelated to all the past positions. A normal distribution is used to select the step size randomly. To illustrate, let the initial position in the input space be $\theta 0$. In the next time step the new position is $\theta 1 = \theta 0 + N(\theta_{t-1}, \sigma)$.

Fig. 11.15 presents a Python implementation of the Metropolis algorithm to search for the appropriate test data covering a specific execution path in the program under test. The Python function "metropolis" uses a normal distribution to calculate random walks based on the current point while simplifying calculating the probability of acceptance. Symmetrical distributions such as normal and uniform distributions make it easy to calculate random walks based on the current point while simplifying calculating the probability of acceptance [32]. It is noteworthy that if θ^* is more probable than θ, the samples are moved in a direction with a higher density distribution.

---

[2] Download the source code from: https://doi.org/10.6084/m9.figshare.21322812

```
def metropolis(sigma, bound):
    # initial config
    T = 400000
    boundary = bound
    # theta is an empty chain of 40000 cells
    theta = np.zeros([T,2])
    theta[0] = [random.uniform(boundary[0], boundary[1]), random.uni
                form(boundary[0], boundary[1])]
    # sampling
    t = 0
    answers = []
    while t+1 < T:
        t = t + 1
        # Executes the random walk by randomly selecting a point
        theta_star = np.random.normal(theta[t-1], sigma)
        print(theta_star)
        if get_fitness(theta[t-1][0], theta[t-1][1]) != 0:
            alpha = min(1, get_fitness(theta_star[0], theta_star[1])
                        /get_ fitness(theta[t-1][0], theta[t-1][1]))
      else:
            alpha = min(1, get_ fitness(theta_star[0], theta_star[1])/1)
    # print(alpha)
    u = random.uniform(0, 1)
    if u > alpha:
      theta[t] = theta_star
      if get_ fitness(theta_star[0], theta_star[1]) == 0:
        answers.append(theta_star)
    else:
        theta[t] = theta[t-1]
ans = list()
x= list()
y = list()
for i in answers:
    x.append(i[0])
    y.append(i[1])
    ans.append(get_ fitness(i[0], i[1]))
plt.plot(x, y, 'bo')
plt.show()
```

Fig. 11.15 Using random walks to select points in a program input space

In the main body of the program, first, the function under test, testfunc, is instrumented, and then the metropolis function is invoked to generate test data covering a
specific execution path of the function:

```
# the main body of the metropolis test data generator

instrument_input_file(testfunc)
metropolis(sigma = 10, bound = [0,100])
```

11.4.3 Examples

The function testfuc and its instrumented version are shown in Fig. 11.7. The program
output is shown in Fig. 11.16. The figure illustrates the program space. The shaded
regions in Fig. 11.16 represent all the points covering the desired execution path
specified in the get_fitness function.

As a second example, consider the function, test_func, and its corresponding
control flow graph in Fig. 11.7.

Fig. 11.16 The Metropolis program execution result and the CFG of the function under test

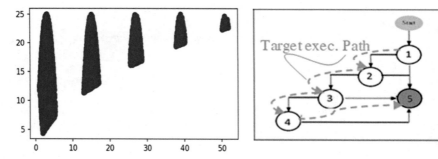

```
def testfunc(x, y):
    if (y - 0.75*abs(x))**2 +
        (0.75 * x)**2 < 1:
        print("ook")
```
(a) Function under test

```
def testfunc_instrumented(x, y):
    if evaluate_condition(1, 'Lt',
        (y - 0.75 * abs(x)) ** 2 +
        (0.75 * x) **2, 1):
        print('ook')
```
(c) Instrumented code

(b) Control flow graph

Fig. 11.17 A sample Python function

Fig. 11.18 The heart-shaped region represents all the points (x,y), satisfying the path constraint

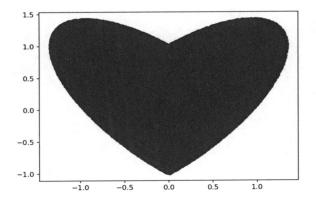

The output of the metropolis test data generator is shown in Fig. 11.18. All the points in the heart-shaped region cover the execution path < start → 1 → 2 → end >. The metropolis function is invoked as follows:

metropolis(sigma = 1, bound = [0,1])

Where 'sigma' is the covariance of the normal distribution, determining the random walks in the input space, and the parameter 'bound' restricts the search space.

11.5 Importance of Domain Coverage

Following the DJkestra argument that testing cannot guarantee the absence of bugs, test adequacy criteria emerged. Nowadays, software test adequacy criteria are considered rules to determine the effectiveness of a test dataset to uncover any possible fault in software under test. For instance, complete path coverage is the most decisive criterion subsuming all the other structural coverage criteria. A test dataset is adequate for the complete path coverage criterion if it includes test data covering each possible combination of the true and false values of the branching conditions in the program under test at least once. For instance, consider the function "foo" in Fig. 11.19. There is a latent fault in line 4 of the "foo" function. The condition "x + y < 40" in line 4 of the code is supposed to be "x + y < 30". This sort of fault may not reveal unless the value of x and y is in the specific domain of "$30 \leq x + y < 40$" .

There are three branching conditions in statements 3, 4, and 8. Therefore, at least 6 test data are required to cover all the execution paths. The execution paths are six sequences of basic block numbers as follows:

Path 1 : < 1, 2, 3, 6, 7, 9 >   Path 5 : < 1, 5, 6, 7, 9 >

Path 2 : < 1, 2, 3, 6, 8, 9 >   Path 6 : < 1, 5, 6, 8, 9 >

```
1.      void foo(int x, int y){
2.        w = 0;
3.        if(x > y)
4.          if(x+y<40) //should be:(x+y<30)
5.            W+ = 1;
6.          else  w+ = 2;
7.        else w+ = 3;
8.        if(x>2*y)
9.          w += 5;
10.       else w = 4;
11.       printf("w = %d", w);
12.     }
```

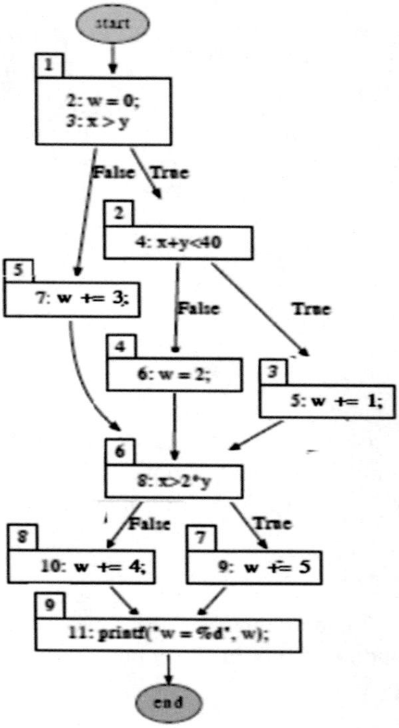

Fig. 11.19 Statement 4 should be if(x + y < 30)

Path 3 : < 1, 2, 4, 6, 7, 9 >
Path 4 : < 1, 2, 4, 6, 8, 9 >

Therefore, by generating six influential test cases, each covering a different execution path, the path coverage will be 100%. However, considering the faulty statement in line 4, the fault reveals only when $30 \leq x + y < 40$. The region of input space that reveals the fault in line 4 of the function "foo" is shown in Fig. 11.20.

Any test data (x,y) such that x > y examines the faulty statement at line 4. However, the fault is uncovered only if "$30 \leq x + y < 40$". Therefore, instead of solely generating a single data covering a test requirement, we are better off computing the domain of inputs satisfying the test requirement.

The problem is that 100% coverage does not mean any bugs. The solution is to modify test data generation algorithms to provide the boundaries rather than the actual test data, satisfying one or more test requirements. The coverage criteria shown in Fig. 11.21 suggest testing the software structure, e.g., branch or path, at least once. This chapter offers search-based genetic algorithms to look for the domain of inputs satisfying the test requirements.

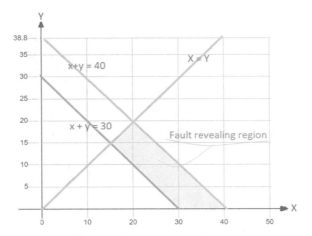

Fig. 11.20 The input space for the function foo in Fig. 11.19

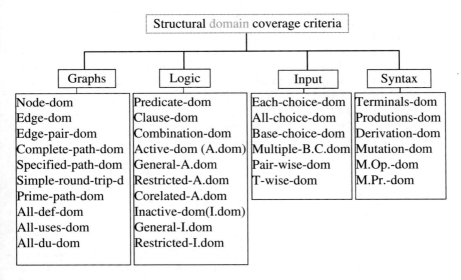

Fig. 11.21 Coverage criteria for the four structure

Structural coverage criteria are used to evaluate the adequacy of test suites when generating test data for software under test. Structural coverage criteria [33] are adequacy measures to qualify if a test objective is met when executing test cases on software under test.

The main reason for the test adequacy criteria is to ensure no faults in the software under test. The adequacy of the test is critical in the case of safety–critical software systems. However, it is observed that even full path coverage does not guarantee fault-free code.

Fig. 11.22 The domain of
inputs covering the path <
1,2,4,6,8,9 >

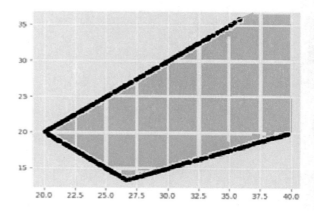

Computing the domain of inputs instead of solely generating test data, satisfying a test requirement, ensures the adequacy of tests. Using a domain partitioning technique facilitates the search for any possible latent faults. For instance, assume the input domain for the function foo, in Fig. 11.19, is $0 \leq x,y \leq 40$. The input domain for the path < 1,2,4,6,8,9 > is illustrated in Fig. 11.22.

Rapid access to the feasible test data satisfying the test requirement is facilitated by providing a decision tree shown in Fig. 11.23. The implementation of the genetic algorithm determining the boundaries of the domain of inputs examining a given path is described in Sect. 12.5.5 (Fig. 11.23).

Figure 6.5 illustrates the domains of the values of the parameters x and y of the function foo. The labels one and zero in the leaves indicate acceptable and unacceptable values.

11.6   Exercises

(1) Use the functions presented in this chapter to write a genetic program in Python to generate test data for the functions and methods with any number of input parameters. Generate coincidentally correct test cases for a given Python function.

(2) Large population size can enhance exploration which influences the ability to search for an optimum solution in the input space if the search space is not small. Use an adaptive approach to determine the population's size dependent on individuals' fitness [30].

(3) Use the Roulette wheel selection operator instead of the Tournament and compare the impact on the test data generation algorithm's performance. Roulette-wheel is the most commonly used selection operator. The probability. P(i), of selecting a chromosome, I, with the roulette wheel operator, is:

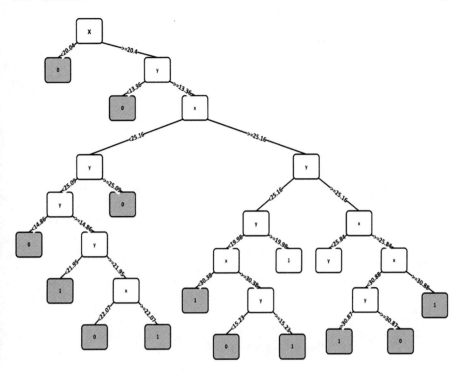

Fig. 11.23 Decision tree domain representing the inputs for x and y

$$P(i) = \frac{F(i)}{\sum_{j=1}^{N} F(j)}$$

where N is the size of the genetic population and F indicates the fitness.

(4) Add a gender field to each individual in the genetic population and restrict the crossover operator to heterosexual individuals. Notice that the mutation rate of male individuals is higher than that of females. Evaluate these modifications' impact on the genetic algorithm's speed and accuracy.

(5) How can the fitness function be modified to conform to this saying by Dave E. Smalley that a fitter individual is the one who is prepared for adaptation, one who can accept and conform to the inevitable, one who can adapt to a changing environment?

(6) Do experiments on genetic differences as the basis of evolution by modifying the selection operator to select the most different rather than fittest individuals.

(7) Complete and run the metropolis program to generate test data. Compare the performance of the metropolis program with that of the genetic. Try to use different distributions instead of the normal distribution to get better results.

(8) Write a program to generate test data covering the boundaries of a given execution path. Table. 11.2 in Sect. 11.2. above is used to compute the fitness of test

data, T_i, covering the path $P^i = (pr_1^i, pr_2^i, ..., pr_n^i)$ as the some of the distances of its predicates pr_k^i from a target execution path [34]. The branch distance for pr_k^i predicate is calculated by by applying the rules defined in Table. 11.2.

For example, for the predicate "$x > 20$," if the value of x for a test case is 15, the branch distance to the true branch distance is.

$$\text{Branch\_distance}(x = 15) = 20--15 = 5 \, (\text{unacceptable}).$$

If the value of x is 26:

$$\text{Branch\_distance}(x = 26) = 20--26 = -6 \, (\text{acceptable}).$$

Therefore, in the case of "$a > b$," the branch distance should be less than or equal to zero. However, when looking for the boundary points, the absolute value of the branch distance should be δ where δ is a tiny number, considered as the width of the boundary. The following table shows how to compute the branch distance in different cases. In this table, K is a constant value set by the user.

Predicate	BD Computation
Boolean	if true, then 0 else K
$a > b$	if $b - a < 0$, then 0 else $(b - a) + K$
$a \geq b$	if $b - a \leq 0$, then 0 else $(b - a) + K$
$a < b$	if $a - b < 0$, then 0 lese $(a - b) + K$
$a \leq b$	if $a - b \leq 0$, then 0 else $(a - b) + K$
$a = b$	if $|a - b| = 0$, then 0 else abs $(a - b) + K$
$a \wedge b$	$BD(a) + BD(b)$
$a \vee b$	Min $[BD(A), BD(b)]$

When looking for all the test cases covering the boundaries for the domain of inputs covering a particular execution path:
$$\forall \in \delta.$$

$$pr_k^i P^i \Rightarrow \sum_{k=1}^{n} |\text{Branch\_distance}(pr_k^i)| \leq \delta$$

(9)   Consider the following Python function:

```python
def equation8 (x: float, y: float):
    k = math.exp (x + y) - 1
    h = math.sin ((x)) + 1.9 * math.cos ((y)) + 1
    if k < 0:
        if h <= 0:
            return True
```

Suppose the domain of input variables x and y is $[-5,5]$.

1 Instrument the code,
2 Generate test data to cover the boundaries of the execution path in which all the conditions, i.e., "k < 0" and "h < = 0," are true.
3 Plot the sub-domain of the input domain covering the execution path. It should be similar to the following:

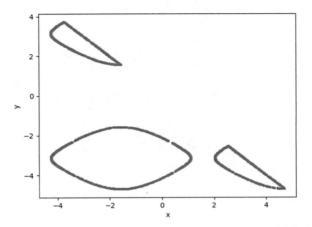

(10) It is claimed that GA sometimes traps in a local optimum or even arbitrary points rather than the global optimum of the problem. Mutation operators are used in GA to bypass local optima. However, the use of mutations slows down the performance by randomly mutating the solutions. Trapping in local optima can be avoided by applying a local search, such as hill climbing in the neighboring particles/points in the search space. Modify your genetic algorithm to use the hill climbing function instead of the mutation operator. Use the Multiswarm algorithm to complete your code. The idea behind this algorithm was inspired by a flock of birds' social, attacking behavior to search for food.

Always Follow the
experienced and enthusiastic
to reach your goal

References

1. Hassan, A., Almohammadi, K., Alkafaween, E., Hammouri, E.A.A., Surya Prasath, V.B.:
   Choosing mutation and crossover ratios for genetic algorithms—a review with a new dynamic
   approach. Information **10**, 390 (2019). doi:https://doi.org/10.3390/info10120390
2. Renders, J., Flasse, S.P.: Hybrid methods using genetic algorithms for global optimization.
   IEEE Trans. Syst. Man, Cybern. B Cybern. **26**, 243–258 (1996)
3. Vavak, F., Fogarty, C.: Comparison of steady-state and generational genetic algorithms for
   use in nonstationary environments. In: Proceedings of the IEEE International Conference on
   Evolutionary Computation, Nagoya, Japan, pp. 192–195 (1996)
4. Chaiyaratana, N., Zalzala, A.: Hybridisation of neural networks and genetic algorithms for time-
   optimal control. In: Proceedings of the 1999 Congress on Evolutionary Computation-CEC99
   (Cat. No. 99TH8406), Washington, DC, USA (1999)
5. Man, F.K., Tang, S.K., Kwong, S.: Genetic Algorithms: Concepts and Designs. Springer,
   Berlin/Heidelberg, Germany (1999)
6. Ammar, H.H., Tao, Y.: Fingerprint registration using genetic algorithms. In: Proceedings of the
   3rd IEEE Symposium on Application-Specific Systems and Software Engineering Technology,
   Richardson, TX, USA, pp. 148–154 (2000)
7. Jareanpon, C., Pensuwon, W., Frank, R.J., Dav, N.: An adaptive RBF network is optimized using
   a genetic algorithm applied to rainfall forecasting. In: Proceedings of the IEEE International
   Symposium on Communications and Information Technology, Sapporo, Japan, pp. 1005–1010
   (2004)
8. Meng, X., Song, B.: Fast genetic algorithms used for PID parameter optimization. In: Proceedings of the IEEE International Conference on Automation and Logistics, Jinan, China,
   pp. 2144–2148 (2007)

9. Krawiec, K.: Generative learning of visual concept using multi-objective genetic programming. Pattern Recognit. Lett. **28**, 2385–2400 (2007)
10. Hamdan, M.: A heterogeneous framework for the global parallelization of genetic algorithms. Int. Arab J. Inf. Technol. **5**, 192–199 (2008)
11. Liu, J.: Application of fuzzy neural networks based on genetic algorithms in the integrated navigation system. In: Proceedings of the IEEE Conference on Intelligent Computation Technology and Automation, Changsha, China, pp. 656–659 (2009)
12. Ka, Y.W., Chi, L.: Positioning weather systems from remote sensing data using genetic algorithms. In: Computational Intelligence for Remote Sensing, pp. 217–243. Springer, Berlin/Heidelberg, Germany (2008)
13. Belevi˘cius, R., Ivanikovas, S., Šešok, D., Valenti, S.: Optimal placement of piles in real grillages: Experimental comparison of optimization algorithms. Inf. Technol. Control **40**, 123–132 (2011)
14. Ai, J., Feng-Wen, H.: Methods for optimizing weights of wavelet neural network based on adaptive annealing genetic algorithms. In: Proceedings of the IEEE International Conference on Industrial Engineering and Engineering Management, Beijing, China, pp. 1744–1748 (2009)
15. Sorsa, A., Peltokangas, R., Leiviska, K.: Real coded genetic algorithms and nonlinear parameter identification. In: Proceedings of the IEEE International Conference on Intelligent Systems, Varna, Bulgaria, pp. 42–47 (2009)
16. Krömer, P., Platoš, J., Snášel, V.: Modeling permutation for genetic algorithms. In: Proceedings of the IEEE International Conference on Soft Computing Pattern Recognition, Malacca, Malaysia, pp. 100–105 (2009)
17. C. An adaptive genetic algorithm based on population diversity strategy. In: Proceedings of the IEEE International Conference on Genetic and Evolutionary Computing, Guilin, China, pp. 93–96 (2009)
18. Lizhe, Y., Bo, X., Xiangjie, W. B.P.: Network model optimized by adaptive genetic algorithms and the application on quality evaluation for class. In: Proceedings of the IEEE International Conference on Future Computer and Communication, Wuhan, China, p. 273 (2010)
19. http://www.macfreek.nl/memory/Trace_in_Python
20. https://www.educative.io/edpresso/what-is-astnodetransformer-in-python
21. Pachauri, A.: Program test data generation for branch coverage with genetic algorithm: comparative evaluation of a maximization and minimization approach. The First International Conference on Information Technology Convergence and Services (2010)
22. Toğan, V., Daloğlu, A.T.: An improved genetic algorithm with initial population strategy and self-adaptive member grouping. Comput. Struct. **86**, 1204–1218 (2008)
23. Ghoreishi S., Clausen A., Joergensen B.: Termination criteria in evolutionary algorithms: a survey. In: Proceedings of the 9th International Joint Conference on Computational Intelligence (IJCCI 2017), pp. 373–384 (2017)
24. Goh, K.S., Lim, A., Rodrigues, B.: Sexual Selection for Genetic Algorithms, Artificial Intelligence Review, Vol. 19,pp. 123–152. Kluwer Academic Publishers (2003)
25. Gangadevi, E.: Study of various selection operators in Genetic Algorithm. Int. J. Res. Sci. Eng. (2018)
26. Goldberg, D.E., Deb, K.: A comparative analysis of selection schemes used in genetic algorithms. In: Rawlins, G.J.E. (ed.) Foundations of Genetic Algorithms, pp. 69–93. Morgan Kaufmann, Los Altos (1991)
27. Blickle, T., Thiele, L.: A Comparison of Selection Schemes used in Genetic Algorithms. TIK-Report, Zurich (1995)
28. Poon, P.W., Carter, J.N.: Genetic algorithm crossover operators for ordering applications. Comput. Oper. Res. **22**(1), 135–147 (1995)
29. Liu, Y., Zhou, A., Zhang, H.: Termination detection strategies in evolutionary algorithms: a survey. In: Proceedings of the Genetic and Evolutionary Computation Conference (2018)
30. Graham, C.: Markov Chains: Analytic and Monte Carlo Computations. Wiley Series in Probability and Statistics, 1st Ed. (2014)
31. https://storopoli.io/Bayesian-Julia/pages/5_MCMC/

32. Robert, C., Casella, G.: Introducing Monte Carlo Methods with R. Springer (2010). https://tcb egley.com/blog/mcmc-part-1
33. Rajakumara, B.R., George, A.: APOGA: an adaptive population pool size based genetic algorithm. AASRI Proc **4**, 288–296 (2013)
34. Gotlieb, A., Petit, M.: A uniform random test data generator for path testing. J. Syst. Softw. **83**(12), 2618–2626 (2010)

Chapter 12
Testing Unknowns

12.1 Introduction

Learning physical principles marks the start of a new era for software testing in cyber-physical systems. Test cases labeled with the reactions of physical systems to the tests could serve as learning samples to model physical principles. Cyber-physical systems should be equipped with test engines to discover principles governing the behaviors of the physical system they control. Behaviors change drastically at the boundaries. Behavioral coverage evaluates the test suite covering expected and unexpected behavioral domains as a test adequacy metric. Test data sets for evaluating a cyber-physical system (CPS), labeled with the reactions of the CPS to tests, could serve as a dataset to learn the known/unknown rules governing the behaviors of the physical systems.

This chapter uses two CPS scenarios, the pedestrian avoidance problem in CARLA autonomous car simulator and autopilot simulator GCAS for F16 aircraft, to demonstrate the application of software testing to reveal expected/unexpected behavioral regions of a program input space. Machine learning techniques are used to determine the rules governing the CPSs behavior.

© The Author(s), under exclusive license to Springer Nature Switzerland AG 2023
S. Parsa, *Software Testing Automation*,
https://doi.org/10.1007/978-3-031-22057-9_12

12.2 Behavior Coverage

Reactions to events form behaviors. Accordingly, I define the behavior coverage of
a test suite as the percentage of the influential events and responses of the program
to the events that the test suite exercises. Using test cases initiating cyber-physical
system activities as learning samples and the reactions of the physical system to the
activities as the labels of the samples, physical rules governing the behavior of the
physical system may be extracted. For instance, CARLA is an Urban driving simulator, providing 3-D environments and valuable tools to easily simulate sensorimotor
control systems in scenarios with complex multi-agent dynamics [2].

Table 12.1 represents the physical rules extracted when testing a CARLA ego
vehicle controller with test cases generated during an evolutionary process, described
in Sect. 12.3.2. The rules describe the unexpected behavior of the vehicle colliding
with a pedestrian crossing the road. It is worth mentioning that the CARLA system
rejects unacceptable test case scenarios. The CARLA scenario_runner.py module
validates and rejects "abnormal" initial states.

As shown in Table 12.1, despite CARLA filtration of invalid initiating samples, test
case samples conducting the autonomous vehicle to crash into a passing pedestrian
have been generated. The genetic algorithm finding I/O intensive paths is described
in Sect. 12.3.1. As shown in Table 12.1, as the number of inputs/outputs increases,

Table 12.1 Rules extracted while searching for unexpected behavior

# I/Os	# C.R.s	**Rules Governing Car Collision with Pedestrians** (Four of the rules extracted by the decision trees model)
8	25	wind_intensity > 43.8146 $\wedge$ sun_altitude_angle $\leq$ -24.4492 $\wedge$ start_distance $\leq$ 12.4088
		walker_yaw > 296.2417 $\wedge$ sun_altitude_angle $\leq$ -21.9091 $\wedge$ start_distance $\leq$ 12.4088
		walker_target_velocity > 1.4312 $\wedge$ walker_yaw > 296.2417 $\wedge$ wind_intensity $\leq$ 66.7154
		walker_target_velocity $\leq$ 3.3770 $\wedge$ friction $\leq$ 0.7842 $\wedge$ x $\leq$ -54.1156
26	14	walker_target_velocity $\leq$ 3.3303 $\wedge$ walker_target_velocity > 1.7494 $\wedge$ start_distance $\leq$ 13.7205
		walker_target_velocity $\leq$ 3.3303 $\wedge$ walker_target_velocity > 1.4602 $\wedge$ walker_yaw $\leq$ 311.3634
		walker_target_velocity $\leq$ 3.2471 $\wedge$ walker_target_velocity > 1.2562 $\wedge$ start_distance $\leq$ 13.6638
		walker_target_velocity $\leq$ 3.2225 $\wedge$ walker_target_velocity > 1.2562 $\wedge$ walker_yaw $\leq$ 313.1714
33	21	walker_target_velocity $\leq$ 3.2683 $\wedge$ x > -96.2315 $\wedge$ start_distance > 7.8806
		walker_target_velocity $\leq$ 3.2683 $\wedge$ friction $\leq$ 0.7881 $\wedge$ x > -96.2315
		walker_target_velocity $\leq$ 3.2690 $\wedge$ x > -96.4950 $\wedge$ start_distance > 7.0573
		walker_target_velocity $\leq$ 3.2690 and y $\leq$ -105.4253 $\wedge$ yaw > 11.6392

the number of critical regions detected by the NSGA II algorithm described in Sect. 12.4.2 increases.

Behaviors are sensitive to boundary conditions. Boundary conditions are divorced from the scientific and mathematical domain in which they properly reside. They are statements about particular events observable through the system reactions. Therefore, behavioral changes are mainly detectable at boundaries. Boundary conditions cause the program to branch its execution towards the conditional statements, ensuring the boundary conditions are satisfied. Domain analysis can highlight the boundary conditions under which the system behavior significantly changes.

Section 12.3.2 offers a multi-objective genetic algorithm, NSGA-II [3], to determine the boundaries of given I/O intensive paths covering an expected/unexpected behavior. Boundary conditions are less important than laws and a variety of dimensions: they are smaller in scope, contingent rather than necessary, and do not come with a background of theoretical support. Most importantly, unlike stable, reliable laws, they are highly variant under intervention. For instance, in the autopilot system of an aircraft, when the inputs are conditioned to $(0.44 < \text{Throttle} \wedge \text{Throttle} \geq 0.57 \wedge \text{Pitchwheel} \geq 5)$, the system manifests an unsafe behavior [4]. Here, PitchWheel is the degree of adjustments applied to the aircraft's pitch, and Throttle is the power applied to the engine. As another example, an advanced driving assistance system, ADAS, reflects an unsafe behavior for the inputs within the range $(\theta \geq 100.52\circ$ and $\text{VP.} \geq 1.17 \text{ m/s})$ [5] where θ indicates the pedestrian orientation, and VP is the speed of the pedestrian.

The testing of cyber-physical systems is different from general-purpose computing systems. The real-time constraints imposed by environmental inputs, and the critical demands of precise monitoring and controlling, require behavior-based testing. Behavior-based testing is when you expect specific interactions to occur.

12.3 Rule Extraction for Autonomous Driving

This Section uses an open-source simulator, CARLA (Car Learning to Act), to demonstrate the use of software testing to extract rules underlying the behavior of the physical system in a CPS. Rules governing the collision of a self-driving car with a passing pedestrian can be extracted while generating test data as inputs to the SenarioRunner component of CARLA. SenarioRunner provides all the required environmental configurations to get the CARLA simulator ready to run the autonomous vehicle for a particular time on a selected road on the map. A CARLA function called srunner_start is invoked to run the simulator. The simulator should be conducted to drive the vehicle in an I/O intensive execution path. Apparently, the more interactions with the environment, the more the behavior of the autonomous vehicle will be revealed. Section 12.3.2 presents a genetic algorithm to look for I/O intensive execution paths.

Once I/O intensive execution paths are detected, a multi-objective genetic algorithm, NSGA II, generates test data covering the selected I/O intensive path while looking for the critical region of the input space where the safety is violated and the autonomous vehicle collides with the crossing pedestrian. The algorithm generates test data in an evolutionary process to cover those regions of the simulator input space that rush the autonomous vehicle into a passing pedestrian.

Fig. 12.1 presents an algorithm, "behavior_detector," to look for unexpected behaviors along a given I/O intensive execution path. The algorithm invokes the NSGA-II to look for subdomains of the input domain, uncovering the desired behavior along a given io-intensive execution path. The objectives of the algorithm are twofold:

$$\text{individual.objectives}[:] = [\ \overbrace{approach\_level, branch\_distance}^{\text{1. Coverage level ojectives}}, \underbrace{min\_distance, speed}_{\text{2. Coverage level objectives}}]$$

1. Coverage level objectives: An objective of the multi-objective evolutionary process is to minimize the normalized ranch distances of the execution path examined by each test data generated as a chromosome in the evolutionary process from the target I/O intensive path.

$$\text{fitness} = approach\_level + normalized\_branch\_distance$$

2. System-level objective: The system-level objective is given as a parameter function to the algorithm. For instance, in the case of the crossing pedestrian problem, the objectives are:

$$\text{Speed} = \text{The vehicle's speed at the time of the collision,}$$
$$\text{Min\_distance} = \text{minimum distance of pedestrian from the ego vehicle}$$

   (a) To minimize the distance between the crossing pedestrian and the ego vehicle.
   (b) To maximize the speed of collision of the vehicle with the crossing pedestrian.

The solutions are, in effect, test scenarios passed to the CARLA ScenarioRunner module. ScenarioRunner allows the CARLA simulator to define and execute traffic scenarios for a limited time slot, e.g., 40 s. Running the simulator is the most time-consuming task. At each time step, e.g., 0.01 per second, of the simulation, the outputs required to calculate the fitness functions are recorded. The NSGAII multi-objective algorithm invokes the scenario runner to evaluate and evolve solutions in the genetic population. It receives two sets of outputs concerned with the coverage and system-level objectives to compute each solution's fitness.

Each solution in the genetic population is labeled feasible or infeasible depending on its fitness. A feasible solution resides on the boundary of a region of the input space, covering the target I/O intensive path. Once the genetic process terminates, all

Algorithm: Behavior_Explorer
_____

Input: time_budget, population_size I/O_intensive_path, inputs_domain, system_level_objectives,
Output: governing_rules
Begin
1.   Current_domain = inputs_domain;
2.   Population ← ∅;   // Population of solutions
3.   **Repeat:**
4.   **For each** region in current_domain **do**:
5.   Initial_population←Generate_initial_population ((population_size, region, Population);
           // each solution is a vector of values for $(v_1, v_2, ..., v_n)$ */
6.   Solutions = NSGA-II (Initial_population,time_budget, source_code_level_objectives,
7.   system_level_objectives, I/O_intensive_path);
           /* the source-code-level-objectives includes approach-level and branch-distance */
8.   Population ← Population ∪ Solutions;
9.   **For each** solution in Solutions **do**:
10.   learning_sample = solution;
11.   (approach_level,branch_distance)=Get_source_cdoe_level_objectives_val (solution);
12.   label = Compute_labels_using_Eq7 (approach_level,branch_distance,
13.   I/O_intensive_path);
14.   learning_samples = learning_samples + (learning_sample, label);
15.   **End for**
16.   **End for**
17.   model = Create_ensemble_rule_learning_model (learning_samples);
18. $region_1, ..., region_x$ ← Extract_rules (model);
19.   Current_domain ← ∅;
20.   **For** j = 1 to x **do**:
21. new_region_j ← Prune_region (initial_domain, $region_j$);
22.   Current_domain ← Current_domain ∪ new_region_j ;
23.   **End for**
24. **Until** the termination_condition is met
25. learning_samples ← ∅;
26. **For each** solution in Population **do**:
27.   learning_sample = solution;
28.   other_objectives = Get_system_level_objectives_val (solution);
29.   label=Compute_labels_using_Eq7_and_Eq9(other_objectives, I/O_intensive_path);
30.   learning_samples = learning_samples + (learning_sample, label);
31. **End for**
32. model = Create_ensemble_rule_learning_model (learning_samples);
33. governing_rules ← Extract_rules (*model*);
   End
_____

Fig. 12.1 An algorithm to extract known/unknown rules governing physical behaviors the population of labeled solutions created in the genetic process is used as a learning set to build a classifier model, such as an ensemble learner.

The model helps to classify feasible and infeasible regions. An infeasible region means a region that does not cover the target I/O critical path. A bagging classifier is preferably used to distinguish homogenous regions of the solutions. The regions may include infeasible subregions. Therefore, the NSGA II function is recursively applied to the feasible subregions to detect denser subregions of feasible solutions. This step does not need to run the simulator and uses all the existing test scenarios (solutions) generated by NSGA-II.

The NSGA II function is recursively applied to the detected subregions as far as the test budget, e.g., the time, is over. Each solution generated during the evolutionary process is accompanied by a label indicating the distance and speed to approach the crossing pedestrian. The labeled solutions are fed into an explainable machine learning model [31] based on the bagging classifier to explain why the intended and unintended behaviors occur. Then, the governing rules are extracted from the classifier (line 33). Below is an example of a rule governing an unintended behavior showing the condition under which the ego vehicle collides with a crossing pedestrian:

$$\text{Walker\_target\_velocity} \leq 6.3218 \text{ and } X > -56.2541 \text{ and Yaw} \leq 79.7086$$

Each solution is a test data representing a vector $V = < v_1, v_2, \ldots, v_n >$, in the program input space. Each solution is initiated with a random value within the range of the corresponding physical feature.

As an example of a solution for the dynamic object crossing scenario in the autonomous vehicle case study, consider the initial configuration shown in Table 12.2. The following instruction is used to copy the domain of the ego vehicle features into the solution variables:

```
_cloudiness = Real(problem.domains[0][2], problem.domains[1][2]).rand()
_precipitation = Real(problem.domains[2][2], problem.domains[3][2]).rand()
...
        solution.variables =

            [_cloudiness, _precipitation, _precipitation_deposits,

            _wind_intensity, _sun_azimuth_angle, _fog_density, _fog_distance,

            _wetness, _start_distance, _other_actor_target_velocity,

           _other_actor_yaw, _friction, _IO_number, spawn_point.location.x,

           spawn_point.location.y,

                    spawn_point.location.z, spawn_point.rotation.yaw,

                    _curvature_id]
```

Table 12.2 Features used to describe the physical environment

Group	Input variables	Initial domain	Description
	Cloudiness	[0, 100]	0 indicates a clear sky
	Precipitation	[0, 100]	0: no rain & 100: shower
	precipitation_deposits	[0, 100]	0: no water & 100: completely flooding. (Area of puddles)
	wind_intensity	[0, 100]	0: no wind & 100, a strong wind
Weather	sun_azimuth_angle	[0, 360]	0: origin in the sphere
	sun_altitude_angle	[-90, 90]	-90: midnight & 90: midday
	fog_density	[0,1]	gray levels ranging from black (low density) to white (high)
	fog_distance	[0, 150]	fog start distance (in meters)
	Wetness	[0, 10]	The water layer thickness in mm Affects the tire traction
	start_distance	[5, 15]	distance between the ego vehicle and the pedestrian
Pedestrian	walker_target_velocity	[1, 18]	The velocity the pedestrian may reach in m/s
	walker_yaw	[270, 320]	the rotation angle of the pedestrian on the Z-axis
Road	Friction	[0.00001, 1]	the friction of the ego wheels
	X	[-99.98,-53.78]	The initial position of the ego vehicle on the X-axis
Ego	Y	[-138.61, -102.61]	The initial position of the ego vehicle on the Y-axis
	Yaw	[0.44, 85.32]	The initial rotation angle of the ego vehicle

The multiobjective algorithm, NSGA II, is taken from a Python library called Platypus. Platypus is a framework for evolutionary computing in Python focusing on multiobjective evolutionary algorithms (MOEAs). It differs from existing optimization libraries, including PyGMO, Inspired, DEAP, and Scipy, by providing optimization algorithms and analysis tools for multiobjective optimization.

12.3.1 Configuring the CARLA Simulator

In fact, the NSGA II algorithm generates test scenarios taken as input by the CARLA ScenarioRunner component to configure the environmental features. Each solution (test scenario) the algorithm generates saved as an XML file in a Carla directory, "examples." ScenarioRunner uses this XML file and the "DynamicObjectCrossing" scenario to configure the CARLA simulator. It is worth mentioning that DynamicObjectCrossing is one of the thirteen scenarios supported by CARLA [1]. In this scenario, a cyclist suddenly drives into the path of the ego vehicle, forcing the vehicle to stop. The genetic algorithm modifies the scenario so that the car does not necessarily stop and passes by the cyclist. Figure 12.2 shows, a pedestrian and cyclist crossing the road while an ego vehicle is approaching. The genetic algorithm generates test data scenarios the ScenarioRunner uses to configure the simulator.

Listed in Table 12.2 are the CARLA simulator environmental features in four groups: weather, pedestrian, road, and ego. The domain of values and a sample value are presented for each feature.

All the features listed in Table 12.2 are used to generate test data for the crossing pedestrian scenario. The features are categorized into four groups: weather, pedestrian, road, and ego vehicle. The features come in four categories described below:

- *Cloudiness*: Cloud cover is based on a 0–100 scale, with 0 representing clear skies and 100 being complete cloud cover..
- *Sun_azimuth_angle:* The s*un-azimuth angle* is the compass direction from which the sunlight comes. The azimuth values: $0°$ = due North, $90°$ = due East, $180°$ = due South, and $270°$ = due West.
- *Fog_density*: Fog density is defined by the degree to which light is attenuated due to scattering and absorption. It is related to visibility distance. Typical values for the absorption coefficient per unit depth of fog are 0.003 m $- 1$ (light fog), 0.006 m^{-1} (medium fog), and 0.03 m^{-1} (heavy fog).

Fig. 12.2 A crossing pedestrian in the front of an autonomous vehicle

- *Fog_distance*: shows the fog start distance (in meters).
- *Wetness*: wetness can quickly endanger driving stability, depending on the driving situation.

  Start_distance: the initial distance between the ego vehicle and the pedestrian.

- *Walker_target_velocity*: the target velocity that the pedestrian should reach (in meters per second).
- *Walker_yaw*: the initial rotation angle of the pedestrian on the Z-axis.
- *Friction*: indicates the friction of the ego wheels.
- *X*: the initial position of the ego vehicle on the X-axis.
- *Y*: is the initial position of the ego vehicle on the Y-axis.
- *Yaw*: is the initial rotation angle of the ego vehicle around the Z-axis.

The values of the above features may vary depending on the environmental conditions and events received via the sensors. The autonomous vehicle's behavior is revealed by altering the value of these features in response to the events received through the sensors.

ScenarioRunner is a CARLA component responsible for running (and repeating) a single scenario or a list of scenarios. CARLA scenario runner takes an XML file to configure the scenarios using the described features.

12.3.2 CARLA Components

During the autonomous vehicle navigation process, a sequence of activities, sensing, perception, planning, and control, is performed repeatedly in fixed time steps. The CARLA processing steps are illustrated in Fig. 12.3.

As shown in Fig. 12.3, the Carla process starts with an event raised by the sensors. *Sensors* are the hardware that gathers environmental data. For instance, three cameras are lined up in a row behind the top of the windshield of Carla's vehicle. Additionally, the vehicle is equipped with a single front-facing radar embedded in its bumper and a 360-degree lidar mounted on the roof. Carla may utilize other sensors, like ultrasonic, IMU, and GPS. The sensor subsystem disseminates the data received from sensors to the components of the perception subsystem.

Carla's *perception* subsystem translates raw sensor data to detect objects such as traffic lights, moving objects, and free space, as well as the vehicle's localization. Using a combination of high-definition maps, Carla's lidar sensor, and sophisticated mathematical algorithms, The Carla vehicle can be localized to within 10 cm or less. A high-definition map of the environment is compared to what Carla's lidar sees during its scans to determine the precise location. The components of the perception subsystem route their output to the planning subsystem.

Carla has a specific *planning* subsystem to build a series of waypoints as spots on the road that it needs to drive over. Carla should match the location and the target velocity associated with each waypoint to pass through. When passing through a

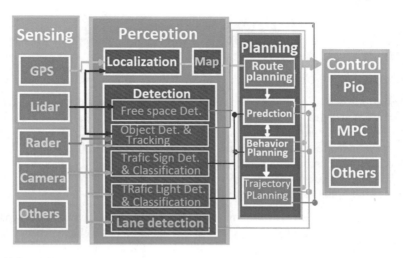

Fig. 12.3 The four subsystems of the CARLA technology

waypoint, Carla should match the location and the target velocity at the waypoint. The planner updates the waypoints accordingly as other vehicles move along the road. For example, if a vehicle in front of Carla slows down, the planning subsystem may advise Carla to decelerate. In this case, the planner slows down the vehicle by generating a new trajectory of waypoints with a lower target velocity. The planning subsystem similarly treats crossing objects, Traffic lights, and traffic signs. Having created a trajectory of new waypoints, the planner passes this trajectory to the final control subsystem.

The input to the *control* subsystem is the list of waypoints and target velocities generated by the planning subsystem. The control subsystem uses an algorithm to calculate how much to steer, accelerate, or brake to hit the target trajectory.

Based on the vehicle's current position detected in the perception layer, the navigation layer finds a path from the vehicle's current position toward its destination and decides on the vehicle's behavior according to driving situations and traffic laws. The navigation system relies on waypoints detected by the GPS to perform each mission. During navigation, the ego-vehicle must adopt some tactical behaviors to reach its destination safely (e.g., avoiding dynamic and static obstacles). However, sometimes mathematical models describing autonomous vehicle behavior cannot consider all environmental conditions. For instance, the parameters such as fog_density, fog_distance, precipitation, cloudiness, and wind_intensity are not directly considered in the mathematical models describing the autonomous vehicle dynamics. Therefore, when running CARLA with test scenarios generated automatically by the software tool described in this section, it was observed that the ego vehicle collides with the passing pedestrian in certain circumstances. Using the test scenarios labeled by the test results as learning samples, the following rules describing

the conditions under which the CARLA vehicle collides with the crossing pedestrian
were extracted:

> precipitation > -03.4141 and walker_target_velocity ≤ 4.4243 and $y \leq -106.0692$
> walker_target_velocity ≤ 4.4193 and walker_target_velocity > 1.4456 and yaw
> ≤ 79.6707
> walker_target_velocity ≤ 5.1355 and fog_distance ≤ 115.82409 and fog_distance
> > 49.2675
> walker_target_velocity ≤ 5.1355 and wind_intensity ≤ 40.4964 and
> wind_intensity > 3.3185

Apparently, considering the number of parameters affecting the behavior, scien-
tists can not easily detect the above rules manually. Genetic programming can define
the above rules more precisely as mathematical models, such as partial differential
equations. Also, the scientist may apply the rules to modify the existing mathematical
models. The question is how to reduce the search space to extract the desired rules.
The detected rules represent the behavior of the autonomous vehicle under certain
circumstances. Behaviors are revealed in reactions to the external events gathered
by the sensor subsystem. Therefore, in search of a particular behavior, it is preferred
to emphasize the execution paths with relatively more interactions with the physical
environment. We call such paths I/O intensive execution paths.

12.3.3   Extracting I/O Intensive Execution Paths

Behavior is a reflexive reaction to antecedent external events. As shown in Fig. 12.4
a cyber-physical system software captures events through sensors, makes decisions
based on its controlling algorithms, and reacts to the events by issuing commands to
actuators. Therefore, I/O intensive paths are probably preferred to the prime paths to
grasp the cyber-physical system behavior rapidly.

An I/O intensive path exhibits relatively more controlling behavior because of
having relatively more interactions with the environment. Due to the exponentially
increasing number of paths in controller software, heuristic-based search is used to

Fig. 12.4 Process in a
cyber-physical system

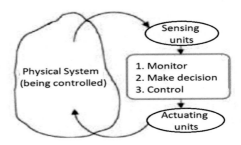

find the execution paths with a relatively higher number of environmental interactions. Figure 12.5 presents a genetic algorithm to generate test data covering execution paths with more I/O function calls. The CARLA controller software interacts with the environment by sending and receiving specific commands to the simulator through API calls (Fig. 12.5).

The chromosomes are defined as input parameters to configure and run the CARLA simulator with the crossing pedestrian scenario details. Figure 12.6 illustrates a sample chromosome structure including all the parameters affecting an autonomous vehicle control software:

For each of the parameters in Fig. 13.3, the initial domain should be specified. For instance, the initial domain of the values for the "start_distance" and "fog_density" parameters are [5.0.15] and [0.0.1], respectively. The initial genetic population is generated by randomly selecting a value in the domain of each gene (parameter) in the chromosome, shown in Table 12.6. Below is a part of the code to generate the initial genetic population randomly.

```
def create_initial_population(self):
    population = []
    for i in range(self.population_size):
        _cloudiness = random.uniform(*self.problem_domains[0])
        _precipitation = random.uniform(*self.problem_domains[1])
        ...


def genetic_algorithm(self):
    generation = 0
    print('Initialize population...')
    # population = list of chromosomes, each representing a test data
    population = self.create_initial_population()
    # population[i] = the ith chromosome
    # chromosome[1] = fitness, chromosome[0] = test data,
    population = self.evaluate_population(population)
    test_data_set = []
    start_time = time.time()
    # the main loop termin
    while (self.test_budget == 'time' and (time.time()-start_time)/60
            < self.time_budget) or
            ( self.test_budget != 'time' and generation < self.iteration):
        print("\n*** Iteration {}  ====> ".format(generation))
        new_population = []
        self.round += 1
        while len(new_population) < len(population):
```

Fig. 12.5 An algorithm to generate test data covering I/O intensive paths

```
    # 1- Selection
    parent1 = self.selection(population, self.tournament_size)
    parent2 = self.selection(population, self.tournament_size)
    # 2- Crossover
    if random.random() < self.crossover_rate:
        offspring1, offspring2 = self.crossover(parent1[0], parent2[0])
    else:
        offspring1 = parent1[0]
        offspring2 = parent2[0]
    # Mutation
    offspring1 = self.mutate(offspring1)
    offspring2 = self.mutate(offspring2)
    offspring1 = self.evaluate_individual(offspring1)
    offspring2 = self.evaluate_individual(offspring2)

    new_population.append(offspring1)
    new_population.append(offspring2)
generation += 1
population += new_population
pos = int(len(population) / 2)
population.sort(key=lambda individual: individual[1])
new_res = []
for pop in population:
    if pop not in new_res:
        new_res.append(pop)
population = new_res
for i in population:
    if i[1] == 0:
        self.invalid_samples += 1
    elif i[0] not in test_data_set:
        test_data_set.append(i[0])
population = population[:pos]
print("no. invalid samples = ", self.invalid_samples)
self.test_data_generation_time = time.time() - start_time
print(" the final population size is:", len(test_data_set))
return test_data_set
```

Fig. 12.5 (continued)

The objective of the genetic algorithm is to look for test cases covering execution paths with a relatively higher number of environmental interactions via inputs and outputs. Therefore, a list of the I/O methods used by the CARLA simulator should be collected in the first stage. Below is a list of the I/O methods in the CARLA simulator Python program:

```
individual = {
    "cloudiness": _cloudiness,
    "precipitation": _precipitation,
    "precipitation_deposits": _precipitation_deposits,
    "wind_intensity": _wind_intensity,
    "sun_azimuth_angle": _sun_azimuth_angle,
    "sun_altitude_angle": _sun_altitude_angle,
    "fog_density": _fog_density,
    "fog_distance": _fog_distance,
    "wetness": _wetness,
    "start_distance": _start_distance,
    "other_actor_target_velocity": _other_actor_target_velocity,
    "other_actor_yaw": _other_actor_yaw,
    "friction": _friction,
    "x": spawn_point.location.x,
    "y": spawn_point.location.y,
    "z": spawn_point.location.z,
    "yaw": spawn_point.rotation.yaw,
    "curvature_id": _curvature_id,
}
```

Fig. 12.6   A chromosome is defined as a set of key-values

```
simulator_config = {
    "python_api_path":  r"C:\CARLA\carla_0.9.9\PythonAPI",
    "exe_path":  "C:\CARLA\carla_0.9.9",
    "scenario_name":  "DynamicObjectCrossing",
    "scenario_config_file": "ObjectCrossing",
    "io_functions":  [
        # waypoint
        'get_left_lane', 'get_right_lane', 'next',
        # vehicle
        'apply_control',  'apply_physics_control',  'get_control', 'get_physics_con
        trol', 'get_speed_limit',  'get_traffic_light', 'get_traffic_light_state',
        'is_at_traffic_light', 'set_autopilot',
        # TrafficLight
        'freeze', 'get_elapsed_time', 'get_green_time', 'get_group_traffic_lights',
        'get_pole_index',  'get_red_time', 'get_state', 'get_yellow_time',
        'is_frozen', 'set_green_time', 'set_red_time', 'set_state', 'set_yelow_time',
        # sensor
        'listen', # 'stop',
        # map
        'generate_waypoints', 'get_spawn_points', 'get_topology', 'get_wayoint',
```

12.3.4 Detecting Unexpected Behavior

In search of the unexpected behavior of the CARLA ego vehicle when approaching a crossing pedestrian, an extension of the multiobjective algorithm, NSGA II, is introduced in this section. The algorithm searches for the challenging solutions used in scenarios where the system fails to satisfy the safety requirements, i.e., the vehicle crashing with a high velocity into or passing at a close lateral distance of the crossing pedestrian. As shown in Table 12.6, the input space of the CARLA simulator includes 19 parameters. Therefore, the search space, which is, in a sense, the input space of the simulator, is large.

Three techniques are applied to reduce the size of the search space:

1. The search space is restricted to the boundaries of the execution paths covering the boundaries of a selected I/O intensive path.
2. NSGA II algorithm is recursively applied to the feasible subregions of input space covering the selected I/O intensive path boundaries. In this way, the search space is limited to the subregions of the input space covering the boundaries of the selected I/O intensive execution path.
3. The search space is restricted to the paths leading the autonomous vehicles to collide or pass nearby the pedestrian.

After configuring the simulator with a test scenario, the simulator runs for a certain amount of time, e.g., 40 s. In each time-step of about 0.01 s, the Carla simulator goes through the sequence of activities: sensing, perception, planning, and control, described in Sect. 12.3.2. After each time step, the part of the execution path covered by the test scenario is recorded. The paths are kept as far as either the ego vehicle collides with or passes very close to the pedestrian. The search space is limited to the paths leading to either colliding with the pedestrian or passing nearby the pedestrian, and the remaining paths are ignored.

After executing each scenario in the genetic population for 40 s, all the solutions in the genetic population are labeled as feasible and infeasible depending on whether or not they cover the target execution path. Figure 12.7 represents the code for determining the labels for each individual.

The labeled samples are used as a learning dataset to build a machine learning classifier, such as an ensemble learner or decision tree (DT). The classifier may distinguish feasible and infeasible regions. By infeasible, I mean those regions of input space, including relatively more data points to cover the target I/O intensive path. For instance, the following Python code fragment shows the features used for building the classifier and different classes of solutions used as test scenarios. A test scenario is either safe or unsafe if it results in the collision of the ego vehicle with the pedestrian. It either resides within the boundaries of the unsafe region or is interior to the input domain. A solution is feasible, as it covers the target I/O intensive execution path.

```
# the objectives are described in detail later in this chapter
individual.objectives[:]=[approach_level,branch_distance,min_distance, speed]
if approach_level == 0 and branch_distance <= 0:
    label = 1                                        # feasible config
else:
    label = -1                                       # infeasible config
if label == 1:                                       # negative delta active point
    feasible_samples.append((fitness, frame_risk))
speed = solution.objectives[3]
if speed > -1:                                       # Collision speed
    statues = 'unsafe'
else:
    statues = 'safe'
```

Fig. 12.7 Defining labels for solutions

```
features = ['cloudiness', 'precipitation', ...] # features in table 13.2
class_name = [ "boundary_feasible_safe","boundary_infeasible_safe",
               "boundary_feasible_unsafe","boundary_infeasible_unsafe",
               "interior_feasible_safe"," interior _infeasible_safe",
               "interior _feasible_unsafe"," interior _infeasible_unsafe"]
```

A learnable evolutionary model [10] using an ensemble learning model comprising 30 decision trees is used to classify test data generated by the multiobjective genetic process into homogeneous regions of feasible solutions. A learnable evolutionary algorithm uses a classifier to restrict the search space to the feasible regions with a higher probability of locating the desired solutions.

The function "extract_rules," in Fig. 12.8, takes the genetic population, labeled_solutions, as input and builds four different classifiers, GradientBoosting, DecisionTrees, RandomForest, and SkopeRules.

The rules generated by the classifiers in Fig. 12.8 are used to determine the dense regions, including more feasible solutions. Figure 12.9a illustrates a sample decision tree representing feasible and infeasible regions of the input space covering an execution path, including eight inputs/outputs with the environment. The number of regions increases as I/Os in the target path increases. For instance, Fig. 12.9b shows a wage image of the decision tree for a path with 21 I/Os.

The field "sample" indicates the total number of feasible and infeasible solutions in each tree node. The field "value = [feasible, infeasible]" indicates the number of feasible and infeasible solutions.

The NSGA-II algorithm was executed ten times (rounds), each time for 2 h. Figure 12.9a shows the decision trees generated after the first round. Compared to the decision tree, relatively more sophisticated rules were obtained using the SkopeRules classifier. SkopeRules finds logical rules by fitting classification and regression trees to sub-samples.

```
def extrat_target(labeled_population, save_path, min_samples, round,
                    class_names, features):
    lines = []
    X_train, X_test, y_train, y_test = train_test_split(
        data.drop([target], axis=1), data[target], test_size=0.25,
                    random_state=42, stratify=data[target])
    feature_names = X_train.columns

    # Train a gradient-boosting classifier for benchmarking
    gradient_boost_clf = GradientBoostingClassifier(random_state=42,
                    n_estimators=30, min_samples_leaf = min_no_samples)
    gradient_boost_clf.fit(X_train, y_train)
    # Train a random forest classifier for benchmarking
    random_forest_clf = RandomForestClassifier(random_state=42,
                    n_estimators=30, min_samples_leaf = min_no_samples)
    random_forest_clf.fit(X_train, y_train)
    # Train a decision tree classifier for benchmark
    decision_tree_clf = DecisionTreeClassifier(random_state=42,
                        min_samples_leaf = min_no_samples)
    decision_tree_clf.fit(X_train, y_train)
    cn = decision_tree_clf.classes_
    fig, axes = plt.subplots(nrows=1, ncols=1, figsize=(4, 4), dpi=300)
    tree.plot_tree(decision_tree_clf,
                feature_names=fea,
                class_names=class_names,
                filled=True)
    date_time = str(datetime.now().strftime('%Y-%m-%d-%H-%M-%S'))
    fname = 'decision_tree_{}_{}'.format(round, date_time)
    fig.savefig('{}/{}.png'.format(save_fig, fname))  # (i + 1) * g

    # Train a skope-rules-boosting classifier
    skope_rules_clf = SkopeRules(max_depth_duplication=2,
                n_estimators=30,
                precision_min=0.7,
                recall_min=0.7,
                feature_names=feature_names)

    skope_rules_clf.fit(X_train, y_train)
    rules = skope_rules_clf.rules_
    return rules
```

Fig. 12.8 SkopeRules classifies feasible and infeasible regions

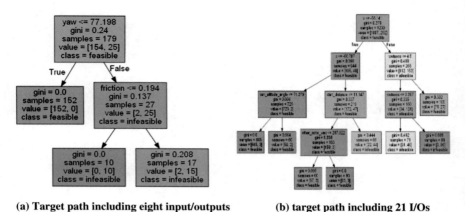

(a) Target path including eight input/outputs   **(b) target path including 21 I/Os**

Fig. 12.9 Decision tree generated after the first round of executing the NSGA II algorithm

Fig. 12.10 illustrates a tree representation of the textual rules generated by the SkopeRules classifier. It is observed that compared to the decision tree resulting from the decision tree classifier, shown in Fig. 12.9, more regions are distinguished.

For instance, walking down the tree, in Fig. 12.10, from the root node toward the leaves, a sample rule defining the boundaries of a feasible subdomain, including 48 feasible samples covering the target path and two infeasible samples, is:

$$-122.94 < y <= -110.272 \text{ and } 13.124 < sun\_attitude \text{ and } fog\_distance <= 57.892$$

The boundaries of each feature can be completed based on the initial domain of its values in Table 12.2. The NSGA II algorithm is recursively applied to the generated sub-regions for 20 h.

The number of samples covering the above feasible regions is 48. Therefore, if the size of the genetic population is more than 48, new solutions within the boundaries of the feasible region should be generated randomly to complete the initial population for the NSGA II algorithm. The final decision tree represents the rules to distinguish feasible from infeasible regions, as shown in Fig. 12.11.

Finally, once the desired regions are detected, the SkopeRules classifier is applied to the final feasible regions to extract the rules governing the collision of the ego vehicle with the crossing pedestrian. Below are two of the rules representing circumstances under which the ego vehicle crashes into the crossing pedestrian:
Rule 1

<--[wind_intensity=43.11..83.44 : 210,279,42%,210,279,42%]
[fog_density=0.03585..0.9913 : 416,555,42%,204,272,42%]
[fog_distance=49.8..140.3 : 277,294,48%,164,122,57%]
[start_distance>=6.411 : 358,463,43%,137,95,59%]
[other_actor_target_velocity=1.609..2.677 : 253,100,71%,91,22,80%]
[other_actor_yaw<=319.4 : 444,558,44%,89,16,84%]

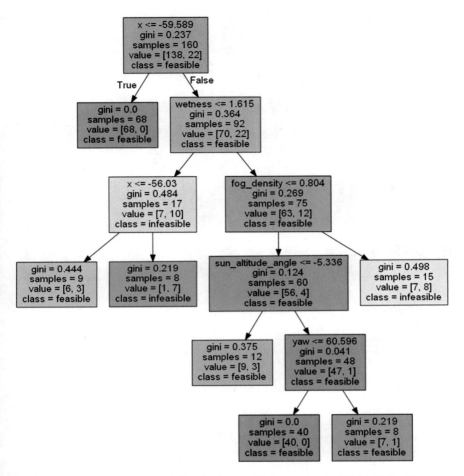

Fig. 12.10 Decision tree generated after the first round of executing the NSGA II algorithm

[ego_location_y=-138.6..-106.3 : 416,450,48%,88,3,96%]
[ego_location_yaw<=79.32 : 416,449,48%,88,0,100%]
: p=88,np=17,u=12,cx=50,c=1,s=88 # 21807

Rule 2

<-[wind_intensity>=63.33 : 180,243,42%,180,243,42%]
[fog_density=0.1535..0.9692 : 358,499,41%,159,220,41%]
[fog_distance=26.74..136.1 : 302,399,43%,139,156,47%]
[wetness>=1.555 : 336,403,45%,101,120,45%]
[start_distance=5.394..9.629 : 263,279,48%,83,61,57%]
[other_actor_target_velocity=1.447..3.985 : 380,165,69%,80,25,76%]
[other_actor_yaw=273.3..307.5 : 292,370,44%,73,17,81%]

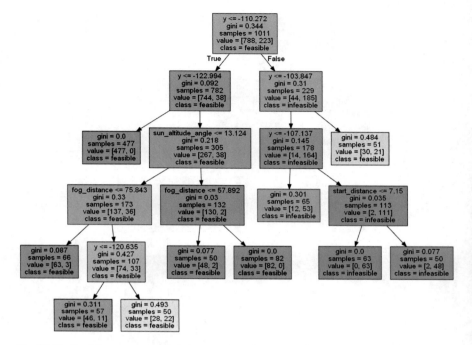

Fig. 12.11 Decision tree generated after the last round of executing the NSGA II algorithm

[ego_location_x=-96.73..-54.04 : 433,552,43%,73,16,82%]
[ego_location_y<=-107.2 : 400,447,47%,73,0,100%]
[ego_location_yaw<=77.2 : 386,443,46%,73,0,100%]
y: p=73,np=73,u=3,cx=62,c=1,s=73 # 21808

12.3.5 Fitness Evaluation and Ranking

As described in Sect. 12.2, the aim is to generate test data covering the boundaries
of an I/O intensive execution path in an evolutionary process. The objectives of the
test data generation algorithm are:

individual.objectives[:] = [approach_level, branch_distance, min_distance, speed]

Approach level indicates the degree to which a candidate solution will likely to
match a given target node within the program's control flow graph. The approach
level is the number of unexecuted nodes on which the target node is transitively
control-dependent [6].

The branch distance indicates how close a branching condition is to being true,
and its calculation varies depending on the corresponding relational conditions.

Figure 12.3 shows a list of branch distance computations for various relational operators [5]. Therefore, the objective to exercise a selected I/O critical execution path is as follows:

$$Target\_path\_objective = Minimize(branch\_distance + approach\_level)$$

The other two objectives are concerned with the behavior of the physical system being controlled by the cyber-physical system:

$$distance\_from\_pedestrain = Minimize(get\_distance\ (ego, pedestrian))$$
$$collision\_velocity = Maximize(get\_velocity(ego, pedestrian))$$

The branch distance, bd, for the path covered by a solution, $solution_{path}$, calculated using the rules shown in Table 12.3. For example, when executing the predicate "R > 12" is 10, the branch distance to the true branch is $12 - 10 + K = 2 + K$, where K is a constant value defined by the user. If there is more than one predicate in an execution path, the maximum value of the branch distances is considered as the distance of that execution path from the target path.

The following equation is used to compute the branch distance, bd, between the target I/O intensive path, tar_{path} and the path covered by a given solution, sol_{path}. In this equation Max_{bd} Indicates the maximum branch distance, and δ is the thickness of the boundary.

$$Bd(tar_{path}, sol_{path}) =$$
$$\begin{cases} 1 & if\ t_{path}\ not\ covered \\ 0 & -\delta \le Max_{bd} \le \delta \\ normalized(|Max_{bd}|) & Max_{bd} \le -\delta\ or\ Max_{bd} > 0 \end{cases} \quad (12.1)$$

Table 12.3 Branch distance evaluation rules suggested by Tracey [5]	Predicate	Branch distance	
		If the true branch is covered	If the false branch is covered
	Boolean	K	K
	$a = b$	K	$|a - b|$
	$a \ne b$	$|a - b|$	K
	$a < b$	$(b - a)$	$(a - b) + K$
	$a \le b$	$(b - a) + K$	$(a - b)$
	$a > b$	$(a - b)$	$(b - a) + K$
	$a \ge b$	$(a - b) + K$	$(b - a)$
	$a \vee b$	Min (d (a), d(b))	
	$a \wedge b$	Max (d(a), d(b))	
	$\neg a$	Negation propagated over a	

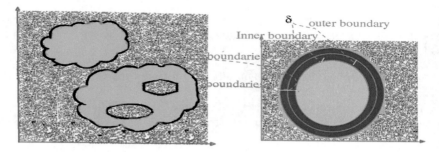

Fig. 12.12 The evolutionary process tends to conduct the population toward the boundaries

In the above equation, the normalized value of $|Max_{bd}|$ is computed as follows:

$$normalized(|Max_{bd}|) = \frac{|Max_{bd}|}{|Max_{bd}| + 1}$$

The multiobjective evolutionary process aims to detect the boundaries of the regions of input space, revealing the desired behavior. Therefore, each individual's fitness (solution) depends on its distance from the boundaries of the feasible regions within the input space. As a result, the fitness of a solution residing within a feasible region may be less than some other solutions outside the region. As shown in Fig. 12.12, the boundaries of the feasible regions are thickened with a constant δ.

The rules extracted from the classifier may recommend several subdomains for the input space. The number of regions (subdomains) sometimes gets very high, and it takes a relatively long time to proceed with all these regions. Therefore, there should be a mechanism to rank the regions and select the ones with higher ranks. The ranking is performed based on the labels assigned to the solutions involved in the regions, as follows:

1. Apparently, a solution, sol, residing within the boundaries of the behavioral region is preferred. In the following equation, 'al' indicates the approach level, 'sol' is a solution.

$$label(Sol) = \begin{cases} 1 & \text{If } -\delta \leq Max_{bd} \leq 0 \text{ and } al(tar_{path}, sol_{path}) = 0 \\ 0 & \text{otherwise} \end{cases} \quad (12.2)$$

2. The status of the solution in terms of satisfying the critical path domain is either "feasible" or "infeasible" see Fig. 12.3. Equation 12.2 returns 1 when the solution is feasible. Otherwise, it returns 0.

$$label(sol) = \begin{cases} 1 & \text{If } Max_{bd} \leq 0 \text{ and } al(t_{path}, S) = 0 \\ 0 & \text{otherwise} \end{cases} \quad (12.3)$$

Table 12.4 The ranking of the behavioral regions

Rank	Boundary (Eq. 3)	Feasible (Eq. 2)	Unsafe (Eq. 4)
1	1	1	1
2	0	1	1
3	1	1	0
4	0	1	0
5	1	0	1
6	0	0	1
7	1	0	0
8	0	0	0

3. Once the solution is executed, the system output can be "safe" or "unsafe" regarding whether or not the vehicle collides with a pedestrian. Equation 12.4 returns 1 when the output is unsafe; Otherwise, it returns 0.

$$\text{label}(Sol) = \begin{cases} 1 & \text{If the Sol.ego hits the pedestrian} \\ 0 & \text{If the Sol.pedestrian passes the road} \end{cases} \tag{12.4}$$

The labeled solutions are then ranked according to the corresponding classes shown in Table 12.4. As a final step, we include the high-ranking solutions within each region in the related initial population up to a given size.

Naturally, test scenarios revealing the same behavior tend to belong to homogeneous regions of the input space [7, 8]. A homogeneous region is an area of input space within which every point shares one or more distinctive characteristics in common. Therefore, all the solutions belonging to each region are evaluated, as shown below. Regions are ranked based on the quality of the solutions they include. The higher-ranked regions are selected for further processing.

Skop-Rules [9], as a bagging classifier, selects all the relatively more effective rules based on Recall and precision thresholds from the bag it creates. Recall measures the model's ability to detect unsafe samples. It is calculated as the ratio of the number of unsafe samples correctly classified as unsafe to the total number of unsafe samples. Precision is the ratio of correctly classified unsafe samples to the total number of unsafe samples.

12.4   Autopilot Software System Behavior[1]

An upgrade to modern F-16 s is an automated ground collision avoidance system (GCAS). The GCAS has saved six aircraft (at least $25 million each) and seven lives. It detects when a ground collision is imminent and performs a recovery maneuver. However, by seeing critical execution paths with a relatively higher number of inputs

---

[1] Download the source code from: https://doi.org/10.6084/m9.figshare.21324330

and outputs, we could witness many unsafe behaviors ending up with the collision of the aircraft with the ground.

12.4.1 Monitoring Component

The monitoring component of the GCAS software is written in Python language. Python's Coverage library has been used to instrument the GCAS software. After instrumenting the software components with the Coverage tool, the initial scenario (test case) is fed to the main() function of the run_GCAS project. The function is shown in Fig. 12.13.

As shown in Fig. 12.13, the main function of the simulator accepts three parameters. The parameters are aircraft engine power, pow_in, altitude, alt_in, and velocity vt_in. These three parameters and 13 other state variables provide the initial state for starting the ground collision avoidance system (GCAS). Safe and unsafe behavioral domains are computed while generating test data with a genetic algorithm and inputting the test data to run the simulator. The GCAS system overrides the pilot with a recovery autopilot to prevent imminent controlled flight into the terrain. The recovery autopilot should adjust the power given to the aircraft's engines to ensure that the aircraft reaches a safe altitude. Dependent on the initial state set by the input test data, GCAS may succeed or fail to reach a safe attitude. As a result, each test data generated by the genetic algorithm is labeled as safe or unsafe.

In fact, the controller uses 16 continuous variables and piecewise nonlinear differential equations. The variables are given in Table 11.1 (Table 12.5).

12.4.2 Test Results

As described above, the genetic algorithm uses only three of the 16 variables, and the autopilot controller software sets the remaining variables by default. The algorithm is described in Sect. 12.4.3. The target of the algorithm is to generate test data making the aircraft collided with the ground. In fact, the algorithm attempts to determine when the recovery is unsuccessful as a function of the state variables.

In this respect, at first, I/O critical execution paths are selected. Due to a relatively higher number of I/Os, the chance of finding unexpected behaviors is higher on critical paths. However, as shown in Fig. 12.14, the GCAS has been capable of determining when a ground collision is imminent and performing a recovery for a certain region of the input domain.

Similarly, the test results by the GCAS developers show that all the test cases generated for certain domains and values of the input variables in Table 13.2 are safe.

However, many unsafe states are detected despite the GCAS software filtering unsafe initial states. Table 12.6 shows the number of unexpected behaviors where

```
'''
round collision avoidance system (GCAS
'''

import math
from numpy import deg2rad
import matplotlib.pyplot as plt
# run_f16_sim function is invoked by the main()
from aerobench.run_f16_sim import run_f16_sim
from aerobench.visualize import plot
# Initiates the autopilot mode
from gcas_autopilot import GcasAutopilot
"""
the main function of GCAS simulator has three inputs used
as test data to run the simulator. The parameters are:
(1)  pow_in: the aircraft power engine
(2)  aalt_in: the altitude of the aircraft,
(3)  vt_in:  the aircraft's initial velocity (ft/sec)
"""
def main(pow_in,alt_in,vt_in):
    'main function'
    ### Initial Conditions ###
    power = pow_in              # engine power level (0-10)
    # Default alpha & beta
    alpha = deg2rad(2.1215)    # Trim Angle of Attack (rad)
    beta = 0                   # Side slip angle (rad)
                               # Initial Attitude
    alt = alt_in               # altitude (ft)
    vt =vt_in                  # initial velocity (ft/sec)
    phi = -math.pi/8           # Roll angle from wings level (rad)
    theta = (-math.pi/2)*0.3   # Pitch angle from nose level (rad)
    psi = 0                    # Yaw angle from North (rad)
    # Build Initial Condition Vectors
    # state = [vt, alpha, beta, phi, theta, psi, P, Q, R, pn, pe, h, pow]
    init = [vt, alpha, beta, phi, theta, psi, 0, 0, 0, 0, 0, alt, power]
    tmax = 3.51                         # simulation time
    # Initialize autopilot mode
    ap = GcasAutopilot(init_mode='roll', stdout=True, gain_str='old')
    step = 1/30
```

Fig. 12.13 The main function of the GCAS software system

```
# the simulator returns the status of the aircraftin a vetor, res.
res = run_f16_sim(init, tmax, ap, step=step, extended_states=True)
print(f"Simulation Completed in {round(res['runtime'], 3)} seconds")
safety_limits = SafetyLimits(
    altitude=(0, 45000),  # ft \
    )  # deg/s/s
verifier = SafetyLimitsVerifier(safety_limits, ap.llc)
res_status = verifier.verify(res)
"""

res_status = 0 implies that the aircraft has been immune
           = 1 implies that the aircraft has collided with the ground
"""

return re_status
```

Fig. 12.13 (continued)

Table 12.5 Aircraft Model and the Inner Loop Controller: State variables

Variable	Symbol	Meaning
x[0]	Vt	Air Speed
x[1]	α	Angle of Attack
x[2]	β	Angle of Sideslip
x[3]	ϕ	Roll
x[4]	Θ	Pitch
x[5]	Ψ	Yaw
x[6]	P	Roll Rate
x[7]	Q	Pitch Rate
x[8]	R	Yaw Rate
x[9]	Pn	Northward Displacement
x[10]	Pe	Eastward Displacement
x[11]	Alt	Altitude
x[12]	Pow	Engine Power Lag
x[13]	Nz (integrator)	Upward Acce
x[14]	Ps (integrator)	Stability Roll Rate
x[15]	Ny + r (integrator)	Side Accel and Yaw Rate

the aircraft collides with the ground and the number of expected behaviors as unsafe and safe tests, respectively.

An observation in Table 12.7 is that as the number of I/Os in the selected prime paths increases, relatively more unsafe tests are detected. Figure 12.15 illustrates the test results in Eclidean space. All the unsafe tests are in red in Figure.

The initial population of the genetic algorithm includes 200 randomly selected individuals. Each individual is a test data representing a triple (Attitude, Power, AirSpeed). Each time a test data is generated, its fitness is measured by actually

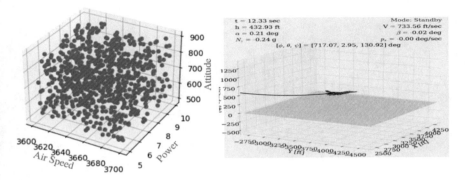

Fig. 12.14 Safe flight domain

Table 12.6 Verification Cases for GCAS system

Case	Initial States
3A-3P	Same as 2A-2P using the 16-dimensional model
3Q	alt = [3600, 3700], xcg _5%
3R	Same as 3Q with _ = [0; _4]
3 T	Same as 3S with xcg _25%
3U	xcg _5%, cxt, cyt, czt, clt, cmt, cnt _40%
3 V	Same as 3U and 3R
3 W	Same as 3U and 3S
3X	cxt, cyt, czt, clt, cmt, cnt _40%
3Y	cxt, cyt, czt, clt, cmt, cnt _45%
3Z	clt, cmt, cnt _55%

Table 12.7 The safe and unsafe test results for different I/o critical paths

Prime paths		Input domain			No unsafe tests	No Safe tests	No test data covering
Path No	No I/Os	Velocity Foot/sec	Altitude Foot	Power level			
1	6	300–900	500–1000	5–10	833	1092	1925
2	9	300–900	500–1000	5–10	787	1109	1886
3	14	300–900	500–1000	5–10	415	1224	1639

running GCAS with the generated test data. The objective of the genetic algorithm is to generate test data covering the safe and unsafe behavioral boundaries of I/O critical prime paths to avoid a ground collision.

The boundaries are represented as a decision tree shown in Fig. 12.16. The decision tree represents the domain of the inputs covering safe and safe regions. The safe region includes the test data set that, when used as the initial state for the autopilot mode,

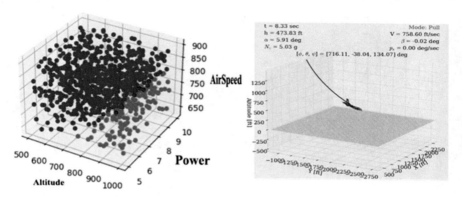

Fig. 12.15 Unsafe test results are red, and the safe ones are green

the GCAS can succeed in the recovery maneuver. As a result, the aircraft rolls until
the wings are level, then begins a 5-G pull up until the nose is above the horizon, at
which point control is returned to the pilot. The physical principles underlying the
safe and unsafe performance of the aircraft are shown in Table 12.3. The principles
are decision rules extracted from the decision tree in Table 12.8.

The decision tree is learned by iteratively partitioning the data sets generated by
the algorithm, described in Sect. 12.4.3.

12.4.3 Test Data Generation

Below is the main function of the domain generator program. It invokes a function,
domain_generaor, to generate test data covering an I/O critical execution prime path.
The genetic algorithm generates a different test data set depending on the value of a
variable, test_mode. The first set includes test data that covers the desired execution
path, set by the user in a configuration file, "config.py." A function called find_points
accepts this data set as input and finds all the test data covering the boundaries of the
regions covering the execution path.

The second set includes test data covering an I/O critical prime path found automatically by the program. Each test data in the second set is labeled as safe or unsafe.
A function called "find_boundary_sim.find_points" accepts the second set and determines the boundaries for the safe and unsafe regions. In other words, the function
finds the behavioral domains for a given I/O critical prime path.

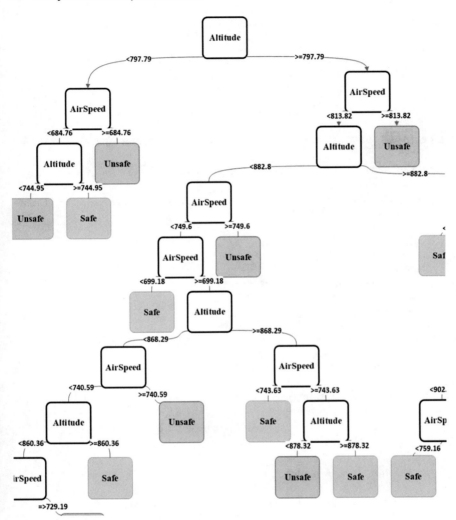

Fig. 12.16 Unsafe and safe domains

Table 12.8 Rules underlying safe and unsafe behaviors of the autopilot system

1	Altitude < 797.79 && AirSpeed > = 684.76	UnSafe
2	Altitude < 797.79 && AirSpeed < 684.76 && Altitude < 744.95	UnSafe
3	Altitude < 797.79 && AirSpeed < 684.76 && Altitude >= 744.95	Safe
4	Altitude >= 797.79 && AirSpeed > = 813.82	UnSafe
5	Altitude >= 797.79 && AirSpeed < 813.82 && Altitude >= 882.8 && AirSpeed < 755.32	Safe
6	Altitude >= 797.79 && AirSpeed < 813.82 && Altitude < 882.8 && AirSpeed > = 749.6	UnSafe
7	Altitude >= 797.79 && AirSpeed < 813.82 && Altitude < 882.8 && AirSpeed < 749.6 && AirSpeed < 699.18	Safe
8	Altitude >= 797.79 && AirSpeed < 813.82 && Altitude >= 882.8 && AirSpeed > = 755.32 && AirSpeed < 890.83	UnSafe

```
import time
import config
from domain_generation import domain_generator
import csv
def main():

    # call genetic to generate test data covering the target path

    # the generated should be labeled with safe and unsafe flags

    self.genetic_algorithm()
    if int(self.test_mode)==1:
        # Find test data covering the boundaries of the target prime path
        p_bound,all_TD=find_boundary.find_points
                    (self.test_data_labeled.gentime, self.time_budget)
    elif int(self. test_mode)==5:
        # Find unsafe & safe boundary points
        labeled_samples,boundarypoint=find_boundary_sim.find_points(
                                rtg,gentime, self.time_budget)
```

The config file includes all the parameters required to configure the genetic algorithm. In addition, some global variables, such as test budget, test mode, and the input domain, are defined in the config file.

```
global_config = {
        'SUT':run_GCAS,        # Software under test = GCAS
        'target_run_io':6,                          # No. I/Os in the taget_path
        'target_path' : tar.t,                      #The prime I/O critical path
        # target_path: {'from statement' -> 'to statement'}
        #eg. " 'target_path' = {'10->13', '4->7', '3->4', '10->14', '4->10'},
        'population_low_up':[[1, 5,10],[1, 3600,3700],[1,500,900]],
        'riemann_delta' : 0.5,
        'test_budget': 'sampling', #[sampling, time]
        'sampling_budget':1000,   #1000000,
        'time_budget':120,        #minutes
        'algorithm':'GA',
        'mod':'5',  #1 for testing Python ordinary functions
                    #2 for detecting I/O critical paths
                    #5 for testing a cyber-physical system
    }
genetic_algorithm_config = {
                        'crossover_rate': 0.7,
                        'mutation_rate': 0.8,
                        'tournament_size': 5,
                        'population':50,
                        'iteration':20,
                    }
```

12.4.4   Domain Generator

The domain_generator function is invoked by the main function to generate test data covering the desired execution path. Each chromosome is a test data represented as an array of lengths three:

$$\text{Chromosome} = <\text{power, attitude, velocity}>$$

All the genetic parameters, such as the population size, crossover rate, mutation rate, and stopping condition, are taken from the config.py file. The Python implementation of the genetic algorithm is given below. The main point of the function is to save each individual generated by the genetic algorithm in a list, decision_tree_samples.

The list is used to compute the domain of inputs covering safe and unsafe behaviors along an I/O critical path.

```python
def genetic_algorithm(self):
    generation = 0
    population = self.create_initial_population()
    evaluated_population = self.evaluate_population(population)
    start_time = time.time()       # Triger the start time for the genetic algorithm
    while (self.test_budget == 'time' and (time.time() - start_time)/60 <
            self.time_budget) or (self.test_budget != 'time'
                                and generation < self.iteration ):
        print("\n*** Iteration  {}  ===> ".format(generation))
        new_population = []
        while len(new_population) < len(population):
            # Selection
            parent1 = self.selection(evaluated_population, self.tournament_size)
            parent2 = self.selection(evaluated_population, self.tournament_size)
            # Crossover
            if random.random() < self.crossover_rate:
                offspring1, offspring2 = self.crossover(parent1[0], parent2[0])
            else:
                offspring1 = parent1[0]
                offspring2 = parent2[0]
            # Mutation
            offspring1 = self.mutate(offspring1)
            offspring2 = self.mutate(offspring2)

            new_population.append(offspring1)
            new_population.append(offspring2)

        generation += 1
        population = new_population
        evaluated_population = self.evaluate_population(population)
        # keeps all the individuals ( test data) in a pool
        for individual in evaluated_population:
            if individual[0] not in test_data_pool:
                test_data_pool.append(individual)
    self.test_data_generation_time = time.time() - start_time
    return test_data_pool,evaluated_population
```

The fitness of each chromosome (test data) is the Jaccard distance between the path examined by the chromosome and the target path. The objective of the genetic algorithm is to obtain the fittest chromosome.

12.4.5   *Fitness Function*

The fitness function operates in two modes depending on the value of the test_mode parameter set by the user in the config file. If the value of test_mode is equal to 1, then the objective of the genetic algorithm will be to generate test data covering a particular path. On the other hand, if test_mode is equal to 5, the objective will be to generate test data covering an I/O critical path. It is worth mentioning that due to relatively more interactions with the environment.

The fitness of each test data, generated by the genetic algorithm, is computed as the Jaccard distance of the path, covered_path, examined by the test data from the target path, target_path:

$$\text{Jaccard} - \text{distance(test data)} = \frac{\text{target\_path} \cap \text{covered\_path}}{\text{target\_path} \cup \text{covered\_path}}$$

According to the above relation, if a test data covers the target path, its fitness will be zero. Considering each test data as a point within the program input space, all the test data covering the targeted I/O critical prime path constitute one or more feasible regions. The objective is to generate test_data covering the borders of the feasible regions.

The fitness of each test data is computed as its minimum distance from the borders of the feasible regions. If the test data resides inside a feasible region, its fitness equals its minimum distance with the points that do not reside there. Conversely, the fitness of test data residing outside the feasible regions is computed as its minimum distance from the nearest point inside a feasible region. This distance is normalized to a value between zero and one concerning the farthest possible distance. The fitness is computed as follows:

$$\text{Fitness(test\_data)} = \alpha^* \text{Jaccard\_distance(test\_data)}$$
$$+ \beta^* \text{distance(test\_data, nearest\_point)}$$

where $\beta = \alpha = 0.5$

```python
# This function computes the fitness of a given chromosome (test data)
# This function computes the fitness of a given chromosome (test data)
    def get_fitness(self, test_data):
        if int(self.test_mode) == 1:   # testing an ordinary function
            covered_path = path_coverage(args) # the path covered by the test data
            if covered_path != self.target_path:    # if the taget_path is not
                                                    # covered by the test
                ar_temp = []
                for i in range(len(test_data)):
                    ar_temp.append(test_data[i])
                ar_temp.append(0)                   # infeasible test data is marked 0
                self.labeled_test_dataappend(ar_temp)
                self.infeasible_points.append(test_data)
                self.jaccard_distance = \
                    (1 - float(len(self.target_path.intersection(covered_path)) / \
                        len(covered_path.union(self.target_path))))
                if self.feasible_found == 0:        # feasible test data is found
                    return self.jaccard_distance
                else:
                    # compute the minimu distance from a feasible point
                    self.distance = min_distans_points(self.feasible_points, test_data)
                    dis_norm = self.distance / self.max_distance # Normalize distance
                    return (((abs(dis_norm) * 0.5) + self.jaccard_distance * 0.5))
            else:
                self.feasible) += 1
                ar_temp = []
                for i in range(len(test_data)):
                    ar_temp.append(test_data[i])
                ar_temp.append(1)                   # Feasible test data is marked 1
                self.labeled_test_dataappend(ar_temp)
                self.feasible_points.append(test_data)

                self.distance = min_distans_points(self.infeasible_points, test_data)
                dis_norm = self.distance / self.max_distance # Normalize distance
                return (((abs(dis_norm) * 0.5) + self.jaccard_distance * 0.5))
        # Runs the modified main function of GCAS and, in return, gets.
    # 1. A flag indicating whether the simulator has accepted the test data.
    # 2. The safety status is "safe" or "unsafe."
    # 3. The fitness of the chromosome (test data).
    if int(self.test_mod) == 5:   # to test the simulator
        unacceptable_init_state, safety_status, fit = run_GCAS.main(*test_data)
```

```
if    unacceptable_init_state == 1:  # GCAS does nt accept the test data
      return 1
if fit != 0:   # The targeted execution path is not covered by the test data
      return fit
else:
      print("found:)")
      labeled_test_dataset_temp = []
if safety_status == "unsafe":
    label = 1;
else:
    label = 0
      for i in range(len(test_data)):
          labeled_test_dataset_temp.append(test_data[i])
          labeled_test_dataset_temp.append(label) # label test data as unsafe
          self.decision_tree_samples.append(labeled_test_dataset_temp)
return fit
```

The main function run_GCAS.main(*test_data), of the simulator invokes a function called run_f16_sim. This function is instrumented by the "coverage" software tool to keep track of the simulator execution.

12.4.6 *Instrumenting the Simulator*

Source code instrumentation techniques often provide coverage reports from the information gathered by the executed probes. The probes are embedded throughout the software at points where runtime information is required. However, in the case of interpreters and just-in-time compilers, runtime information can be obtained directly from the runtime environment that runs the program or the interpreter. Therefore, there will be no need to insert probes inside the code.

Coverage.py is a tool for monitoring Python programs, noting which parts of the code have been executed. The following instruction creates coverage_object as an object of the class Coverage:

```
coverage_object = coverage.Coverage(branch=True)
```

The coverage_object keeps records of all the functions invoked between the coverage_object.start() and coverage_object.stop() hooks. Since the "branch" parameter is "True," branch coverage will be measured in addition to the usual statement coverage. Coverage measurement only occurs in functions invoked after start() is called. Measurements do not apply to statements within the start() scope. The measurement applies to all the functions invoked before the stop() method is invoked. The coverage_object.save() command saves the collected coverage data to a file. The coverage_object.start(), coverage_object.stop(), and coverage_object.save(), hooks are set inside the main loop of the run_ run_f16_sim() method, shown below. The coverage_object.xml_report() call saves the coverage results in an XML file. It is possible to omit some functions from the XML report through the "omit" argument of the xml_report() function.

```
Import coverage.py
    def run_f16_sim(...):
      try:
        #Simulates and analyzes autonomous F-16 maneuvers
        ...

        # the main loop of the simulator
        while integrator.status == 'running':
          #start with computing the coverage
          cov = coverage.Coverage(branch=True)
          cov.start()
          ...
          cov.stop()
          cov.save()
          # Create XML report including the line and branch coverage data
            cov.xml_report(omit=['aerobench/util.py', 'aero
                  bench/run_f16_sim.py', 'aerobench/lowlevel/thrust.py'],
                  outfile="Temp")
          # parse the XML file to determine the execution path
          executed_path = determine_exec_path(outfile)
```

The xml_report function creates an XML report of the coverage results in a file named "coverage.xml." Each line number examined by the test data is marked with hits "n," where "n" is the number of times the line has been executed. The line marked with 0 hits is thus the line never tested.

Aside from reporting statement coverage, xml_reprt() also measures branch coverage. Where a line in the program could jump to more than one next line, xml_report() tracks the visited destinations and flags lines that have not visited all of their possible destinations. Each branch destination in the file is an execution opportunity, as is a line. The report provides information about which lines had missing branches. The XML report gives information about which lines had missing branches. The XML report also contains branch information, including separate statements and branch coverage percentages. Below is a sample report generated by the xml_report function:

```xml
<?xml version="1.0" ?>
<coverage version="5.5" timestamp="1644989963491" lines-valid="11" lines-covered="6" line-
rate="0.5455" branches-valid="8" branches-covered="4" branch-rate="0.5" complexity="0">
  <!-- Generated by coverage.py: https://coverage.readthedocs.io -->
  <!-- Based on https://raw.githubusercontent.com/cobertura/web/master/htdocs/xml/coverage-
04.dtd -->
    <sources>
      <source>C:\Users\pythonProject\ST-Project</source>
    </sources>
    <packages>
      <package name="sut_convert" line-rate="0.5455" branch-rate="0.5"
                    complexity="0">
      <classes>
        <class name="BMI.py" filename="sut_convert/BMI.py" complexity="0"
          line-rate="0.5455" branch-rate="0.5">
          <methods/>
          <lines>
            <line number="1" hits="0"/>
            <line number="2" hits="1"/>
            <line number="3" hits="1" branch="true" condition-coverage="50%
              (1/2)" missing-branches="4"/>
            <line number="4" hits="0"/>
            <line number="5" hits="1" branch="true" condition-coverage="50%
              (1/2)" missing-branches="7"/>
            <line number="6" hits="1"/>
            <line number="7" hits="1" branch="true" condition-coverage="0%
              (1/2)" missing-branches="8,9"/>
            <line number="8" hits="0"/>
            <line number="9" hits="1" branch="true" condition-coverage="0%
              (1/2)" missing-branches="10,12"/>
            <line number="10" hits="0"/>
            <line number="12" hits="0"/>
          </lines>
        </class>
      </classes>
      </package>
    </packages>
</coverage>
```

```
Timport xml.etree.ElementTree as ET                                          rt to
detedef determine_exec_path():
    exec_path = {None}
    """
    XML.tree, or in short ET, is a Python package to generate a parse tree for XML
    file coverage.xml.
    """
    tree = ET.parse(r"coverage.xml")
    # for each line in XML tree
    for instance in tree.findall(".//line"):
        # if the executed line is a branching condition and the condition is true
        if (f"{instance.get('hits')}") == '1'and (f"{instance.get('branch')}") \
            == "true":
            # if condition coverage is 100^ then
            if (f"{instance.get('condition coverage)}")=="100% (2/2)":
                # The true and false branches are covered
                exec_path.add((f"{instance.get('number')}")+"->one")
                exec_path.add((f"{instance.get('number')}")+"->two")
            else: # the else part is going to be executed
                path.add((f"{instance.get('number')}")+"->
                        "+(f"{instance.get('missing branches)}")+"?")
    return path
```

12.4.7   Calculating Boundary Points

Each test_data generated by the genetic algorithm is saved as a pair (test_data, label) in a list named labeled_test_data. The label of feasible test data covering the target execution path is 1, while infeasible test data are labeled 0. The following algorithm is used to calculate boundary points:

For.

Step 1. Choose a feasible point a and an infeasible point b.
Step 2. Calculate the middle point c between a and b.
Step 3. If c is feasible, a = c; otherwise b = c.
Step 4. Calculate the distance between a and b.
Step 5. If the distance $\leq \varepsilon$, terminate; otherwise, go to Step2,

The points in the feasible and infeasible regions are selected as follows:

```
for i in range(0,len(feasible)):
  if test_budget ==1: # if no test budget left
    break
  #call boundary search algorithm for find all boundary tast data
  for j in range(0,len(infeasible)):
    boundary_point = locate_boundary_point(feasible[i], infeasible[j])
    boundary_points_pool.append(boundary_point)
    plot_boundary.plot(end_of_boundary)
```

The locate_boundary_point function uses a binary search to find the boundary point residing between a given feasible and infeasible point:

```
#binary search algorithm for finding boundary points
def generate_boundary_two_point(feasible_point, infeasible_point)
  while(True):
    #calculate middle point beetween a feasible and an infeasible point
    mid_point =[]
    for dimension in range(len(feasible)):
    mid_point.append(float(feasible_point[dimension]+
                    infeasible_point[dimension])/2)
  # take the target path from the config file
  target_path=config.global_config['target_path']
  #run c(middle) on SUT and get path covered
  executed_path = coverage_eval.execute_sut_with(mid_point)
  if executed_path == target_path :
    feasible_point = mid_point
  else:
    infeasible_point = mid_point
  dist= math.dist (feasible_point, infeasible_point)
  # if  the distance between feasible and infeaible point is < 0.001
  # then the mid_point is a boundary point
  if dist<0.001:
    break
return feasible_point
```

Example 1: BMI

Consider the body mass index (BMI) category example. The BMI is the body mass in kilograms divided by the square of body height in meters. This is expressed as:

$$\text{BMI} = \frac{weight}{height^2}$$

The following function calculates the Body Mass Index (BMI):

```
def BMI(weight, height):
    bmi=( weight / (weight * weight))
    if bmi < 18.5:
      print ("underweight")
    elif bmi < 25:
      return ("normal")
    elif bmi < 30:
      return ("overweight")
    elif bmi < 40:
      return ("obese")
    else:
      return ("very obese")
```

Fig. 12.17 illustrates the normal BMI region in green.

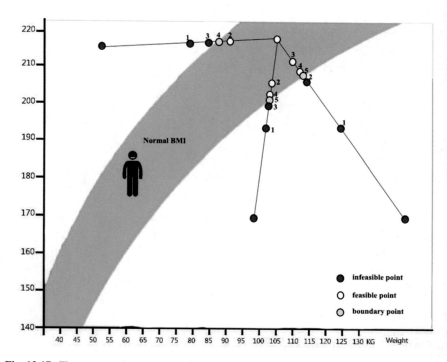

Fig. 12.17   The green region represents the normal BMI

Fig. 12.18 The boundaries
between the safe and unsafe
regions

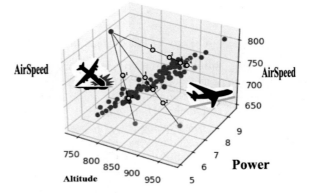

The above example shows how to determine the boundary points for a given prime path of the BMI function. In this example, the normal BMI is examined. Figure 12.18 shows how to determine the aircraft control system's boundaries between unsafe and safe regions.

The following function uses a binary search to find the boundary point residing between safe and unsafe points:

```
#binary search algorithm for finding boundary points
def generate_boundary_two_point(safe_point, unsafe_point)
    while(True):
        #calculate the middle point between a safe and an unsafe point
        mid_point =[]
        for dimension in range(len(safe_point)):
            mid_point.append(float(safe_point[dimension]+
                             unsafe_point[dimension])/2
        #run simulator with mid_point
        safety_state = run_GCAS.main(*mid_point)
        if sfety_satate == 1 :
            unsafe_point = mid_point
        else:
            safe_point = mid_point
        dist = math.dist (safe_point, unsafe_point)
        # if the distance between safe and unsafe point is < 1
        # then the mid_point is a boundary point
        if dist<1:
            break
    return unsafe_point
```

12.5   Summary

Finally, the key messages of this chapter are summarized as follows:

- The use of software testing to uncover unexpected behavior in a cyber-physical system opens new horizons for future research.
- Design and implementation of an algorithm to extract boundary conditions for unintended and intended behaviors.
- Introducing a test adequacy criterion, behavioral domain coverage, to evaluate the adequacy of the test data set used to determine the behavioral boundaries.
- Applying a multi-objective evolutionary approach to determine boundaries for intended and unintended behaviors while using ensemble learning models to determine regions representing specific behaviors.
- Instead of prime path coverage, a new test requirement called critical path coverage should be applied for testing cyber-physical systems, where the change of behaviors and behavioral boundaries are of primary concern.

References

1. Ben Abdessalem, R., Nejati, S., Briand, L.C., Stifter, T.: Testing vision-based control systems using learnable evolutionary algorithms. In: Proceedings of the 40th International Conference on Software Engineering – ICSE'18, pp. 1016–1026 (2018)
2. Dosovitskiy, A., Ros, G., Codevilla, F., Lopez, A., Koltun, V.: Carla: An open urban driving simulator. In: Conference on Robot Learning, pp. 1–16. PMLR (2017)
3. Deb, K., Pratap, A., Agarwal, S., Meyarivan, T.: A fast and elitist multi-objective genetic algorithm: Nsga-ii. IEEE Trans. Evol. Comput. **6**(2), 182–197 (2002)
4. Heidlauf, P., Collins, A., Bolender, M.S.B.: Verification challenges in f-16 ground collision avoidance and other automated maneuvers. EPiC Series Computing **54**(Arch18), 208–217 (2018)
5. Korel, B.: Automated software test data generation. IEEE Trans. Software Eng. **16**(8), 870–879 (1990)
6. Wegener, J., Baresel, A., Steamer, H.: Evolutionary test environment for automatic structural testing. Inf. Softw. Technol. **43**(14), 841–854 (2001)
7. Gotlieb, A., Petit, M.: A uniform random test data generator for path testing. J. Syst. Softw. **83**(12), 2618–2626 (2010)
8. Huang, R., Sun, W., Chen, T.Y., Ng, S., Chen, J.: Identification of failure regions for programs with numeric inputs. IEEE Trans. Emerg. Top. Comput. Intell. **5**(4), 651–667 (2020)
9. "SKOPE-Rules," [Online]. Available: https://github.com/scikit-learn-contrib/skope-rules. Accessed 23 Apr 2020
10. "CARLA Scenario Runner 0.9.8". [Online]. https://github.com/carla-simulator/scenario_run ner/releases/tag/v0.9.8. Accessed 10 Mar 2020
11. "CARLA 0.9.9". [Online]. https: //github.com/carla-simulator/carla/releases/tag/0.9.9. Accessed 23 Apr 2020

Chapter 13
Automatic Test Data Generation
Symbolic and Concolic Executions

13.1 Introduction

Dynamic-symbolic execution is a fantastic area of research from a software industry point of view. It relieves the ability to use the benefits of both static and dynamic analysis of programs simultaneously while avoiding their weaknesses. Symbolic execution is an effective automated static analysis technique. It is a crucial element for code-driven test data generation. As described in Sect. 13.3, symbolic execution works by running the program under test on symbolic inputs and computing over the expressions expressed in terms of the inputs. However, as described in Sect. 13.4.2, it may get stuck by calls to binary library functions whose source code is unavailable and loops with many iterations, pointers, strings, and large-scale or very complex programs. The use of dynamic-symbolic techniques could mitigate such difficulties.

Dynamic-symbolic execution, also known as Concolic execution, is a hybrid method for collecting path constraints and test data for unit testing. Cocncolic execution is the subject of Sect. 13.4. In these hybrid methods, each program instruction is executed simultaneously in concrete and symbolic modes. Concolic execution also faces several challenges, one of which is the problem of path explosion. A possible solution is to consider basis execution paths solely. Basis execution paths are described further in Sect. 13.5.

Concolic testing begins with randomly selected input data. After the execution, a branching condition addressed by the program's symbolic execution tree is selected and negated to obtain the following path constraint. The path constraint is given as input to a solver to generate test data to execute the path. Constraint solvers are popular tools for solving a path constraint and creating test data to run the path. Along with the dynamic-symbolic execution of a path, a symbolic execution tree representing all the possible executions through the program is formed gradually. Tree search strategies are conventionally applied to select and flip appropriate branching conditions considering the search objectives. The question is how to use the vital information obtained from the symbolic expressions and run time state of the program, to reduce

© The Author(s), under exclusive license to Springer Nature Switzerland AG 2023
S. Parsa, *Software Testing Automation*,
https://doi.org/10.1007/978-3-031-22057-9_13

the search space. This question is discussed and partially answered in Sects. 13.6.3 and 13.7.

Most potential search strategies have their strength and weaknesses. To take advantage of the power of different search strategies, and avoid their flaws, the use of meta-strategies, taking advantage of different search strategies is proposed. The question is which combination of search strategies in what order and dependent on what program execution states could be most preferably applied. This is an open-ended area of research. For a detailed description, refer to Sect. 13.7.

13.2   Dynamic Symbolic Execution

In a dynamic-symbolic or, in other words, Concolic execution, program execution starts with both symbolic and concrete (real) input values. For instance, in line 4 of Fig. 13.1, the two input variables *i* and *j* are initially assigned the symbolic values "*i*0" and "*j*0" as well as the concrete values 8 and 2, respectively. As shown in line 5, the execution continues computing both concrete and symbolic values.

	Sval = Symbolic value, Cval = Concolic value				
		Sval-1	Cval–1	Sval-2	Cval–2
	void main()				
	{ double i, j;				
	double * p = &i, q = p;				
1.	p = &i;			p ="address i"	
2.	q = p;			q ="address i"	
4.	scanf("%f %f", i, j);	i = "i0", j = "j0"	i = 8, 2	i = i = "i0", j = "j0"	i = 2, j = 4
5.	if (i > j*2)	P1="i0>j0*2"	True	P1="i0 >j0*2"	
6.	*p = i*j – 4;	i = "i0*j0 – 4"	i = 12		
7.	else			P2 = not P1	True
8.	{ j = sqrt(i * 2);			j = ?	j = 2
9.	*q = fabs(j - i);}			i = ?	i = 0
10.	If(*q > j)	P3=("i0*j0 – 4)>j0"	True	P3: ?	
11.	printf(" j = %f", i);				
12.	else			P4 = not P3	True
13.	printf(" i = %f", i);				
14.	return(); }				

Fig. 13.1 Dynamic-symbolic execution of a sample C program

To this aim, two columns to keep concrete and symbolic values are added to the symbol tables, and the symbol table is kept and used at runtime. Symbolic expressions are constructed by replacing all the variables appearing in the program under test with their symbolic values. For instance, by substituting the symbolic values of i and j in the predicate, $i > j*2$, in line 5, the symbolic predicate "$i0 > j0 + 1$" will be obtained. At the same time, the concrete values $i = 2$ and $j = 4$ are substituted into the predicate $i > j*2$, resulting in the predicate returning true. In line 6, the symbolic and concrete values of i are "$i0*j - 4$" and 12, respectively. After each execution is completed, one of the conditions along the execution path is selected and negated to cover more paths. As described in Sect. 13.7, different search strategies may be applied to choose a condition, dependent on the target of the dynamic-symbolic execution. Suppose the condition at line 5 is selected. The condition is then negated, and its symbolic representation is passed as a string to SMT solver to yield the input values, e.g., $i = 2$ and $j = 4$, satisfying the condition.

After executing the program, in Fig. 13.1, with inputs $i = 2$ and $j = 4$, the predicate at line 6 returns false. This time the else-part of the if-statement is executed. However, the symbolic execution gets stuck since the source codes of the sqrt and fabs functions are unavailable. In general, black-box functions whose source codes are not available may cause a problem for symbolic execution. However, as described in Sect. 13.3.4, using a dynamic disassembler, dynamic-symbolic execution can be continued inside a black box function as its code is disassembled.

13.3 Symbolic Execution

The Symbolic theory was introduced by King in 1976 [1]. Symbolic execution, as a static program analysis technique, is aimed at exploring execution paths. Symbolic execution is the program implementation using symbolic inputs in place of actual (concrete) input values and symbolically performing all the program operations. In symbolic execution, the value of each variable is a string representing a symbolic expression in terms of the values of the program input variables. For example, consider the code fragment presented in Fig. 13.2.

With the symbolic implementation of this function, the condition $C == 100$ turns into the symbolic expression "$x0 * x0 - y0 * y0 == 100$". After the symbolic execution of the first line of the program, the symbolic value of 'a' will be "$x0 + y0$". After executing the second line, the symbolic value of 'b' will be "$x0 - y0$". By executing the third line of the program, the symbolic value of 'c' will be equal to the product of the symbolic values of a and b. As a result, the symbolic value of the variable 'c' will be equal to "$(x0 + y0) * (x0 - y0)$" which can be simplified to "$x0 * x0 - y0 * y0$". Symbolically, the condition of the program at line 4 will be "$x0 * x0 - y0*y0 == 100$".

Symbolic execution is the usual way to infer path constraints. Program execution with this method means that the program is being manipulated symbolically, as far as, all the branching conditions along the execution path include only input variables

```
void main( ) //                                    Symbolic Execution
{
1.      int a = x + y ; //                            a = x0 + y0;
2.      int b = x - y ; //                            b = x0 - y0 ;
3.      int c = a * b ;        //                     c = (x0 + y0)*( x0 -
y0);
4.      if(C = = 100)  //                        if( (x*x -y*y) == 100)
5.          Print(C) ;  //                            Print(100);
}
```

Fig. 13.2 A sample code and its corresponding symbolic representation

as their operands. For example, suppose 'a' and 'b' are input variables, and we have the following code fragment:

1. $c = a + b$;
2. $d = a - b$;
3. $e = c * d$;
4. if $(e > 5)$
5. println(e);

Using symbolic execution, the condition '$e > 5$' in this code is converted to "$a * a - b * b > 5$". At first, in the statement '$e = c * d$', c and d are replaced with "$a + b$" and "$a - b$," respectively. Therefore, we have '$e = (a + b) * (a - b)$', which is simplified and converted to "$e = a * a - b * b$". By using this value instead of 'e' at '$e > 5$', we have "$a * a - b * b > 5$". If the constraint solver finds a solution to "$a * a - b * b > 5$", then there will be test data that executes the corresponding path. The constraint solver function is to find the input values for the program so that a specific path can be executed [7]. The logical conjunction of all the constraints along an execution path is called the path constraint.

For a better understanding of the path constraints and the symbolic state of the program, consider the code fragment in Fig. 13.3. The parameters x and y are initiated with the symbolic values "$x0$" and "$y0$", respectively. The symbolic execution of the if-statement at line number 1, results in two symbolic expressions "$x0 <= 2$" and "$!(x0 <= 2)$". Therefore, from this point on, the execution path is branched into two directions, and accordingly, the symbolic execution proceeds in two branches.

Table 13.1 shows the path conditions and symbolic states for each sentence in program E. The symbolic state column represents the values of the variables in the sentence, addressed in the last column, in terms of the program input variables. For example, after executing the statement in line 2, the symbolic value of the variable y will be "$y0 + 1$".

As shown in Table 13.1, all possible branches are followed simultaneously after each branching condition. Symbolic execution trees are constructed to characterize execution paths explored during symbolic executions. For each symbolically

Fig. 13.3 A sample
program to show states in
symbolic execution

Program E(int x, int y
1.   **if** (x <= 2)
2.   ++y
3.   **else**
4.   --y
5.   **if** (y > 2)
6.   print 1
7.   **else**
8.   print 0

Table 13.1 Symbolic
execution of the program in
Fig. 13.3

Symbolic state	Path conditions	Inst.#
$x = x_0, y = y_0$	True	1
$x = x_0, y = y_0$	$x_0 \leq 2$	2
$x = x_0, y = y_0$	$x_0 > 2$	4
$x = x_0, y = y_0 + 1$	$x_0 \leq 2$	5
$x = x_0, y = y_0 - 1$	$x_0 > 2$	
$x = x_0, y = y_0 + 1$	$(x_0 \leq 2) \wedge (y_0 > 1)$	6
$x = x_0, y = y_0 - 1$	$(x_0 > 2) \wedge (y_0 > 3)$	
$x = x_0, y = y_0 + 1$	$(x_0 \leq 2) \wedge (y_0 \leq 1)$	8
$x = x_0, y = y_0 - 1$	$(x_0 > 2) \wedge (y_0 \leq 3)$	

executed program path, a path condition is maintained. The path condition represents the condition on the inputs for the execution to follow that path. A symbolic execution tree characterizes the explored execution paths.

13.3.1 Symbolic Execution Tree

A symbolic execution tree characterizes all the paths traversed during the symbolic execution. Each node in the tree represents a symbolic state of the program, and each edge indicates a transition between the two states. The symbolic state is a picture of the current execution and consists of one or more real states. For example, the symbolic execution tree in Fig. 11.1 contains seven symbolic states. The branching in the symbolic execution tree occurs whenever the symbolic execution cannot be accurately determined during the symbolic execution.

A path in the tree is feasible if and only if the logical expression representing the path constraint is solvable. Otherwise, the path is infeasible. The symbolic execution tree in Fig. 13.4 represents two feasible or practical execution paths in the m() function.

```
Int     m(int
y){
    1:  if (y > 0)
    2:      y++;
    3:  else
    4:      y--;
    5:  return y;
    }
```

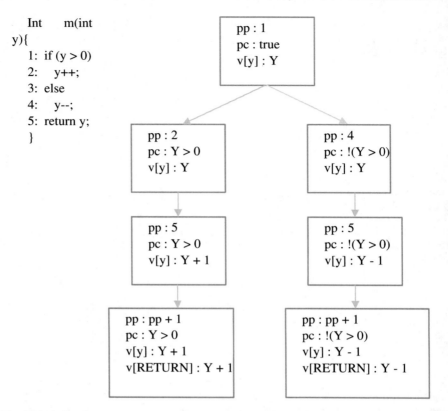

Fig. 13.4 A function m() and its corresponding symbolic execution tree

Each state of the program is symbolically defined as a triple (pp, pc, v). The program position, pp, addresses the next instruction to be executed. The path constraint, pc, is the conjunction of the branching clauses that are marked as constraints along the current execution path. The corresponding path constraint should be solvable for the program to follow a specific path. Symbolic value store, v, is a mapping from the memory locations or the program variables to the symbolic expressions, representing the path constraint.

Figure 13.5 illustrates a sample program and its corresponding symbolic execution tree. The program has four different execution paths.

The four leaves of the symbol tree represent the final states of the program. The two final states of the program are in line 6, and the other two final states are in line 8 of the program. For example, consider the rightmost leaf of the tree. The path conditions for this leaf are $x > 2$ && $y > 2$; when x and y are greater than 2, the program will execute instructions 1, 4, 5, and 8 in sequence, and the output will be zero. As another example, consider the leftmost leaf of the tree. The path conditions

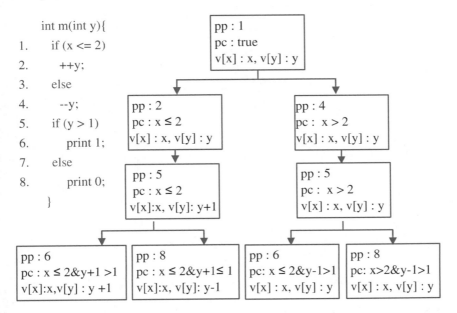

```
        int m(int y){
1.        if (x <= 2)
2.          ++y;
3.        else
4.          --y;
5.        if (y > 1)
6.          print 1;
7.        else
8.          print 0;
        }
```

Fig. 13.5 A sample symbolic execution tree

of this node are $X \leq 2$ && $Y > 1$. When the condition is true, the program executes instructions 1, 2, 5, and 6 in sequence, and '1' will be printed in the output.

13.3.2 Path Constraint

When executing a program, an execution trace is generated as a sequence of states along the execution path. The conjunction of all the conditional expressions along the trace will form a path constraint. The difficulty is how to modify the conditions such that they could be brought together as a whole representing the path constraint. As an example, consider the following function f(). The inputs are 'a' and 'b' and the symbolic execution for the path condition, $((a + 1) > b)$ && $((a\_b) - (a * b)) > (a + 2)$, is shown in Fig. 13.6. The path is highlighted in the control flow graph for f(), shown on the right side of the Figure.

The function f(), presented in Fig. 13.6, has two contradicting conditions '$a > b$' and '$b > a$.' However, it should be noted that the value of 'a' and 'b' differs in these two expressions. The symbolic expression corresponding to the condition '$a > b$' is '$(a + 1) > b$' and for '$b > a$' is '$((a + b) - (a * b)) > (a + 2)$'. Therefore, the path constraint to follow the trace, highlighted in the CFG in Fig. 13.7, will be:

$$Pc = ((a + 1) > b) \,\&\& \, (((a + b) - (a*b)) > (a + 2)).$$

where 'a' and 'b' are input to the function $f()$.

```
Int f(int a, int b);
{
  int c;                              // Symbolic execution
  c = a + b;                // c = a+b;
  a = a + 1;                // a = a + 1;
  if (a > b)                // if((a+1) > b)
      { a = a + 1;          //    { a = (a + 1) + 1
        b = c*a;}           //      b = (a*b); }
  b = c - b;                // b = (a+b) - (a*b)
  if ( b > a)               //  if (((a+b) - (a*b)) > (a+2))

      b = a*c;              //    b = (a+2)*(a+b)
  return ( c * a - b)       // return ((a+b)*((a+2) - ((a+2)*(a+b)))
}
```

Fig. 13.6 A sample function and its corresponding symbolic execution trace

A trace of all symbolic expressions along the execution paths will be generated upon a program's symbolic execution. The conjunction of all symbolic expressions, representing branching conditions, along an execution path forms the path constraint. Constraint solvers, such as Satisfiability Modulo Theories (SMT) solvers, can be applied to assign a set of inputs that satisfy the current path constraint. This assignment is useful for test data generation as they set the input values to exercise target paths in the program. If the solver cannot determine valid inputs that can simultaneously satisfy all constraints along the given path, then it may be either the path being infeasible or the solver's inability to solve the path constraint.

13.3.3 Path Explosion

The number of execution paths of a program may be huge, especially if there are branching conditions within loops. We commonly have loops with many branching insides. So many different paths can be created by taking a different branch in each iteration of the loops. The high number of executable paths represents a severe barrier to symbolic execution, denoted as the path explosion problem. Program loops, recursion, and branching sequences are the three essential sources of barriers to symbolic execution. These sources often appear together in programs. In practice, especially in numerical and scientific computations, programs often contain much more complex loops with loop nesting and internal branching.

Another important source of the barrier is recursion. Recursions can be translated into loops. Conversely, loops can be translated into recursions. Therefore, a loop and a recursion are two different syntactical notations for the same barrier source,

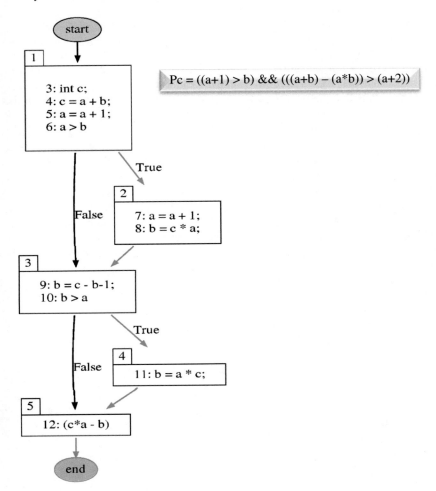

Fig. 13.7 CFG for the function *f* in Fig. 13.6

a notion of repetition. Even without any loop and recursion, we can get a sizeable symbolic execution tree relative to the program size.

Scaling symbolic execution to real systems remains challenging despite recent algorithmic and technological advances. A practical approach to addressing scalability is reducing the analysis's scope. For example, in regression analysis, differences between two related program versions are used to guide the analysis. While such an approach is intuitive, finding efficient and precise ways to identify program differences and characterize their impact on how the program executes has proved challenging in practice.

Programs with loops and recursion may generate infinitely long execution paths. Testers may specify depth bounds as input to symbolic execution to guarantee execution termination in such cases. In the following, an example is given to show a symbolic execution tree for execution paths, including loops. The program in Fig. 13.8 contains only a single very simple loop without internal branching. The program accepts two parameters, x and y, and returns x to the power of y. On the right side, the executable tree of the program is shown.

Instructions 1 and 2 execute first, and if the value of y is greater than or equal to 1, it enters the loop; otherwise, line 6 runs. If the loop does not execute, instructions 1, 2, 3, and 6 execute, and value one is returned. The second path is taken when the loop executes once, instructions 1, 2, 3, 4, 5, 3, and six execute, and the value of x is returned. All the other paths of the program are obtained in the same way.

Let us look further at Fig. 13.9, where a small fraction of the symbolic execution tree of a function, Sample, is depicted on the right-hand side of the Figure. This program has a single loop with an if-else statement inside the loop. The loop repeats 30 times. Generally, long loop iterations are the leading cause of path explosion in symbolic executions. Such loops may cause path explosion, resulting in incompleteness and scalability challenges.

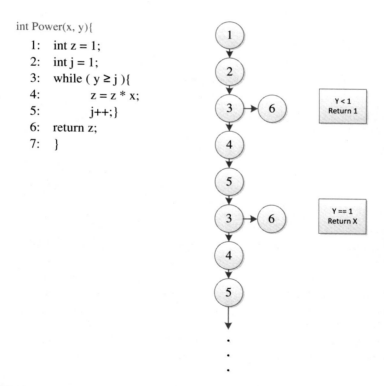

Fig. 13.8 A sample program and its symbolic execution tree

```
int Sample(int[] a)

{

01.  int sum = 0;

02.  int i = 0;

03.  while( i< 30){

04:      if(a[i]=45)

05:          sum ++;

06:      else

07:          sum --;

08:      i++;

09: }

10: return sum;

    }
```

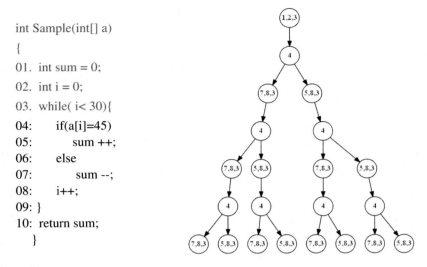

Fig. 13.9 An example of a path explosion

Figure 13.9 shows part of the symbolic execution tree, only for three repetitions of the loop. The condition on line 4 of the program results in a path explosion. As shown in Fig. 13.9, with each loop repetition, the number of paths of the symbolic executable tree doubles. This way, the number of execution paths after 30 iterations of the loop will be about 2^{30} paths (more than 1 billion).

13.3.4 Black Box Functions

A function whose source code is not available, or in other words, black-box functions, cannot be executed symbolically straight away. Where applicable, black box functions are often replaced with their return value in a branching condition. The underlying assumption is that the value returned by a function remains the same over successive invocations of the function with similar inputs. However, this assumption is not always true. The function code could be disassembled or possibly decompiled to resolve the problem.

At first glance, it seems to be a difficult job to walk through a binary code step by step and execute it both concretely and symbolically. For example, consider the dynamic-symbolic execution steps of the code fragment presented in Fig. 13.2.

As shown in Table 13.2, the Concolic execution of the function, f, starts with randomly selected input values, $a = 16$ and $b = 4$, as a result of which the path constraint p1 $= $ "$\neg (c > d) \wedge (d > c)$" is satisfied. Afterward, the branching condition $\neg(c > d)$ is negated, and the path constraint p2 $= $ "$(c > d) \wedge (d > c)$" is passed as a parameter to a constraint solver to generate test data satisfying the constraint. The

Table 13.2 Concolic execution of a sample function

Cvals = Concrete values		Svals = Symbolic values			
		Cvals-1	Svals-1	Cvals-2	Svals-2
int f(int a, int b)		a = 16	a = "a0"	a = 10	a = "a0"
		b = 4	b = "b0"	b = 10	b = "b0"
{int c, d;					
1.	c = a*b;	c =16*4 = 64	c = "a0 * b0"	c = 10*10 = 100	c = "a0 * b0"
2.	d = a*5;	d =16*5 = 80	d = "a0*5"	d = 10*5 = 50	d = "a0*5"
3.	if(c > d)	64>80=false	"¬(a0*b0> a0*5)"	100 > 50 = true	"¬(a0*b0> a0*5)"
4.	d =sqrt(a*b)*20;			d = 200	d = ?
5.	else				
6.	d = d*2;	D=80*2=160	d = "a0*10"		
7.	c = c+a;	c=64+16=80	c="a0*b0+a0"	c=100+10=110	c="a0* b0+a0"
8.	if (d > c)	160>80=true	"a0*10>a0*b0+a0"	200>110 = true	?>a0* b0+a0"
9.	a = a + c;	a=16+80=98	c="a0*b0+2*a0"	a=101+10=120	c="a0 *b0+2*a0"
10.	else				
11.	a = a – d;				
12.	return a;	return 98		return 120	
Path constraint-1 : p1 = "¬ (c > d) ∧ (d>c)"					
Path constraint-2 : p2 = " (c > d) ∧ (d>c)"					
Path constraint-3 : p3 = "¬ (c > d) ∧¬ (d>c)"					
Path constraint-4 : p4 = " (c > d) ∧ ¬ (d>c)"					

resultant test data is a = ten and b = 10. This time, the function d = sqrt($a * b$) is invoked, for which the source code is unavailable, and therefore, it is impossible to carry on with the symbolic execution.

Symbolic execution of black-box functions, whose source code is unavailable, seems impractical. However, a binary code can be disassembled step by step and in each step while executing symbolically and dynamically. Assembly source code for a function can be easily obtained by applying a dynamic disassembler. Dynamic disassemblers are used to bypass any obfuscations that may be applied to a binary code.

To build a dynamic disassembler, an open-source 64-bit disassembler engine, such as BeaEngine, can be used [20]. BeaEngine.lib is a disassembler library for intel64 and IA-32 architectures. It is in C. BeaEngine.lib contains a function, Disasm, to decode instructions from 32 and 64 bits Intel architectures. The library includes a standard instruction set from FPU, MMX, SSE, SSE2, SSE3, SSSE3, SSE4.1, SSE4.2, and VMX technologies. The Disasm function of BeaEngine can be invoked to disassemble a hex code instruction by instruction.

A debugger engine such as ArchDasm can be used to disassemble and execute a binary code. Arkham is an open-source 64-bit debugger and disassembler for

Windows. To build a debugger, one can use windows debugging APIs. Windows has provided an event-driven process debugging model similar to the Linux signals. For instance, an API function called DebugActiveProcess is invoked to attach a debugger to the debuggee process. The debuggee process is suspended upon the debugger attachment. To implement kernel debuggers with relatively more robust features, hardware-assisted virtualization technology supported through Intel processor extensions, Intel VT-x, can be used.

13.4 Concolic Execution

The term Concolic stems from concrete and symbolic. Concolic execution is also known as dynamic-symbolic execution because it combines concrete and symbolic executions. Concolic execution is implemented by instrumenting the code under test so that the program execution pauses after each instruction, and the symbolic execution of the same instruction proceeds before the subsequent instruction execution. Then, the next instruction is executed, and this process is repeated as far as the execution is completed. Instrumentation tools such as CIL [1], PIN [2], VALGRIND [3], and DYNAMORIO [4] can be used to both instrument and generate path constraints.

Concolic execution is a solution to alleviate the obstacles of symbolic execution. It starts the simultaneous concrete and symbolic execution over random concrete and symbolic inputs. While proceeding with the execution of each instruction along a path, if an invoked function cannot be executed symbolically, its return value, obtained by the concrete execution, will be used to enable the symbolic execution.

Concolic execution is aimed at path exploration while collecting path constraints. During a concolic execution, the path constraint is gradually formed as the conjunction of branching conditions encountered along the path. Each time a branching condition and the conjunction of its negation with the previous conditions is given to a constraint solver to generate test data. The solver will attempt to find the range of input values that satisfy the negated condition. As a result, the solver will generate input data covering the path.

It should be noted that path constraints are collected while executing the program in symbolic mode. As an example, consider the code fragment and the results of its Concolic execution in Fig. 13.10.

As shown in Fig. 13.10, the initial input to the program is (0,0,0). While running the program over this input, the path constraint will be as shown in row 1 of the table in Fig. 13.10. To exploit a second execution path, the branching condition, "$a! = 1$", is negated. The resultant path constraint ($a! = 1$) is then passed to a constraint solver. The solver returns the values 1, 0, and 0 for the input variables a, b, and c, respectively. The concolic execution restarts with the new test data (1,0,0). This time the path constraint will be "$(a == 1) \wedge (b! = 2)$". Again the branch condition "$b! = 2$)" is negated, and the constraint "$(a == 1) \wedge (b == 2)$" is passed to the solver. This time the solver returns the values 1, 2, and 0 for the input variables a, b, and c, respectively. Calling f with these arguments, the symbolic path constraint will be

```
// Test input a, b, c
void f(int a, int b, int c)
      S1. { if (a == 1)
      S2.     if (b == 2)
      S3.         if (c == 3*a+b)
      S4.             Error(); }
```

	(a,b,c)	Path constraint	Negated condition
S1	(0,0,0)	(a!=1	a==1
S2	(1,0,0)	(a==1)∧(b!=2)	(b==2)
S3	(1,2,0)	(a==1)∧(b==2)∧(c!=3*a+b)	(c==3*a+b)
S4	(1,2,5)	(a==1)∧(b==2)∧(c==3*a+b)	

Fig. 13.10   A function, f, and its concolic execution results

"$(a == 1)\wedge(b == 2)\wedge(c! = 3 * a + b)$". Negating the branch condition $(c! = 3 * a + b)$", and passing the constraint "$(a == 1)\wedge(b == 2)\wedge(c == 3 * a + b)$" to the solver, this time the solver will return the values, 1, 2 and 5 for the input variables a, b and c, respectively. Calling the function, f, with these actual argument values, the ERROR() function, in line 4, will be executed.

Test data for the function f in Fig. 13.10 were generated by using random inputs for the program's first run; four other input data were automatically generated while running the program with the test data generated by the solver. However, using a random method for test data generation, it could happen that even after 1000 trial executions, all the execution paths could not be explored. An overall algorithm for the concolic execution is given in Fig. 13.11.

Below are a few questions to bear in mind when designing a new Concolic execution tool or selecting an existing one:

1. What representation is used for transforming the program under test to a suitable form for instrumentation?
2. Which functions are inserted by the instrumentation tool into the program to keep track of program execution both in symbolic and concrete modes?
3. Which strategy is applied by the tool to select the next execution path?
4. Which solvers are used by the tool?
5. Are the solvers nonlinear?

13.4.1   Instrumentation for Concolic Execution

Concolic execution begins with the instrumentation of the code under test. The instrumentation is aimed at inserting additional code into the program under test to allow

Algorithm: Concolic Execution
Input: instrumented program
Output: test data
Steps:

1. Instrument the program
2. Select input variables for symbolic execution.
3. Generate random concrete input values.
4. Concretely run the program over the generated input values.
5. Collect the symbolic path constraint along with the concrete execution of the path, P.
6. Select and negate a branch condition to obtain a new path, Q, constraint.
7. Invoke a solver to solve the constraint, Q, and generate new concrete inputs.
8. If a solution is found, go to step 4 with the newly generated concrete inputs.
9. If not found (or could not solve) declare an infeasible path and go to 3.

Fig. 13.11 Concolic execution algorithm

simultaneous dynamic and symbolic execution of the program. The most straightforward approach to instrumentation is to equip the source or intermediate representation of the code under test with function calls, to work out symbolic expressions, and link symbolic and concrete states while executing the program. For instance, CUTE [9] is a concolic testing environment. It uses the CIL [10] framework to translate and monitor the execution of C programs.

Concolic testing of binary code is supported by the dynamic instrumentation of the binary code when it is loaded into the memory. Thus, there will be no need for the source code, and binary code can be automatically tested. This approach can detect failures caused by executable code. SAGE [14] is a concolic testing tool that has been mainly applied to identify security bugs in ×86-binaries. A limited number of concolic execution engines can work on binary programs, and Triton is one of the strongest. Other engines include S2E, SAGE, Mayhem, and Bitblaze. Triton is faster because its uses C++. Triton provides APIs both for use in C++ and Python. It supports ×86 and ×64 instructions. The Triton implementation is based on Intel Pin. Other projects like S2E, SAGE, Mayhem, and Bitblaze like Triton do binary analysis, but Triton has a higher level language and, more importantly, uses Pin in the background.

Instead of direct instrumentation of source, intermediate or binary code, virtual machines can be instrumented as runtime environments for executing binary code. This approach to instrumentation modifies virtual machines on which the target programs are supposed to run. One advantage of this approach is that it can access runtime information provided by the virtual machine. Java PathFinder (JPF) implements software model checking by executing the program under test in its own virtual machine (VM) that can backtrack the entire state of the program [23]. JPF combines runtime analysis with model checking so that warnings produced from the runtime analysis are used to guide a model checker.

The instrumentation task can be simplified by transforming the code under test into a simplified intermediate form, such as the three address codes. A C program is preferably translated into C Intermediate Language (CIL) [10], a simplified subset of the C programming language. CIL comprises fewer programming constructs than C. For instance, in CIL, all loop constructs, such as *while, for,* and *do-while,* are converted into the while loop. The while loop takes advantage of explicit break statements to terminate. A sample C program and its corresponding code in CIL are presented in Table 13.3.

It should be noted that Microsoft Common Intermediate Language, MSIL, also known as CIL, differs from C intermediate language. All the compilers in the .Net framework translate their source code into MSIL.

Many concolic test data generation tools, such as CROWN [24], CREST [12], and CUTE [13], are instrumented by CIL. These tools operate on C programs only because CIL requires C source code. The CIL compiler translates C programs into CIL. After translating the program under test into CIL, function calls are inserted throughout the CIL code to enable the program's simultaneous concrete and symbolic execution. The calls allow symbolic execution of the program under test by a symbolic stack and a symbolic memory map. The stack facilitates the 'postfix' symbolic evaluation of the expressions. The functions should be written in OCaml programming language before they can be accessed through the CIL code because it is developed in OCaml.

13.4.2   Solving the Path Explosion Problem

Theoretically, dynamic symbol execution can proceed as far as the program's paths are covered. However, in practice, realistic programs may have many execution paths, and as the program size increases, the number of execution paths could increase exponentially. This is called the path explosion. For example, consider a simple program consisting of an iteration loop with a branch condition inside it. If the program loop is repeated 20 times, the number of program execution paths will be 2^{20}.

The strategy used to search the execution tree for appropriate constraints will significantly impact the speed of test data generation to achieve the maximum possible code coverage. Dynamic symbolic execution is inevitable to face an important problem of how to choose a proper path in the huge path space. Some search strategies are adopted for the dynamic symbol execution. These strategies improve

Table 13.3 A sample C program and its translation in CIL

	Program C	CIL translation	
1	int main(void)	/* By CIL v. 1.3.7 */	break;
2	{	//print_CIL_Input is true	}
3	int x, y;		#line 8
4	for(int i = 0; i < 5; i++)	#line 1 "ex2.c"	i += 2;
5	{	int main(void)	Cont: // CIL Label
6	if(i == 5) continue;	{	#line 4
7	if(i == 4) break;	int x;	i++;
8	i += 2;	int i;	}
9	}	{	#line 10
10	while(x < 5)	#line 4	while (x < 5)
11	{	i = 0;	{
12	if(x == 3) continue;	#line 4	#line 12
13	x++;	while (i < 5)	if (x == 3)
14	}	{	{
15	}	#line 6	#line 12
16		if (i == 5)	continue;
17		{	}
18		goto __Cont;	#line 13
19		} #line 7 if (i == 4) { #line 7 }	X++; } #line 15 return (0); }

The #line instruction in the CIL source indicates the line number of the relevant instruction in the C source code

the coverage of symbolic execution, which enables dynamic symbol execution to detect bugs more efficiently. The search strategies are described in Sect. 13.6.

There could be a massive number of execution paths in a program due to many loops containing many branching instructions or an enormous number of instructions that need to be executed. Therefore, generating test data to cover all the execution paths may take a long time. The time spent solving path constraints by a solver is the main portion of the time taken to use a concolic testing method to generate test data. One way to reduce the difficulty is to select the basis paths. Basis paths are described in Sect. 13.5.

13.5   Basis Paths

As described in Sect. 3.7, all possible execution paths in a program are a combination of atomic level, loop-free, and unique paths called basis paths. Path constraint of a basis path covers a relatively higher number of branches. In many real applications, it is preferred to cover all program branches rather than the execution paths. For instance, the following path constraint consists of four branching conditions:

$$(a < 1) \wedge (b < 2) \wedge (c < 3) \wedge (d < 4)$$

In this case, if each time one of the branching conditions is negated, there would be four basis paths constraints as follows:

$$(a \geq= 1) \wedge (b < 2) \wedge (c < 3) \wedge (d, < 4)$$
$$(a < 1) \wedge (b \geq= 2) \wedge (c < 3) \wedge (d < 4)$$
$$(a < 1) \wedge (b < 2) \wedge (c \geq= 3) \wedge (d < 4)$$
$$(a < 1) \wedge (b < 2) \wedge (c < 3) \wedge (d \geq= 4)$$

Basis paths are linearly independent vectors called basis path sets [7]. This core set is the most miniature set of paths that can be combined to create any other possible paths within the function under test. Here, an execution path is represented as a vector $v = <v_1, v_2, ..., v_n>$, in which v_i is the number of times that the edge $e_i \in$ CFG (control flow graph) of the function under test appears in the path. A path p_1 is linearly dependent on $p_2 =< a_1, a_2, ..., a_n>$ if and only if there exist coefficients β_0, $\beta_1, \beta_2, ..., \beta_n$ such that:

$$P_1 = \beta_0 + \beta_1 * a_1 + \beta_2 * a_2 + ... + \beta_n * a_n$$

The total number of linearly dependent vectors in the basis set is considered as the cyclomatic complexity, CC(G) of the CFG G where:

$$CC(G) = no. \ edges \ (G) + no. \ vertices \ (G) - 2 * no. \ disconnected \ (G)$$

Natural loops within the program control flow graph (CFG) are collected instead of looking for linearly independent paths. A natural loop has a single entry point that dominates all the nodes in the loop and a back edge that enters the loop entry point. Two natural loops are distinct if one is not a cyclic permutation of the other. The two back edges connecting the loop entry to the CFG entry point and the CFG exit to the loop exit points are added to the CFG to complete paths generated within a natural loop. As an example, consider the CFG in Fig. 13.12.

$1{\rightarrow}2{\rightarrow}3{\rightarrow}4{\rightarrow}6{\rightarrow}7$
$1{\rightarrow}2{\rightarrow}3{\rightarrow}5{\rightarrow}6{\rightarrow}7$

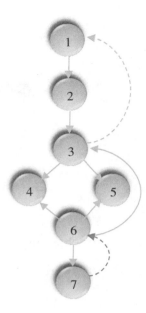

Fig. 13.12 A CFG with two added back edges $(7 \rightarrow 6, 3 \rightarrow 1)$

13.6 Search Strategies

Starting with random input data, all the symbolic branching conditions, bc_i, encountered along the execution path and their concrete execution results (true or false) are kept, and the path constraint, pc, is gradually built as follows:

$$pc_1 = bc_1 \wedge bc_2 \wedge \ldots \wedge bc_n$$

For the next run of the program under test, a branching condition, bc_j, is negated, and as a result, a new path constraint pc_2 is generated as follows:

$$pc_1 = bc_1 \wedge bc_2 \wedge \ldots \wedge \text{not } bc_j$$

A concrete input vector for another execution path could be generated by solving the new path constraint, pc_2, with a constraint solver. If the solver cannot solve the path constraint, the path will be marked as infeasible. Afterward, the program is executed with the generated input vector. This process is repeated as far as acceptable coverage is achieved.

As described above, in dynamic symbolic (concolic) testing, along with the concrete and symbolic execution of the program, the corresponding symbolic tree is

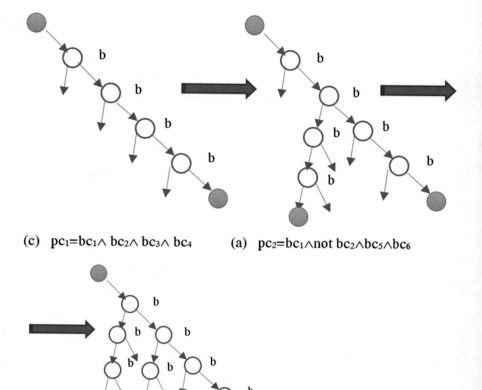

(c) $pc_1=bc_1\wedge bc_2\wedge bc_3\wedge bc_4$ (a) $pc_2=bc_1\wedge not\ bc_2\wedge bc_5\wedge bc_6$

(b) $pc_3=not\ bc_1\wedge bc_7\wedge bc_8$

Fig. 13.13 Selecting and negating a branch condition

gradually completed. Therefore, an execution tree may include only feasible execution paths. Figure 13.13, illustrates the process of completing an execution tree by adding execution paths to the tree.

As shown in Fig. 13.13, after the execution of the program under test is terminated, a branch condition addressed by a node in the execution tree is selected and negated to generate the next path constraint. A key issue is which search strategy should be applied to select the next node in the execution tree. The appropriate choice of strategy will highly improve code coverage. To this aim, I suggest combining search strategies to achieve higher code coverage. The evaluations show that the combined strategy could provide relatively higher coverage. Search strategies are further described in the following section. A generic search algorithm is presented in Fig. 13.14.

Algorithm Generic-Search
  Input: A program, Prog;
 Output: inputDataSet,
          Coverage-information,
          ExecTree;

    Select input-vector, randomly;
    inputDataSet ← inputVector;
    ExecPath ← ConcolicExec(Prog, inputVector);
    ExecTree ← ExecPath;
  While Coverage-rate **is not** adequate **do**
        brCond ← **Select** a branch condition addressed by a node **in** ExecTree
        ExecPath ← **Find** an execution path, including the branch condition, **in**
               ExecTree
      PathConstraint ← true;
      Forall branchCondition in ExecPath **before** brCond **do**
           PathConstraint ← PathConstraint ∧ branchCondition;
         PathConstraint ← PathConstraint ∧ not brCond;
      InputVector ← SMTSolver(PathConstraint);
      If InputVector **is** Empty, **then**
        Mark ExecPath as infeasible-path;
      else
        ExecPath = ConcolicExec(Prog, inputVector);
        Add ExecPath to ExecTree;
        Add inputVec to inputDataSet;
      end if;
  end while;

Fig. 13.14 A generic search strategy algorithm

Three search strategies are used for selecting branch conditions: classical search, heuristic search, and pruning strategies [11].

13.6.1 Classical Search Strategies

The classical search strategies are based on graph traversal algorithms such as burst-first (BFS), depth-first (DFS), and random search. DFS algorithms are applied by concolic execution tools, such as DART [12] and CUTE [13], in the early days. It is the earliest search algorithm used in concolic execution. While traversing a path from the root to a leaf of the symbolic execution tree, DFS builds an incremental path constraint as the conjunction of the branch condition comes across on its way.

On the other hand, BFS algorithms tend to choose a completely different execution path every time. In a random branch search, every time, a branch condition within the current execution path is randomly selected and negated. When applying a DFS algorithm, care should be taken to ensure that input vectors are not generated repeatedly to execute loops or recursive calls.

On the other hand, the BFS strategy gives little chance of being selected to the branches that only appear later in the execution path. In a queue-based search strategy, the tree nodes are added to the end of a queue. When a node is requested, the first node in the front of the queue is returned. Stack-based search strategies push nodes in a stack. An algorithm for DFS is given in Fig. 13.15

As an example, consider the C code fragment shown in Fig. 13.16.

Consider the C program presented in Fig. 13.16. Suppose the program initial execution is on $m = 3$ and $n = 4$, yielding the path (1, 3, 4, 11, 15, 16, 5, 6) with the path constraint $\neg (m > n) \wedge \neg (m = 5)$. Considering the DFS algorithm, given in Fig. 13.15, the depth-first search first forces branch $11 \rightarrow 12$. Solving the corresponding path constraint, $\neg (m > n) \wedge (m = 5)$, results in, e.g. $m = 5$ and $n = 7$. Executing the program with these inputs, will cause it to go through the path (1, 3, 4, 11, 12, 14, 16, 5, 6) with the path constraint $\neg (m > n) \wedge (m = 5)$.

This time the first branch condition is negated as next path constraint will be $m > n$. Solving the path constraint, $m > n$, results in, e.g. $m = 30$ and $n = 2$.

Algorithm **DFS**
    Input: program-under-test Prog,
            Array-of-anded-predicates PathConstraint[ ],
            Last-negated-branch-condition LastNegatedBrCond;
    Output: Set-of-input-vectors InputDataSet;
            brCondNo ← Length(PathConstraint);
   while brCondNo >= LastNegatedBrCond **do**
     brCond ← PathConstraint[brCondNo];
     nextPathConstraint ← true;
     Forall branchCondition **in** ExecPath **before** brCond **do**
         nextPathConstraint ← nextPathConstraint ∧ branchCondition;
     End Forall;
     nextPathConstraint ← nextPathConstraint ∧ not brCond;
     InputVector ← SMTSolver(nextPathConstraint);
     If InputVector ± Empty then
       nextPathConstraint ← ConcolicExec(Prog, inputVector);
      Add inputVec **to** inputDataSet;
      DFS(Prog, nextPathConstraint, , brCondNo + 1);
     end if;
     brCondNo ← brCondNo – 1;
   end while;

Fig. 13.15 DFS algorithm for concolic execution

void main(int m, int n)	Int f1(int m, int n)	int f2(int m, int n)
{	{	{
1. if (m > n)	7. If(m < 0)	11. If(m == 5)
2. n = f1(m, n+2);	8. exit(1);	12. If(m*n > 40)
Else	Else	13. exit(2);
3.. m = m+1;	9. m = m +n;	Else
4. m = f2(m, n);	10. return n - m;	14. m = m * n;
5.	. }	else
writeln("%d",m);		
6. exit(0);		15. Writeln("%d",m);
		16. return m+n;

Fig. 13.16 A simple C program

Executing the program on these inputs yields the path (1, 2, 7, 9, 10, 11, 15, 16, 5, 6). The search explores the three aforementioned feasible paths and their corresponding input vectors. The number of branch conditions to be negated along each execution path can be restricted.

DFS attempts to navigate toward the most recently discovered nodes in the search tree. It tends to rapidly deep into the search tree to reach depth. Opposite to DFS, the breadth-first search (BFS) traverses the execution tree level by level towards the leaves. DFS may get stuck in loops, while BFS does not show any loop-related dilemma. However, in symbolic executions with large search space, it seems to be hard for BFS to reach certain depths of the execution trees.

13.6.2 Heuristic Search Strategies

Heuristic search strategies, applied for dynamic-symbolic executions, consider the amount of code coverage that may be achieved by selecting a branch. In greedy approaches, the branch leading to relatively higher code coverage is selected in each branching step. Generational strategy [14], CarFast [15], Control-Flow Directed Search [16], and Fitness-guided Search Strategy [17] are known greedy search approaches applied by dynamic-symbolic execution tools.

Carfast is a greedy strategy [14] that looks for those branch conditions, leading the program execution toward a branch or basic block with a relatively higher number of statements. The number of statements in each branch, transitively control-dependent on the branch condition, is calculated to this aim. Care should be taken when applying a greedy strategy such as CarFast, to ensure that the search space is not limited to those branches whose opposite branches are already covered.

The generational search strategy [15] generates input vectors for all the branch conditions along an execution path, and the execution path of the input vector with the highest coverage is selected as the next execution path. This search strategy also begins with randomly selected input data. After each symbolic-concrete execution is terminated, for each branch condition, bc_j, in the current execution path constraint, pc, a path constraint, pc_j, is generated as follows:

$$\forall j \in [1..|pc|] \Rightarrow pcc_i = bc_1 \wedge bcc_2 \wedge \ldots \wedge \neg bcc_i$$

Each newly generated path constraint is subsequently sent to a constraint solver to generate a concrete input vector, satisfying the constraint. While executing the program under test with each input vector, the amount of block coverage obtained by executing the path is computed. The input vector for the path with the highest block coverage is selected and added to a list, sorted by the amount of code coverage. The input vector with the highest code coverage is then picked off the list, and the process is repeated as far as the list is not empty.

A control-flow-directed search strategy [16] starts with a random input as the first data and executes the program under test with the selected test data. Then uses a DFS search to walk down the CFG along the execution path, path-1, while using the next uncovered branch condition, br_k, to form a new path constraint as the conjunction of $Br_1 \wedge Br_2, \ldots \wedge \neg Br_k$. Here, the branch conditions br_1 to br_{k-1} are assumed to be before Br_k along the execution path. The newly formed path constraint is given to a constraint to generate new test data. The process is continued recursively by running the code under test with the new test data.

For example, consider the C program given in Fig. 13.16. Suppose the program initial execution is on $m = 4$ and $n = 2$, yielding the execution path (1, 2, 7, 9, 10, 4, 11, 15, 16, 5, 6) with the path constraint $(m > n) \wedge \neg (m < 0) \wedge \neg (m = 5)$. This time the CFG-directed algorithm prefers to force branch 12 $\rightarrow$ 15 with the distance 1, rather than the other alternate branches, 3 and 7. Solving the path constraint $(m > n) \wedge \neg (m < 0) \wedge (m = 5)$, results in, e.g. $m = 5$ and $n = 7$. The intuition underlying CFG-directed search is that selecting an uncovered branch close to the current execution path may result in a relatively higher branch coverage.

Fitness-guided search strategies [17] use a fitness function to measure the fitness gain that could be achieved by exploring a specific path or branch. To compute the fitness gained by flipping a branching condition, the values to be added to the variables in the condition to get the flipped condition true can be considered. For instance, the fitness gain of the predicate $x > 100$, if the actual value of x is 90, will be 11. The fitness of the predicate $x > 100 \wedge y > 102$, if the value of both x and y is 90 will be 13. A path's fitness is considered the worst fitness value of the branch conditions along the path. Here, fitness value 0 is the optimal case.

Classical and heuristic approaches do not take full advantage of the vital information provided by dynamic-symbolic executions to reduce the search space. The following sections propose the use of integrated and combined search strategies to make use of the strength of the search strategies and avoid their weaknesses, as well as take full advantage of the available information provided.

13.6.3 Integration of Search Strategies

Most potential search strategies have their own merits and pitfalls. To take advantage of the benefits of different search strategies and avoid the pitfalls, the use of meta-strategies, taking advantage of different search strategies, is proposed. The question is which combination of search strategies in what order and dependent on what program execution states could be most preferably applied.

Search strategies applying certain heuristics or structural attributes to select branches could be biased toward certain parts of the execution tree. For instance, BFS favors the first branches in the program paths, while the depth-first search is biased toward the final branches. The bias could be reduced by applying a combination of different search strategies. For instance, Pex [18] provides various basic strategies and combines them into a meta-strategy. This meta-strategy starts with a root node in the execution tree and, before committing to provide the next node, selects an appropriate strategy and asks for the strategy for the node. The main strategy of Pex can be defined as follows [19]:

ShortCircuit(CodeLocationPrioritized[Shortest], DefaultRoundRobin)

A search frontier decides in which order the search is performed. To provide and integrate sets of standard search frontiers, Pex has offered an interface class called IPexBuiltinSearchFrontiers. This interface class provides several search strategies, each implemented as a separate method. For instance, a method called *Shortest* implements a sort of depth-first strategy. To this aim, it orders the nodes ascendingly by their depth in the execution tree. The strategy begins with the first node in its ordered list and provides the next node in the list each time it is asked for a node. Random is another method that provides a random search strategy frontier.

The CodeLocationPrioritized[Shortest] method creates code location prioritized frontiers. This is achieved by partitioning all the nodes in an execution tree into equivalence classes based on the control-flow locations (branches) of which the nodes are instances. Here, an inner frontier of type Shortest is considered for each equivalent class. When a node is requested, an appropriate choice is made among all the classes. Then the inner frontier of the selected class is asked to provide the node.

Short-circuit is a meta-search strategy applied to combine several basic search strategies. The *ShortCircuit* method of the IPexBuiltinSearchFrontiers interface class integrates search frontiers. ShortCircuit $(y_0,..,y_i, ..., y_j, ..., y_n)$ is a strategy combinatory that does not ask for a later strategy y_j, as far as an earlier strategy y_i, such that $i < j$, provides more execution tree nodes to flip.

Round-robin is another meta-search strategy applied to select a strategy, in turn, one after the other in a loop. DefaultRoundRobin is a method implementing a search frontier that always processes less visited branches first. An implementation of this method is presented in [19].

Pex facilitates the implementation of new strategies. A new strategy can be implemented as a subclass of a Pex class, PexSearchFrontierAttributeBase. The new search strategy subclass should implement an interface class, IPexSearchFrontier. This interface class has the following methods:

```
bool HasContent { get; }
bool IsProgressing { get; }
void Add(IExecutionNode node);
void Remove(IExecutionNode node, bool used);
void LocationUpdated(IExecutionNode node);
bool Contains(IExecutionNode node);
bool TryRemove(out IExecutionNode node);
int SkipMany();
void DumpStatistics( DumpTreeWriter writer, int depth);
```

The HasContent property determines whether the search frontier does not have memory or is empty. If the frontier is not empty, then the IsProcessing method can be invoked to determine whether the search frontier is in progress and should be given a higher priority. SkipMany() skip many nodes to speed up the search process for the price of incompleteness. Contains() checks whether a node is already contained. The Remove() method is invoked to remove a node from a search strategy. When a node is added, it can be accessed by all the included strategies in a combined strategy.

13.6.4 Combined Search Strategies

The basis of the integrated search strategy [18, 19] is to partition an execution tree node into equivalent classes based on specific parameters, such as offset, depth, and level of nodes, applied by a tree search strategy. Also, an internal search strategy is applied to each class. When asking for a node, a fair selection is made among all the equivalent classes, and then an internal search strategy is applied to the matching class to find the appropriate node. Another equivalence class will be selected if the internal strategy does not return any node. If no other class is left, no nodes will be returned.

It should be noted that the internal search strategy itself can also be applied as a basis to partition each of the classes based on parameters and heuristics applied by the strategy. For example, an execution tree node may be partitioned based on the depth of the nodes into several classes, and then each class itself is randomly partitioned into several classes. To name such combined strategies, we prefix the name of the internal strategy with the name of the feature used for partitioning the execution tree. For instance, the aforementioned combined strategy is named [DepthFirst [Random]]. As another example, a search strategy named OffsetInMethod [DepthFirsy [Random]], first partitions the execution tree nodes into equivalent classes based on the offset of the corresponding branching conditions in the methods of the program under test. Then, the nodes in the same offset are partitioned into equivalence classes based on their depth in the execution tree. Finally, the nodes in each class are selected randomly.

New nodes are created and added to the program execution tree by running and monitoring a program execution. At the same time, the new nodes are added to the equivalent class to which they should belong. Each equivalence class can be partitioned into three categories: *new*, *active*, and *inactive, to* provide a better basis for selecting the nodes. Those branching nodes not selected yet are categorized as 'new'. An active partition includes all the branching nodes selected at least once and can be selected again. Finally, a branching node is moved to an *inactive* partition if the node and its negation are selected.

The most fundamental parts of any search strategy are two functions, Add and GetNode. As the program under test is being executed and monitored, the Add function is called to insert a node addressing a branch condition into the program execution tree.

13.6.5 A Function to ADD Nodes to an Execution Tree

Figure 13.17 presents a pseudocode description of the ADD function. As mentioned earlier, this function is invoked to add a new node to execution trees.

This function consists of three steps. In the first step, the matching equivalent class of the branching node to be inserted into the execution tree is obtained (line 1). A node class may indicate its depth in the execution tree, offset in the source code, or distance from a target or any other attribute. In the second step, the input node is added to the internal search strategy corresponding to the matching class (line 2). The third step (lines 3 to 9) determines the matching class of the branching node in terms of being new, active, or inactive.

Both the functions Add and GetNode are implemented in Python to be invoked by a dynamic-symbolic execution environment, PyExZ3 [20]. PyExZ3 [20] is a simple open source for dynamic-symbolic execution in Python.

The source code is available on GitHub. PyExZ3 has a main class, ExplorationEngine, that ties everything together. A class, Loader, takes as its input the name of a Python file to import and finds a given function name as the starting point for symbolic execution. This starting point is wrapped via an object of the type FunctionInvocation. Symbolic expressions are built via symbolicType and its subclasses. While executing a path, callbacks collect the path-constraint into PathToConstraint.

Fig. 13.17 Pseudo code description of the ADD function

```
0.   void Add(node)
1.   C = Class of node
2.   Add node to Inner Search Strategy of C
3.   if( C is in Active Classes)
4.       return;
5.   if (C is in Inactive Classes)
6.           Add C to Active Classes
7.           Remove C from Inactive Classes
8.   else
9.           Add C to New Classes
```

```
def Add(node):
    #1. Compute the depth of node in the execu-tion tree.
    c=find_depth(node)
    for i in "{c}".format(c=c)list ["active"]:
    #2. Do nothing if node belongs to an active class
        if node==i:
            return 0
    #3.if node belongs to an inactive class remove it from
    #    inactive class and add it to the active class
    for i in "{c}".format(c=c)list ["Inactive"]:
        if node==i:
                "{c}".format(c=c)list ["active"].add(node)
                "{c}".format(c=c)list ["Inactive"].delete(node)
        # make a new class and add node to the class if the node
        # does not belong to neither an active nor an inactive class
        else:
            "new{m}".format(m=i)list.add(node)
            count=count+1
```

Fig. 13.18 Implementation of the ADD function in Python

The created path constraint is represented by Constraint and Predicate classes. All the discovered path constraints are added to the end of a double-ended queue maintained by ExplorationEngine. A method, ADD, is invoked to add branch conditions encountered, across the execution path, to the execution tree. Each time a method, called WhichPath, of the class PathToConstraint is invoked to get a node from a function, GetNode, and negate it. Figure 13.18 presents a Python implementation of the Add method.

13.6.6   A Function to Look for a Node in Execution Trees

Figure 13.19 presents a pseudocode description of the GetNode function. As mentioned earlier, this function is invoked to look for a new node in the execution trees.

This function consists of two phases. In the first phase (lines 1–7), one of the classes from the *new* class set is randomly selected and removed from the set (lines 2 and 3). Then, the internal strategy of the class is asked for a node. If successful, the returned node is assigned as the output node, and the selected class is added to the active classes (lines 5 to 7). Otherwise, another class is selected.

In the second phase, a class is randomly selected from active classes (line 9). Then, the internal strategy of the class is activated to get a node. In the event of success, the returned node will be the output node (lines 9 to 12). Otherwise, the class is added to inactive classes, and another active class s selected (lines 14 and 15).

```
bool GetNode(out node)
    1-      while (NewClasses.Count > 0)
    2-                  C ← Choose a Random Class in NewClasses
    3-                  Remove C from NewClasses
    4-                  innerSF ← GetInnerSearchFrontier(C)
    5-                  if (innerSF.Getnode(out node))
    6-                              Add C to Active Classes
    7-                              return true;
    8-      while (ActiveClasses.Count > 0)
    9-                  C← Random C in ActiveClasses
    10-                 innerSF ← GetInnerSearchFrontier(C)
    11-                 if (innerSF.Getnode(out node))
    12-                             return true;
    13-                 else
    14-                             Remove C from Active Classes
    15-     Add C to Inactive Classes
    16-     node = null;
    17-     return false;
```

Fig. 13.19 Pseudo code description of GetNode function

13.7 Tools and Techniques

Recent tools for manual and automated testing, including instrumentation techniques used in Crown, Microsoft testing tools, and Microsoft Intellitest for concolic testing and test data generation, are the subjects of this Section.

13.7.1 Crown

Crown is a concolic test generation tool for C. The crown project is based on the crest project and is a branch of the project at GitHub. The project's source code is written in C++ but only supports C-language syntaxes. Many vital structures, such as classes, threads, inheritance, and vector functions, cannot be compiled with this program.

To run the Crown tool, you must first transfer the source of the Crown program to Linux ubuntu16.04. After that, you can compile the Crown program using the GCC compiler. Crown translates the source of the program under test into CIL. It accepts a C program like Test_1.c as its input, compiles it, and outputs the file Test_1.cil.

Crown uses a birth-first search strategy to select and negate constraints and invokes the Z3 solver to solve path constraints and generate test data to examine the paths. Symbolic execution is supported by a symbolic stack and a symbolic memory map data structures to keep symbolic expressions and their concrete values. For instance, crown translates the conditional expression '$a*b > 3 + c$' into the following code:

```
Load(&a, a)
Load(&b, b)
ApplyBinOp(MULTIPLY,  a*b)
Load(0, 3)
Load(&c, c)
ApplyBinOp(ADD,  3+c)
ApplyBinOp(GREATER_THAN,  a*b > 3+c)
```

The instruction Load(&a, a) pushes the symbolic and concrete (real) values of the variable "a" onto the stack. The ApplyBinOp(MULTIPLY, $a*b$) instruction pops the two top stack cells, multiplies them symbolically and concretely, and stacks the results. When executing the above sequence of instructions, the concrete and symbolic values of the expression $a*b > 3 + c$ will be kept on the top of the stack. If the conditional expression was in an if statement:

if $(a*b > 3 + c)$ then ...

Crown uses the following branch instructions to decide whether to execute the then- or else-part of the "if" statements. This depends on the real value of the conditional expression at the top of the stack:

Branch(true_id, 1)

Branch(false_id, 0)

Crown assigns an id to each branch of the program. For instance, true_id and false_id could be 6 and 8, respectively.

Crown uses the instruction, Store(&a), to pop the concrete and symbolic values at the top of the stack and assigns the poped values to the variable "a." For instance, Crown translates the assignment statement $a = 3 + b$ into the following sequence of instructions:

Load(0, 3)

Load(&a, b)

ApplyBinOP(ADD, 3+b)

Store(&a)

The instrumentation operation for function calls is slightly more complicated than assignment statements. The function's parameters are pushed onto the stack for each function call. The values are assigned to the formal parameters in the function's body. For instance, consider the statement $z = max(a, 7)$. The return value of the function "max" is popped off the stack and stored in z by the following instructions:

HandleReturn([concrete returned value])

Store(&z)

13.7.2 Microsoft Testing Tools

Visual Studio provides testing tools that help to adopt manual and automated functional testing practices. These testing tools are accessible within the Visual Studio development environment so that quality can be assured throughout the development process. The unit testing capabilities of Visual Studio help developers and testers to look for errors in C#, Visual Basic, and C++ methods of classes. The unit test tools include:

1. Test Explorer: Visual Studio Test Explorer is a tool for following and managing unit tests. It acts as a hub for unit testing activities by providing a unified environment to run unit tests and visualize the results. Any unit test framework with an adapter for Test Explorer can be used. To open the Test Explorer window, go to Test⇒Windows⇒Test Explorer. The window includes a list of test cases provided by different projects or test suites.
2. Code coverage: The Visual Studio code coverage tools analyze code coverage results. These tools report the amount of the code covered by the unit tests and determine which parts of the program the tests have covered. Many coverage metrics include file, method, path, branch and statement, and MC/DC. Software safety standards necessitate code coverage analysis reports. With one command in Test Explorer, the amount of code covered by the unit tests can be observed.
3. Microsoft Fakes: Microsoft Fakes helps isolate certain parts of code for testing. Isolation of the code is provided by replacing each class out of the scope of the code with small empty classes, implementing the same interface. These empty classes are called stubs.
4. Coded UI test: Tests controlling applications via their user interfaces (UI) are the coded UI tests. For example, one may record a scenario to borrow a book at a website and then edit the generated code to add a loop that lends many books. The Coded UI Test Builder may record any manual UI test. In addition to recording manual test actions, the test builder generates code to replay the actions. The Coded UI Test Editor can be used to edit the generated code.
5. Live unit testing: Live unit testing provides an online facility to automatically depict test coverage and the status of the tests via dynamically executing tests whenever a given code is modified. It works with the three unit-testing tools of xUnit.net, NUnit, and MSTest.
6. Intellitest: Intellitest is a concolic testing tool. It automatically generates test data for the program under test. When running IntelliTest, it is possible to observe failing tests and apply any necessary modifications to fix them. The generated test data can also be saved.

As an example, consider the sample code, SampleCode, shown in Fig. 13.20. SampleCode is a console App (.NET Framework) project that includes a class called clsGrades.cs. The clsGrades.cs class itself consists of a method called Grade-Convertor. This method receives a decimal number as its input and returns a character.

Fig. 13.20 A sample method to be tested both manually and automatically

```
namespace SampleCode
{  public class clsGrades
    {  public string GradeConvertor(decimal score)
        {  if (score >= 90) return "A";
            if (score >= 80) return "B";
            if (score >= 70) return "C";
            if (score >= 60) return "D";
            return "F"; }
    }
}
```

After creating an object from a class built into the Main function, the GradeConvertor method is invoked to convert the grade.

```
namespace SampleCode
{public class Program
    {static void Main(string[] args)
        {  clsGrades ObjGrade = new clsGrades();
            var TheGrade = ObjGrade.GradeConvertor(75);
            Console.WriteLine(TheGrade);
            Console.ReadKey(); }}
```

The unit test class in Fig. 13.21 aims to verify the correctness of the code written for the GradeConvertor function and compare the amount of code coverage provided by manual and automated unit testing methods.

1. **Manual unit test**

   As shown in Fig. 13.22, a unit test project called SampleUnitTest is added to the current solution to test the method manually (Fig. 13.22).

Fig. 13.21 A class to test the GradeConvertor method

```
namespace SampleUnitTest
{  [TestClass]
    public class UnitTest
    {
        [TestMethod]
        public void Try_90_A()
        { clsGrades objGrade = new clsGrades();
            var TheGrade = objGrade.GradeConvertor(90);
            Assert.AreEqual("A", TheGrade); } }
}
```

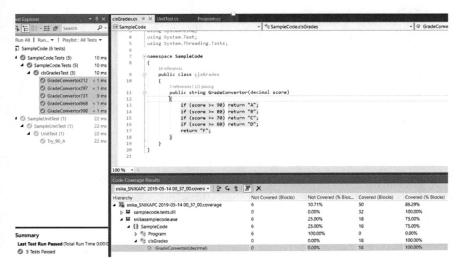

Fig. 13.22 A class to test the gradeconvertor method

Now by running the unit test created in the Test Explorer window, we can see that the expected result is met. In other words, the code written in the GradeConvertor function with the given input has produced the expected output. Nevertheless, remember that the function is only tested with a single value, and there is no guarantee that it will achieve the expected results with other inputs. It is undoubtedly challenging to produce unit tests manually for complex codes.

2. **Automatic unit test using IntelliTest**

To run the unit test automatically, right-click the GradeConvertor method code and select Run IntelliTest from the popup menu. In this case, the test data will be automatically generated and executed for all the execution paths of the method. By selecting the tests produced in the IntelliTest Exploration Results window, all data generated for testing the GradeConvertor can be stored and accessed later when needed.

Intellitest is an automated white-box test-data generation and unit testing tool for Dot NET. It uses concolic execution to collect path constraints. Pex is a famous unit testing tool that automatically generates test data targeted at branch coverage. In practice, IntelliTest and Pex are automatically adapted to generate test data for the unit test of C#. IntelliTest generates test data to cover the conditional branches. To avoid getting trapped in a particular part of the program under test by a fixed search order, Pex combines different fair strategies into a meta-strategy to make a reasonable choice amongst the strategies. Pex uses the Z3 solver to efficiently solve linear and nonlinear path constraints and generate their input values.

The generated input values are passed to parameterized unit tests to provide input for the symbolic execution engine of Pex [10]. A method with a predefined list of parameters is a concrete representation of a parametrized unit test. Therefore, it can be considered as a wrapper interface for the test inputs. Parametrized unit tests not only contain the calls to the unit under test but also look after the initialization, isolation, and assertion logic.

13.7.3   *Parameterized Unit Testing*

A unit is the smallest logically separatable piece of a program. Method, function, procedure, and subprogram are all units of the programs. Unit testing is a way to test a unit of the program under test. A unit test is a self-contained program checking an aspect of the program under test. Below, a typical unit test of a method, *Add*, is shown. The method adds an element to a list.

```
public void AddToListTest ()
    {
        // exemplary data
        object item = new object();
        ArrayList list = new ArrayList (1);
        // method sequence
        list.Add( item );
        // assertions
        Assert.IsTrue(list [0] == item );
    }
```

In the above unit test, an object is added to an empty array list and becomes the first element of the list. Below, a parameterized version of the array list unit test is presented. This time two parameters, List and item, are passed as parameters to AddToListTest. Provided that a given array list is not null, the AddToListTest method asserts that after adding an item to the list, the item is presented at the end of the list:

```
public void AddToListTest (ArrayList List, object item)
    {
    // assumptions
    PexAssume.IsTrue( List != null );
    // method sequence
    int len = List. Count;
    List.Add( item );
    // assertions
    PexAssert.IsTrue( List[len] == item );
    }
```

The above test is more suitable than the original test. It states that the item is contained in the end after adding an item to the list. The above unit testing method,

AddToListTest, can be invoked with various input values. Parameterized unit testing sometimes is referred to as data-driven tests.

13.8 Summary

Manual software testing is laborious and subject to human error. Yet, it is the most prevalent method for quality assurance. Parametrized manual testing opens the way for automatic testing. Automatic testing is achieved by frequent steering of the symbolic execution towards the branches that are most likely to lead to the target program statement and trimming the search space that is unlikely to reach the target. In this respect, automatic test data generation promises better effectiveness, especially for exposing latent bugs. Static approaches to automatic test data generation mainly use symbolic execution to define branching conditions in terms of the input variables.

The main idea behind symbolic execution is to use symbolic values rather than specific concrete values as program inputs to simulate program execution. When executing a program symbolically, all the program variables will be represented as symbolic expressions in terms of the program inputs. In this way, all the branching conditions in a program execution path will also be defined in terms of the program inputs. A solver software is then applied to find input data, satisfying the path constraint. A path constraint is the conjunction of all the branching conditions appearing along the path. The benefits of symbolic execution to software testing could be summarized as follows:

- To establish paths constraints,
- To generate input test data,
- To detect program faults and vulnerabilities,
- To distinguish infeasible paths.

Some benefits of symbolic execution to software testing are as follows:

1. Detect infeasible paths of the program
2. Generate test data for program input
3. Detect errors and vulnerabilities in the program
4. Repair programs.

The main difficulties faced with the symbolic execution are:

1. Complex control flow graphs could cause difficulty in symbolic execution and enhance the complexity of symbolic expressions.
2. Loops and recursions could cause a phenomenon called path explosions.
3. Function calls where the source code for the function is inaccessible can not be executed symbolically.
4. Complex data types such as pointers, sockets, and files may cause difficulty in symbolic execution.

5. Non-linear branch conditions and path constraints could cause difficulty for constraint solvers.

The concolic execution technique suggests simultaneous concrete (real) and symbolic execution of programs. In this way, when not possible, symbolic values are substituted with real ones to proceed with symbolic execution. Below, the discrepancies between symbolic and concrete executions are listed:

1. In symbolic execution, we use symbolic values for variables instead of the actual values of the inputs.
2. In symbolic execution, you can simultaneously run multiple paths from the program.
3. When running a path, we simulate many experiments by examining all the inputs that can run the same path.

Dynamic-symbolic execution often starts with random concrete test data. During the execution, the symbolic state of the program under test is updated. Also, each branching condition met along the execution path is inserted into an execution tree. All the branching conditions met along the execution path are conjunct to build the path constraint. Search strategies are applied to find appropriate branching conditions to negate and create test data to resume the execution along a different path. The difficulty is that most of these search strategies do not consider the vital information collected during symbolic-dynamic executions. To this end, the use of combined search strategies is proposed. Detailed descriptions of the search strategies are given in the next chapter.

13.9   Exercises

1. In practical terms, unit test development comprises an iterative workflow informed by code coverage. Develop a unit test, find the parts of the code that are not covered by the test, develop more tests to cover the parts, repeat as far as all of the code is covered.
2. Download and execute a concolic execution tool and answer the following questions:

   – Does it support composite data types such as strings, pointers, structures, and objects?
   – Does it handle black-box functions?

3. Use the Pex interface class, IPexSearchFrontier, to implement and evaluate the combined search strategy OffsetInMethod [DepthFirsy [Random]], described in Sect. 5.6.1.
4. Apply an Ant-colony algorithm to search through execution trees in such a way that if a node is selected and negated, the amount of pheromone deposited on the corresponding edge is increased.

5. How can the search space for selecting a branch in the program execution tree be restricted by using the information collected during dynamic-symbolic execution?

6. An effort-intensive task during test data generation is isolating the unit under test from external units and libraries. Develop an extension of Intellitest that uses Microsoft Fakes to isolate the unit under test from its environment.

7. QucikCheck for Java, Jcheck, and Feed4Junit tools are applied to generate test data for parametrized unit testing in Java. Use an example to demonstrate their differences.

8. Create a test plan, with IEEE 829 standard, for a method that tests whether an array of characters is in ascending sorted order. Define a class named SortSpot and define a method with the following signature:

    public static boolean isSorted(int [] arr)

    In your Test Plan:

    – Consider arrays up to size 10,
    – Make sure all possible inputs are covered,
    – Test your method on an array with a size greater than 10.

9. Use NUnit to test the method, CalcInvoice(int PurchasedItens[4]), of a class, Invoice. This method invokes the method, MyTax.TaxCalc(invoice_amount), where MyTax is an instance of the class, Tax. To test the method in isolation, the call to Tax.Calc, should be replaced with a stub method. A stub method just returns a simple but valid result. To this end:

    1. Define an interface class, ITax, including an empty method, with the same interface as TaxCalc,

        ```
        public interface ITax
            { public int TaxCalc( int amount );}
        ```

    2. Define the class Tax to implement ITax,

        ```
        Public class Tax : ITax { ... }
        ```

    3. Add two constructors to the class Invoice, as follows:

        ```
        public class Invoice
            { private Tax MyTax;
            public Invoice() { MyTax = new Tax; }
            public Invoice(ITax tax) { MyTax = tax; }
            ... }
        ```

4. Use NUnit to test the method TaxCalc as follows:

```
[TestFixture]
public class InvoiceTests
  {  [Test]
public void
TestTaxCalc()
    { MyTax = new StubTax();
    Invoice MyInvoice = new Invoive (MyTax);
    int result = MyInvoice.CalculateInvoice(100,105,223,142);
        Assert.AreEqual(result,450000);); }  }
internal class StubTax : ITax
  {   public int TaxCalc(int  amount) { return 0; } }
```

10. Unit testing of object-oriented programs could be a hard, if not impossible, job due to the tight coupling of classes because it will not be easily possible to test classes in isolation. Give an example of two tightly coupled C# classes and then show how to refactor them to use loose coupling through the Dependency Injection Pattern, also known as Inversion of Control (IoC). Use Mock framework and NUnit to test the loosely coupled classes. For more information, refer to [26].

11. Maven is a project management tool that can execute unit testing and integration testing. Besides, it provides a superset of features found in a build tool such as Ant. Apply Marven to Test a program of your choice, and uncover the program errors and provide a comprehensive report on your findings. For detailed information on how to install and use Marven refer to:

    https://examples.javacodegeeks.com/enterprise-java/maven/junit-maven-example/

12. Apply the test-driven development techniques and use JUnit to develop a Java program of your choice. The test-driven development technique begins with writing a failing test as if the program is already working. Afterward, the code is completed to pass the test. Finally, before moving to the next unit test, make sure that the code is refactored and optimized. Test-driven development technique suggests repeating the following three steps until the desired result is achieved.

1. Write a failing test as if the program is already working.
2. Complete the code to pass the test.
3. Refactor both the test and the code.

As an example, consider the following test for a function that prints 'Hello World!':

```
public void test(){ Greeting greeting = new Greeting();
assertThat(greeting.getMessage(), is("Hello world!")); }
```

For more details, refer to "https://javacodehouse.com/blog/test-driven-development-tutorial/".

References

1. Necula, G. C., McPeak, S., Rahul, S. P., and Weimer, W.: CIL: Intermediate Language and Tools for Analysis and Transformation of C Programs. In: CC'02: Proceedings of the 11th International Conference on Compiler Construction, (London, UK), pp. 213–228. Springer-Verlag (2002)
2. Necula, G.C.: CIL: Infrastructure for C Program Analysis and Transformation (v. 1.3.7). April 24, 2009, http://www.cs.berkeley.edu/~necula/cil
3. Luk, C., Cohn, R.S., Muth, R., Patil, H., Klauser, A., Lowney, P.G., Wallace, S., Reddi, V.J., Hazelwood, K.M.: Pin: building customized program analysis tools with dynamic instrumentation. In PLDI'05: Proceedings of the 2005 ACM SIGPLAN conference on Programming language design and implementation, pp. 190–200.
4. DynamoRio: Dynamic Instrumentation Tool Platform. http://code.google.com/p/dynamorio/
5. Nethercote, N., Seward, J.: Valgrind: a framework for heavyweight dynamic binary instrumentation. In Ferrante, J., McKinley, K.S. (eds.). PLDI, ACM, 2007, pp. 89–100
6. STP Constraint Solver. http://sites.google.com/site/stpfastprover/STP-FastProver
7. Wang, X., Jiang, Y., Tian, W.: An efficient method for automatic generation of linearly independent paths in white-box testing. Int. J. Eng. Technol. Innov. 5(2), 108–120 (2015)
8. Aho, A., Lam, M.S.: Compilers: Principles, Techniques and tools, 2nd ed. Addison Wesley (2007)
9. Sen, K.: CUTE : A concolic unit testing engine for C. In: Proceedings of the 10th European software engineering conference held jointly with 13th ACM SIGSOFT international symposium on Foundations of software engineering, pp. 263–272 (2005)
10. Necula, G.C., McPeak, S., Rahul, SP., Weimer, W.: CIL: intermediate language and tools for analysis and transformation of C programs. In: Proceedings of Conference on compiler Construction, pp. 213–228 (2002)
11. Liu, Y., Zhou, X., Wei-Wei, G.: A survey of search strategies in the dynamic symbolic execution. In: ITM Web Conference, pp. 3–25 (2017)
12. Godefroid, P., Klarlund, N., Sen, K.: DART: directed automated random testing. ACM Sigplan Notices Vol. 40. No. 6. ACM (2005).
13. Sen, K., Marinov, D., Agha, G.: CUTE: a concolic unit testing engine for C. In: ACM, SIGSOFT Software Engineering Notes. Vol. 30., No. 5. ACM (2005)
14. Godefroid, P., Levin, M., Molnar, D., et al.: Automated white box fuzz testing. In: NDSS (2008)
15. Park, S., Hossain, B.M.M., Hussain, I., Csallner, C., Grechanik, M., Taneja, K., Fu, C., Xie, Q.: CarFast: achieving higher statement coverage faster. In: Proceedings of the ACM SIGSOFT 20th International Symposium on the Foundations of Software Engineering, FSE'12 pages 35:1{35:11, New York, NY, USA (2012)
16. Cadar, C., Engler, D.: Execution generated test cases: how to make systems code crash itself. In: Proceedings of the 12th International Conference on Model Checking Software, Berlin, Heidelberg, (2005)
17. Kaiser, C.: Quantitative analysis of exploration schedules for symbolic execution, degree project, Kth royal institute of technology, school of computer science and communication Stockholm, Sweden (2017)
18. Halleux, N.T.J.: Parameterized Unit Testing with Microsoft Pex. In: Proc. The/FSE, pp. 253–262 (2005)
19. Tillmann, T.X.N.: Fitness-guided path exploration in dynamic symbolic execution. In: IEEE/IFIP International Conference on Dependable Systems & Networks (2009)
20. Ball, T., Daniel, J.: Deconstructing dynamic symbolic execution, Technical report, Jan. (2015)
21. BeaEngine Sweet Home, http://beatrix2004.free.fr/BeaEngine/index1.php
22. Bergmann, V.: Feed4JUnit Homepage. http://databene.org/feed4junit (2011)
23. Pathfinder, J.: A Model Checker for Java Programs, [Online]. http://ase.arc.nasa.gov/visser/jpf/
24. CROWN: Concolic testing for Real-wOrld, https://github.com/swtv-kaist/CROWN

25. Downey, A., Monje, N.: Think Ocaml: how to think like a (functional) computer scientist. Green Tea Press (2008)
26. Oliveira, M. R.: How design patterns can help you in developing unit testing-enabled applications (2007)

Chapter 14
Automatic Test Data Generation for Domain Coverage

14.1 Introduction

Test data generation techniques optimally seek to provide test data sets that, when applied, ensure the accuracy of the program under test, or expose any existing error. In 1972, Dijkstra made this alarming statement that software testing can show the presence of bugs but never their absence [11]. Following this alarming statement, test adequacy criteria emerged. Ideally, an adequate test suite provides enough test data to ensure the correctness of the program under test. The question is, how much test data is enough to test? Sometimes, one has to run a certain path, branch, or statement several times with different input data until the fault is located.

Buffer overflow, null-pointer-assignment, divide-by-**zero**, array boundary violation, data race, and many other latent faults do not reveal themselves unless a certain combination of input data is used to run the faulty statements. Even after detecting a faulty path, branch, or even statement, sometimes it is required to execute the program several times with different test data to determine the cause of the fault and locate it. Sometimes finding a bug is like finding a needle in a haystack. So how can we ensure a program's accuracy?

Existing coverage criteria confirm the adequacy of a test data set if the data set includes enough data to examine a targeted part, such as execution paths, only once. So the question is, what criterion can ensure the adequacy of a given test suite? My answer is, first of all, to try to separate the data generation task from the fault localization. No matter what fault localization technique such as black box, white box, automated or manual, is going to be used, the fact is that if the subdomains of the input domain to cover a given test target are available, then the tester can generate as many test data as required. For instance, if the target is to examine a particular path, and the subdomains are I:[1..20] and J:[1..10], then 20*10 test data can be produced manually or automatically.

A sophisticated test data generation technique should provide subdomains of the input domain to execute certain parts such as paths, branches, nodes of the control flow graph, and statements of the program under test. We have proposed a new

© The Author(s), under exclusive license to Springer Nature Switzerland AG 2023
S. Parsa, *Software Testing Automation*,
https://doi.org/10.1007/978-3-031-22057-9_14

criterion, domain coverage [1]. Domain coverage is the percentage of a program input domain that covers its segments, such as path, branch, and statement.

For instance, suppose the actual subdomain covering a particular program branch is i:[1..15] and j:[4..25]. If a certain algorithm detects the subdomain i:[1..6] and j:[1..11], then according to the domain coverage criterion, the coverage of the domain provided by this algorithm is:

$$\text{Doman} - \text{coverage} = (6 * 10)/(15 * 20) * 100 = 20\%$$

Domain partitioning is a black-box technique that partitions the input space into regions that map the input sub-domains into desired outputs. The primary characteristic of domain partitioning strategies [12] is partitioning the input domain and selecting test data from each partition. All these strategies can be applied to a subdomain covering desired parts of the code under test to locate the root cause of the failure or determine whether there is a fault.

Data flow testing criteria require path segments determined by the definitions and uses [13, 14]. In this respect, the input domain is partitioned so that there is a subdomain corresponding to each slice associated with a statement in a suspicious fault region or safety–critical or accuracy-critical regions along a certain execution path. Similarly, mutation testing strategies can be further improved by seeding faults in critical code regions and detecting subdomains to exercise the regions. Domain detection algorithms are described in the following sections.

14.2   Random Testing

Random Testing (RT) is to select test data at random, with a uniform distribution over the input space of the program under test. A program's input space is the Cartesian product of bounded intervals of its input variables. Such a bounded Cartesian product in an n-dimensional space is called a hyper-cuboid. Random testing over a hyper-cuboid space is implemented by randomly selecting points within the hyper-cuboid. Path-oriented Random Testing (PRT) generates uniformly spread test data that covers a single control flow path of the program under test [2]. PRT is a type of random testing.

14.2.1   Path Random Testing

In search of the sub-spaces of n-dimensional program space covering a certain execution path, PRT suggests partitioning the hyper-cuboid n-dimensional space into 2^k, $k \leq n$, subdomains and invokes a constant propagation function [2] to reject those randomly selected subdomains that do not satisfy the path constraint. Here, the

Fig. 14.1 Input domain space partitioned into 24 subdomains by PRT

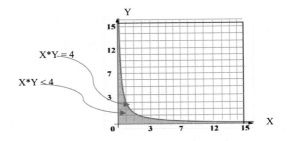

constraint propagation function checks whether the domain of the variables is consistent with the path constraint. Each input variable of a program has its own range [min..Max] of values. The constraint propagation function removes those values that do not satisfy the path constraint. The steps taken by PRT are as follows:

1. Partition the domain of each input variable into K sub-domains, where $K \leq n$, and n is the number of input variables.
2. Call a constraint-propagation function to examine the validity of each sub-domain against the path constraint.
3. Refute invalid sub-domains.
4. If no sub-domain is left, then declare the path as infeasible
5. Otherwise, draw uniformly at random test data from the remaining sub-domains

For instance, as shown in Fig. 14.1, applying the constraint propagation algorithm on the path constraint, $X*Y < 4$ where X and Y range [0..15], only the shaded squares, satisfying the path constraint, remain, and the rest are removed.

In Fig. 14.1, the input domain is a 16×16 plain, including 256 points. The gray region contains 39 valid points, satisfying the path constraint, $X*Y <= 4$. Therefore, about 15% of the whole input domain is valid, and the probability of random selection of an invalid point not satisfying the path constraint is 85%.

14.2.2 The IP-PRT Algorithm

Improved path random testing or IP-PRT [6] improves the PRT [2] algorithm. The improvement is achieved by reducing the number of subdomains to be examined against a path constraint. To this aim, we have suggested successively dividing the input variables domain into two halves, one after the other, and repeating this process until each input domain is divided into K sub-domains [2]. Our proposed method is called Improved PRT or, in short, IP-PRT [6]. IP-PRT, unlike PRT, does not partition all the input variable domains into K equal-sized subdomains at once. Instead, it divides the domains into two equal subdomains, one at a time, and this operation is repeated so that, eventually, the domain of each of the input variables is divided into k equal subdomains. For instance, the process of partitioning the input space of the example given in Fig. 14.1 is shown in Fig. 14.2.

As shown in Fig. 14.2, the number of times the subdomains of the input domain are supposed to be validated by the IP-PRT algorithm is:

$$1 + 2 + 4 + 6 + 10 + 14 + 22 + 30 + 48 = 137$$

Hence, it is observed that using the IP-PRT approach, the number of times required to validate the sub-domains of the input space is reduced from 256 to 137. The PRT algorithm partitions an n-dimensional input space into K^n, regions where the user defines the value of the division parameter, K. PRT invokes a constraint propagation function, K^n times, to check the validity of each region. For instance, in the above example, the input domain is divided into $16*16 = 256$ cells.

The IP-PRT algorithm is presented in Fig. 14.3. The algorithm uses a binary search technique to locate and remove those parts of the program input space which are inconsistent concerning a given path constraint. In search of sub-spaces of a

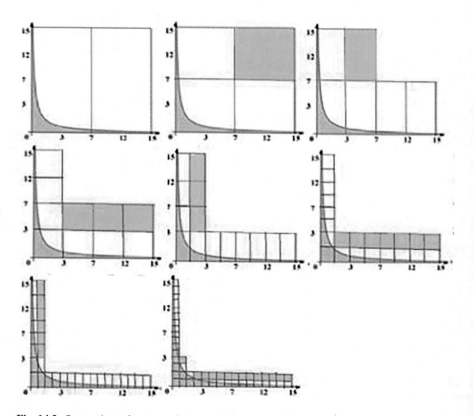

Fig. 14.2 Successive refinement of the input domain in IP-PRT

Algorithm : IP-PRT
_____

Input:
      Inputs-domain: a set of input variables domain,
      PC: path constraint,
      N: maximum number of test data to be returned by the algorithm
      K: division parameter
Output:
      $t_1, \ldots, t_N$ or ϕ (non-feasible path)
Method:
    1.   T:= ϕ;
    2.   D := in-domains := (dom-1, dom-2, .. , dom-n)
    3.   if not (D is inconsistent concerning PC), then Add d to Cur_Set
    4.   For no-divides := 1 to log(K)*n do
    5.   if isEmpty(Cur_Set) then return ϕ; -- Infeasible path
    6.   For dim:= 1 to n do
    7.   Forall D in Cur_Set do
    8.   LowerBound := D.dom[dim].lower;
    9.   UpperBound := D.dom[dim].upper;
   10.   Mid := (D.dom[dim].lower + D.dom[dim].upper) / 2;
   11.   D1 := D2 := D;
   12.   D1.dom[dim].upper := Mid;
   13.   D2.dom[dim].lower := Mid + 1;
   14.   if not (D1 is inconsistent concerning PC), then Add D1 to new_set;
   15.   If not(LowerBOund = UpperBound) then
   16.   if not (D2 is inconsistent concerning PC), then Add D2 to new_set;
   17.   End Forall
   18.   Cur_Set := New_Set;
   19.   New_Set := { }; //Empty Set
   20.   End For;
   21.   End For;
   22.   // Select sample test data from the feasible regions
   23.   Let D1', . . . , Dp' be the list of domains in cur_set;
   24.   if p≥1 then
   25.   while N>0 do // N is initially the number of required test data
   26.   Pick up uniformly D at random from D1', . . . , Dp'
   27.   Pick up uniformly t at random from D;
   28.   if PC is satisfied by t then
   29.   add t to T;
   30.   N:=N−1;
   31.   end if
   32.   end while
   33. end if
      return T;

Fig. 14.3 The IP-PRT algorithm

program input space, satisfying a given path constraint, the use of binary search seems promising.

The binary search is fulfilled by iteratively dividing each part of the input space into 2^n equal parts, for $\log_2(K)$ times, where K is the division parameter and n indicates the number of inputs. The input space is a hyper-cuboid in n-dimensional space, where each dimension represents the values of an input variable. Every part of the input space is divided into two parts in the direction considered for each input variable in each iteration.

In this way, since there are n input variables, each part will be divided into 2^n equal parts. In this way after $\log_2(K)$ Iterations of the input space will be eventually divided into 2^k equal parts. Before dividing each part, it is validated against the path constraint by calling a constraint propagation function. If a part of the input space is inconsistent with the path constraint, it is removed; otherwise, it is divided into two equal parts.

A significant critique of the IP-PRT algorithm is that it validates sub-domains without considering whether their enclosing domain is consistent with the path constraint. The IP-PRT algorithm can be further improved by applying diagrammatic reasoning to reduce the number of times required to call the constraint propagation function to locate those regions of the input domain of the program under test, which is inconsistent with the path constraint.

As shown in Fig. 14.4a, the whole input space, D, is initially examined against the path constraint, $X*Y > 4$, and then it is sub-divided into two equal-size parts, D1 and D2, shown in Fig. 14.4b, This time both the D1 and D2 are examined, by calling the constraint propagation function. In Fig. 14.4(c), D2 = D2-1 $\cup$ D2_2 is valid, and since D2_1 is invalid, it is deduced that D2_2 is also valid, and there will be no need to check it. This deduction is readily possible by the rules of set theory. If we show the values in regions D2, D2_1, and D2_2 with sets A, B, and C, then we will have the following:

$$\begin{cases} \boldsymbol{B \cup C = A} \\ \exists x | x \in A \text{ and } valid(x) \\ \nexists x | x \in C \text{ and } valid(x) \end{cases} \implies \exists x | x \in B \text{ and } valid(x)$$

The above relation states that if set A is acceptable and set C is unacceptable, set B is acceptable and does not need to be examined. This feature, called Free-ride, is an essential feature of Euler/Venn's reasoning. Euler/Venn's reasoning is the subject of the next section. In this way, considering the process of dividing the input space, shown in Fig. 14.4, the number of times the subdomains must be examined against the path constraint will be reduced from 127 to 70.

Fig. 14.4 Number of consistency checks concerning the path constraint

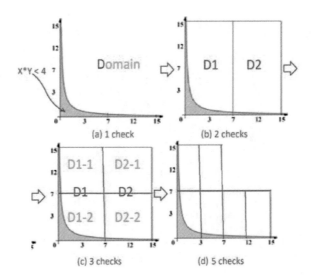

(a) 1 check

(b) 2 checks

(c) 3 checks

(d) 5 checks

14.3   Euler/Venn Reasoning for Test Data Generation

By expressing the problem of reducing the input domain in the language of the theory of sets and Euler/Venn diagrams [9, 15] and using their properties and principles, it is possible to accelerate the process of discovering acceptable sub-domains as well as the production of efficient and effective test data. The main question here is how to illustrate the partitioning of input domains with the Euler/Venn diagram and how an Euler/Venn diagram is implemented and applied. The answers are given in the following subsections.

14.3.1   Euler/Venn Representation of Input Domains

Euler/Venn diagrams provide observational representations of sets and operations on sets such as intersection, union, complement, and difference. Complex set-theoretic expressions can be visualized by drawing Euler/Venn diagrams, representing each set as an oval or combination of overlapping ovals. Euler//Venn diagrams benefit from substantial observational features over set-theoretic expressions. They help to understand what a complex set-theoretic expression means to say. Relationships between sub-domains, such as overlapping and containment, are best represented in Euler-Venn diagrams, shown in Fig. 14.5.

As shown in Fig. 14.5a at the beginning, the Euler/Venn diagram is an empty rectangle, X:<0..15> and Y:<0..15>, representing the input domain, D. In Fig. 14.5b, the input domain is divided vertically into two equal sub-domains, D1 and D2. In the

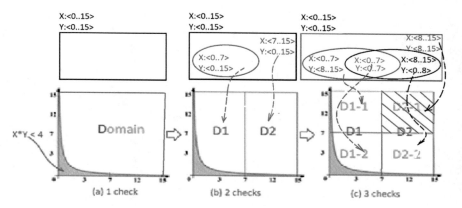

Fig. 14.5 A partitioned input domain and its corresponding Euler/Venn diagram

corresponding Euler/Venn diagram, D1 is represented by an oval in the rectangle, representing the whole input domain, D.

The rest of the rectangle is assigned to D2 as the complement of D1. In Fig. 14.5c, the input domain is horizontally divided into two equal parts. The lower part is represented by a second oval in the Euler/Venn diagram. The rest of the rectangle is considered the complement of the lower part.

If the domain of input variables in a path constraint expression is represented with an Euler/Venn diagram, then finding suitable values to satisfy the path constraint can be considered as finding the corresponding region in the Euler/Venn diagram. Also, the process of dividing the input domain can be illustrated by adding appropriate ovals to the Euler/Venn diagram. For example, as shown in Fig. 14.5c, partitioning the domain of each of the input variables, X and Y, in the path constraint, $X*Y \leq 4$, four distinct sub-domains D1_1, D1_2, D1_3, and D1_4 with the following ranges, are obtained:

$$D1_1 : (X \in 0 \ldots 7, Y \in 8 \ldots 15), D1_2 : (X \in 0 \ldots 7, Y \in 0 \ldots 7),$$
$$D2_1 : (X \in 8 \ldots 15, Y \in 8 \ldots 15), D2 \ldots 2 : (X \in 8 \ldots 15, Y \in 0 \ldots 7)$$

The Euler/Venn diagram in Fig. 14.5c shows these sub-domains and their relations. When adding an oval to an Euler/Venn diagram, its intersection with existing ovals may result in many sub-regions. With the set-theoretic principles, it is possible to determine whether or not the resultant sub-regions are inconsistent with the path constraint without examining them using the constraint propagation algorithm. In the continuation, the use of DAGs to implement Euler/Venn diagrams is described.

14.3.2 DAG Implementation of Euler/Venn Diagrams

It is proven in [47] that the implementation of an Euler/Venn diagram with a Directed
Acyclic Graph (DAG) does not affect the features provided by the diagram. All
principles of the set theory, such as subscription, community, difference, subset, etc.,
are also applicable to DAGs. DAGs can be used interchangeably with Euler/Venn
diagrams in reasoning processes.

That is why we use a DAG data structure to implement Euler/Venn diagrams and
reason about them. Below in, Fig. 14.6, it is shown how a DAG corresponding to
an Euler/Venn diagram can be constructed. Here, the union and intersection of two
domains, A and B, are shown as A + B and AB, respectively. The complement of a
domain, A, in the input domain space, is shown by A'.

1. Base: An empty Euler/Venn diagram, E, representing the input space, S, as an
   empty rectangle we built. The Euler/Diagram is also implemented as a DAG,
   consisting of a single node V as its root.

2. Induction: When dividing the input space into two regions, an oval representing
   one of the regions is inserted into the corresponding Euler/Venn diagram. The
   second region will be considered the complement of the first region in the
   Euler/Venn diagram. To insert an oval, A, representing a new sub-domain into
   the Euler/Venn diagram, E, one of the following cases may apply, in practice,
   dependent on the location of the sub-domain in the input space:

Fig. 14.6 Translating
Euler/Venn diagram into
DAGs

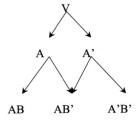

a. The new oval, A, does not intersect with the existing ones in the Euler/Venn diagram, E. A and its complement, A′, are added as children to the root node, V, of the DAG.

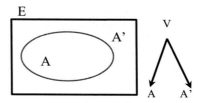

b. The new oval, A, intersects with an existing oval, B, in E. Two leaf nodes labeled B and B′ are added as the children of the root node, V, to the DAG.

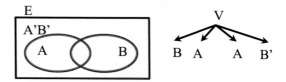

Each of the two overlapping ovals, A and B, are divided into two sub-regions, as shown below:

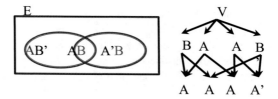

A is subdivided into AB and AB′, where AB = A ∩ B and AB′ = A ∩ B′. Region B is subdivided into A′B and AB.

c. The new oval, A, completely encloses the existing oval, B, in E.

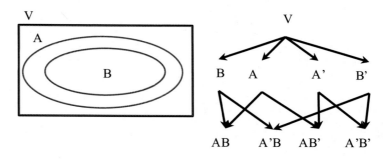

Since, B is enclosed by A, then $AB = A \cap B = B$, $A'B' = A' \cap B' = B'$, $AB' = A \cap B' = \varphi$ and finally $A'B = A' \cap B \neq \varphi$. Therefore, the DAG could be simplified, as shown in Fig. 14.6.

Figure 14.7 illustrates the input space for two input variables, x, and y, together with their corresponding Euler/Venn diagram and DAG representations (Fig. 14.7).

14.4 Domain Coverage Algorithm

In Sect. 14.2.2 above, the IP-PRT algorithm is introduced as an improvement to the PRT algorithm. As described above, the IP-PRT algorithm can be further improved by applying set-theoretic principles to avoid unnecessary calls to the constraint propagation function [1]. The improvement is achieved by applying the Euler/Venn reasoning. To this end, the use of Euler/Venn diagrams to represent the input domain of the program under test is suggested in Sect. 6.3. As shown in Fig. 14.7, Euler/Venn diagrams can be implemented as DAGs. To this end, a new algorithm called DomainCoverage is introduced in this section.

The algorithm uses Euler/Venn reasoning to determine the consistency of sub-domains of the input domain concerning a given path constraint. A sub-domain is labeled as valid if it contains at least a point, and when its coordinates are used as input data, the desired path constraint will be satisfied, and the path will be executed. The DomainCoverage algorithm is targeted at finding sub-domains that fully satisfy the path constraint. The algorithm uses a DAG data structure to keep track of and validate sub-domains generated while breaking down the input domain to find regions that are entirely consistent with a given path constraint or vice versa. A sub-domain is considered valid if it contains at least a point, satisfying the path constraint. The IP-PRT algorithm only proceeds with dividing valid sub-domains. The DomainCoverage algorithm removes invalid nodes from the DAG representation of the input domain.

A worked example is presented in the following sub-section to get the reader ready for an in-depth understanding of the algorithm.

14.4.1 Example

As an example, consider the code fragment presented in Fig. 14.8. Suppose the domain of values for the variables x and y, to examine the execution path leading to the printf statement, is required.

Considering the conditions in statements 1 and 3, the path constraint, PC, for executing the printf statement is:

$$PC = ((x > y) \text{and} (x + y < 30))$$

1. **Base:** An Euler/Venn diagram representing a given input domain, D, is initially shown as an empty rectangle. The corresponding DAG is shown as a single node labeled D

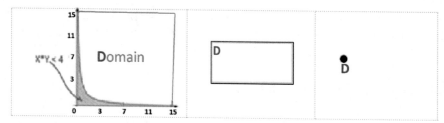

2. Induction: The input domain is broken down into two subregions. Correspondingly, an oval, A, representing one of the sub-regions, is inserted into the Euler/Venn diagram. The second sub-region is represented as the complement of A called A'. The DAG representation of the diagram contains two leaves, A and A'.

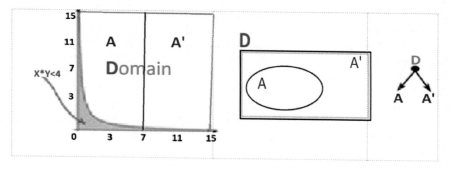

3. The input domain is further broken down horizontally into two parts, as a result of which the input domain is partitioned into four regions. Correspondingly, a new oval, B, crossing the previous one, is inserted into the Euler/Venn diagram, resulting in four sub-domains, AB', AB, A'B, and A'B'. In the corresponding DAG, nodes representing the newly generated subdomains are added to the DAG.

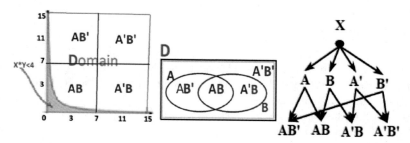

Fig. 14.7 DAG and Euler/Venn representations

Fig. 14.8 A sample code fragment

```
void test1(int x, int y)
{
  1.  if (x>y) {              //Condition1
  2.      s=x+y;
  3.      if (s<30)           //Condition2
  4.          printf ("Solved !"); }
}
```

Fig. 14.9 The valid subdomain is the rectangular region

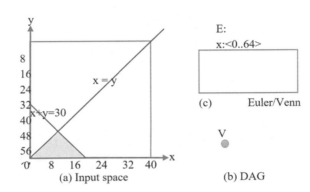

(a) Input space

(b) DAG

(c)   Euler/Venn

As shown in Fig. 14.9, the program input domain is a 2-dimensional Euclidian space. It is assumed that the domain of both variables x and y lie between zero and 64. Initially, the Euler/Venn representation for the input space is an empty rectangle, and the corresponding DAG is a single node labeled V.

As shown in Fig. 14.10, the input space is first vertically divided into two equal parts, A and A′. An oval, labeled A, is successfully inserted into the corresponding Euler/Venn diagram. After adding the node A: (x:<0..32>, y:<0..64>) and its complement A′: (x:<32..64>, y:<0..64>) to the DAG, the constraint propagation function is invoked to validate A and its complement, A′, against the path constraint, PC. Since A′ is invalid, it is removed from the input space and its DAG representation. Removing A′ from the DAG, only one child, A, will be left for node V. This means that V is restricted to A, and there will be no points for keeping node V. Therefore, as shown in Fig. 14.10b, the node V is replaced with its child A.

As shown in Fig. 14.11, in the next three steps, the input space is reduced as far as only a 16 × 16 rectangle surrounding the triangle, representing the valid region, remains.

As shown in step 3 of Fig. 14.11, before a sub-domain can be divided, it is validated against the path constraint. Therefore, it is observed that D' is removed from the input space. As a result, its complement D is moved to the root of the DAG. Movement of a node such as D to the root of the DAG means that the whole input space is now reduced to D. Therefore, the extents of C along each axis must be such that they are less than or equal to the extent of D. Figure 14.11, shows step by step refinement of the input space towards the sub-domains, entirely consistent concerning the given path constraint, PC.

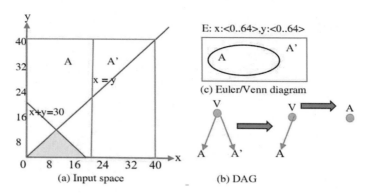

Fig. 14.10 The valid subdomain is the rectangular region

In Fig. 14.12, the rest of the refinements applied to the input space are illustrated. It is observed that the more we refine the input space, the more we get close to the boundaries of the rectangular region, satisfying the path constraint.

14.5   The Domain-Coverage Algorithm

The main body of the DomainCoverage algorithm is presented in Fig. 14.13. The inputs to this algorithm are: Path Condition (PC), the domain of each of the input variables, v1 to vn, represented by an array named Input_intervals, and finally, the division parameter, K. This algorithm returns either an array, Valid_Intervals, of almost fully valid domains or Null to indicate that the path is infeasible. A valid subdomain contains at least one point that satisfies the path constraint, PC. A subdomain is fully valid if all the points residing in it satisfy the path constraint.

The algorithm comprises three main steps. In the first step, A DAG, including a single node representing the whole input domain, is created. A function called isValid() is invoked to make sure that the input domain contains at least one point, satisfying the path constraint, PC. The second and third steps are repeated in a 2-level loop where the outer loops K times and the inner loop iterates for the number of input variables. In each iteration of the inner loop, the input space is partitioned by cutting along the sub-domains of a different input variable. The number of partitions will be two times the number of sub-domains of the input variable. After each sub-domain is partitioned, a function called Add2DAG is invoked to insert the partitions into the DAG. In step 3 of the algorithm, a function called isValid() is invoked to validate the leaves of the DAG. If a leaf is inconsistent concerning the path constraint, PC, a function called DeleteNode(), is invoked to remove the leaf from the DAG. A pseudocode description of the DeleteNode algorithm is presented in Fig. 14.14.

The DeleteNode() function removes a given node, A, from the DAG. In the first step of the algorithm, node A and its children's hierarchy are removed from the DAG.

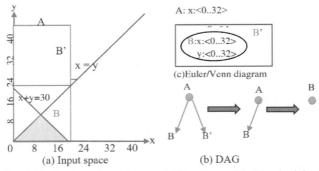

Step 1. The input space is restricted to a 64*32 region, and the B' region is invalid

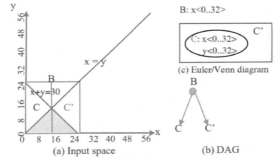

Step 2. The input space is restricted to a 32*32 region, and C and C' are valid.

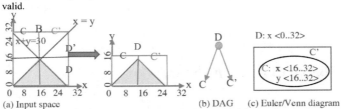

Step 3. The input space is restricted to a 16*32 region, and C and C' are valid

Fig. 14.11 Finding regions satisfying PC $= ((x > y)$ and $(x + y < 30))$

In the second step, all the single child nodes are replaced with their child. To this end, the domain of the single child node is reduced to its child's.

In Fig. 14.15, an algorithm named Add2DAG is presented. Add2DAG is invoked to add a given node, A, to the DAG. The algorithm comprises two steps. First, the smallest DAG node that covers node A is selected as the parent of nodes A and A'. The DAG is then populated with the intersections of the modes A and A' with its leaves.

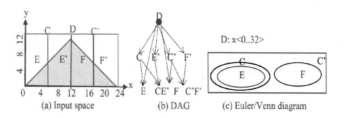

(a) Input space   (b) DAG   (c) Euler/Venn diagram

Step 1: Bisecting C and C' vertically, the two new ovals E and F will be
inside C and C', respectively

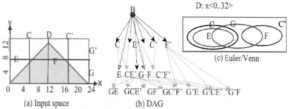

(a) Input space   (b) DAG

Step 2: Bisecting the existing sub-domains horizontally

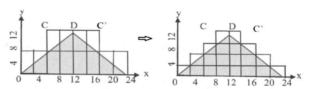

Steps 3 and 4: Bisecting the existing sub-domains vertically and horizontally.

Fig. 14.12   Finding regions satisfying PC

14.6   A Simple Constraint Propagation Algorithm

The abovementioned algorithms, PRT, IP-PRT, and CodeCoverage, aim at finding
a test suite that uniformly exercises a selected path. All these algorithms use a sort
of constraint reasoning to build such a test suite efficiently. Constraint reasoning
consists of constraint propagation and variable labeling [2].

The constraint propagation algorithm removes inconsistent values and labelings
from the domains of variables involved in a constraint and infers solutions. Constraint
propagation algorithms are typical of polynomial time [8]. However, it cannot decide
alone whether a constraint satisfaction problem is satisfiable [2]. That is why they
are typically combined with a search strategy. In continuation, I propose a simple
algorithm to decide whether a constraint is satisfiable.

14.6.1 *Constraint Satisfaction*

Divide and conquer is a valuable technique to open ways to overcome unknowns and complexities. The difficulty is how to break down complex problems into simpler ones that can be solved one at a time. Adopting the divide and conquer principles to tackle problems, we break down a constraint, as a logical expression, into relational expressions, shown below in Fig. 14.16.

Therefore, without loss of generality, we assume each constraint as a relational expression LeftExpr RelOp RightExpr, where LeftExpr and RightExpr are two regular expressions, and RelOP is a relational operator such as $<$, $<=$, $>=$, or $>$. We apply concolic execution to symbolically compute the expressions, LeftExpr, and RightExpr in terms of the input variables. Basic interval arithmetics can be applied to evaluate the domain of the constraint in terms of the domain of the input variables involved.

To compute the domains, LeftDom of LeftExp and RightDom, of RightExp, an algorithm, ExprDom, is introduced. The boundaries of the domains are compared to determine whether the domains include any value satisfying the constraint. As shown in Fig. 14.17a, If the LeftDom and RightDom overlap, then there are values belonging to these domains that satisfy the constraint. Otherwise, as shown in Fig. 14.17b, either

```
Algorithm  DomainCoverage (PC, Input_Intervals) returns Valid_intervals;
   Input:
        1- PC: path-constraint;
        2- input_Intervals: array[1..no.inputs] of Input-variables-domains;
        3- K: Division parameter;
   Output:
        1- Valid_Intervals: List of valid input sub-domains;
   Method:
      Step 1:
        -- Create a DAG with a single node containing the initial intervals of input
           DAG := CreateEmptyDAG(input_intervals, no.inputs );
           Valid := DAG.Root.isValid( );
           If not Valid then return Null  - infeasible path – EndIf;
      Step 2:
        while no.iterations < K  do
           -- Partition input domain by bisecting the input variable domains
           -- one at a time.
           For dimension-no := 1 to no.inputs do
        Step 2.1: Keep non-overlapping (disjoin) intervals in an array, Dom.
              Foreach leaf of DAG do
                  Dom[i] := leaf.dim[dimension-no];
              End foreach;
              Sort(Dom[i]);
              //Keep just one domain and remove the rest
              removeOverlappingIntervals( Dom );
```

Fig. 14.13 The main body of the domain coverage algorithm

Step 2.2:

Divide the input domain in the direction of the axis assigned
    to one of the n input variables into two parts, at a time and
add the resultant parts as ew nodes to the DAG.
For all the DAG leaves addressing the same interval, intv,
    of the input variable v_i do
1. determine the $min_{i,j}$ and $Max_{i,j}$ values of the lower and
    upper bounds of the other input variables, v_j intervals.
2. Insert a new node into DAG, representing a new partition
    that is enclosed within $min_{i,j}$, $Max_{i,j}$.
For i := 1 to Dom.Length do
  NewNode := New DAG;
  Foreah leaf of DAG do
    If (overlaps(leaf.dim[dimension-no], Dom[i])
      For d := 1 to no-dimensions do
        NewNode.dim[d].lowerBound :=
            min(NewNode.dim[d].lowerBound,
                    leaf.dim[d].lowerBound);
        NewNode.dim[d].upperBound :=
            min(NewNode.dim[d].upperBound,
                leaf.dim[d].upperBound)
      End For
    End If
  End Foreah
  NewNode.dim[dimension-no] := randomlyBisect(DOM[i]);
  If NewNode.isValid(PC)
    then Add2DAG(DAG, NewNode); End IF

Step 3

    Call constraint propagation function, isValid(), to detect and
    remove leaves inconsistent
    --   concerning the path constraint, PC.
  Foreach leaf of DAG do
    Valid := leaf.isValid(PC);
    if Not Valid
        then DeleteNode(leaf, DAG)
        else leaf.addToList(validIntervals);
            EndIf;
  End Foreach
 End for ;
End For; - End for dimension.no
End while; - End while K
  Return DAG.leaves;
End algorithm;

Fig. 14.13 (continued)

```
Algorithm DeleteNode (A, DAG)
  Input    A: An invalid node to be removed from the DAG
  Output   DAG: reconstructed DAG
  Method:
  - Step 1: Remove A and all its children hierarchy from the DAG;
          Queue.AddRear ( A );
          While Queue.IsNotEmpty( ) do
                node := Queue.GetFrontt();
                Foeach child of node.GetNextChild() do
                      Queue.AddRear ( node );
                End Foreach;
                DAG.Remove_Node( node );
          End While;
  - Step 2: Remove all single child nodes from the DAG, and add the child to the
          - children's list of each of the node's parents provided that no parent
          - is left for the child.
             while ( SingleChildNode := DAG.FindNextSingleChildNode()) do
                Child = SingleChildNode.GetChild( );
                ParrentsSet = SingleChildNode.GetParrents();
                DAG.RemoveNode( SingleChildNode );
                if Child.HasNoParrents( ) then
                      DAG.AddChild(Child, ParrentsSet );
                End if;
             end while;
  end algorithm;
```

Fig. 14.14 The DeleteNode algorithm

the region of input space circumvented by LeftDom and RightDom fully covers the constraint, or no input values in this region satisfy the constraint.

Now, considering the relation $\Re$ in the constraint LeftExp $\Re$ RightExp, it is observed that the decision on whether the non-overlapping parts of the domains, LeftDom and RightDom, satisfy the constraint is straightforward. The non-overlapping parts are shaded in Fig. 14.17. For instance, if the constraint is $x*2 > y*y$ and $x \in [0.0.4]$ and $y \in [2.0.8]$, then we will have:

LeftDom $= [0*2..4*2] = [0..8]$
RightDom $= [2*2..4*4] = [4..16]$
OverlappingRegion $=$ LeftDom $\cap$ RightDom $= [4..8]$
NonOverlappingRegions $= [0.4] \in$ LeftDom $\cup [8..16] \in$ RightDom

The subdomain $[0..4] \in$ LeftDom is not acceptable because all the values in this subdomain are less than the values in RightDom. Conversely, all the values in $[8..16] \in$ RightDom are acceptable. Therefore, it is observed that the decision about these two non-overlapping parts is straightforward.

To determine whether an area of the input space includes acceptable values to satisfy a given constraint, I offer a new algorithm called Simplified Constraint Propagation (SCP), shown in Fig. 14.18.

Algorithm Add2DAG (A, DAG)
Inputs :
  1- A: A new node to be inserted into the DAG;
  2- DAG: Directed Acyclic Graph
Output :
       1- DAG: the reconstructed DAG
Method :
- Step 1: Find the parent of the node, A, and add A and its complement
A' to the DAG
       Foreach Node in birthFirstTraversal(DAG) do
        If Node.Covers(A) then X := Node; End If;
       End Foreach;
       X.AddChild(A);
       A' = X – A;
       X.AddChild(A');
       If (not A.isValid( )) then DAG.DeleteNode(A);
       Else If (not A'.isValid( )) then DAG.DeleteNode(A'); Endif;
       Endif;
- Step 2: Add intersection of A and its complement A' with leaves to
the DAG
       ForAll Node in DAG.birthFirstTraversal( ) do
        If (Node.isLeaf( )) then
          X := Node $\cap$ A;
          If (X $\neq$ Null $\wedge$ X $\neq$ A) then
            Node.AddChild(X); A.AddChild(X); End If;
          Y := Node $\cap$ A';
          If (Y $\neq$ Null $\wedge$ Y $\neq$ A') then
            Node.AddChild(Y); A.AddChild(Y); End If;
        End IF;
       End ForAll;
End algorithm;

Fig. 14.15 The Add2DAG algorithm

LogicalExpr	$\rightarrow$ not Boolean | Boolean
Boolean	$\rightarrow$ Boolean and Bool | Bool
Bool	$\rightarrow$ Bool or RelationalExpr | RelationalExpr
RelationalExpr	$\rightarrow$ Expr RelOp Expr | Expr
RelOP	$\rightarrow$ < | <= | = | != | > | >=

Fig. 14.16 Logical expressions syntax

Fig. 14.17 Comparison of the LeftDom and RightDom domains

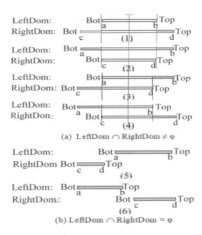

(a) LeftDom ∩ RightDom ≠ φ

(b) LeftDom ∩ RightDom = φ

Fig. 14.18 An algorithm to decide whether a given constraint is satisfiable

Algorithm SCP
Input:
  Expression$_{left}$ Relop Expression$_{right}$: constraint;
  (D1, D2, ..., Dm) : input variables domains;
Output:
  Satisfies: Boolean;
Method:
  Step 1-- Evaluate the domains LeftDom and RightDom using
    [D1,D2, ..., Dm].
    -- ExprDom evaluates the domain of a given expression in
    terms of the domains of its operands while traversing
    the expression tree in a post-order manner.

    LeftAST := CreateAbstractSyntaxTree(Expression$_{left}$);
    RightAST := CreateAbstractSyntaxTree(Expression$_{right}$);
    LeftDom := ExprDom(AST, (D1, D2, ... , Dm));
    RightDom := ExprDom(AST, (D1, D2, ... , Dm));

  Step 2-- Determine whether any values satisfy the constraint.
    if(LeftDom ∩ RightDom ≠ φ) **then** Satisfies := true
    else Satisfies := (LeftDom.Top < RightDom.Bottom) and
    (RelOp = '<') or
    (LeftDom.Bottom > RightDom.Top) and (RelOp = '>');
  End SCP;

14.7 Interval Arithmetic[1]

The ExprDom algorithm accepts an expression tree as input and uses a post-order traversal of the tree to evaluate the domain of the expression. In Fig. 14.19, a regular expression and its corresponding Abstract Syntax Tree (AST) are shown.

Figure 14.19 shows a sample expression and its domain computation steps. As shown in Fig. 14.19(c), the expression domain is computed while traversing the AST

---

[1] Download the source code from: https://doi.org/10.6084/m9.figshare.21324522

$$z = x^2 * f(x, y+2) - \cos(x) * (x - 2*y)$$

$$x \in [0..180], \ y \in [0..100]$$

(a) A sample expression

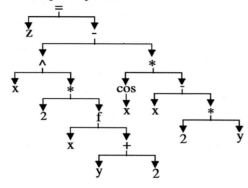

(b) Abstract syntax tree

$$D[1] = \text{Dom}(x^2) = [0..180^2]$$
$$D[2] = \text{Dom}(y+2) = [2..102]$$
$$Lb = \text{searchForMin}(\text{``f''}, \text{Dom}(x), D[2])$$
$$Ub = \text{searchForMax}(\text{``f''}, \text{Dom}(x), D[2])$$
$$D[2] = [Lb..Ub] = [0..1]$$
$$D[1] = [Lb*0..180^2*Ub]=[0..180^2]$$
$$Lb = \text{searchForMin}(\text{``cos''}, \text{Dom}[x])$$
$$Ub = \text{searchForMax}(\text{``cos''}, \text{Dom}[x])$$
$$D[2] = [Lb..Ub] = [-1..1]$$
$$D[3] = \text{Dom}(x*2) = [0..360]$$
$$D[3] = \text{Dom}(x) - D[2] = [-180..0]$$
$$D[2] = D[2] * D[3] = [0..180]$$
$$D[1] = D[1] - D[2] = [0..180^2 - 180] = [0..32220]$$

(c) Steps of computing the domain

Fig. 14.19 A sample expression and its corresponding AST

in postorder. The basic arithmetic for the domain intervals [a..b] and [c..d] are as follows:

$$[a..b] + [c..d] = [a+c..b+d]$$
$$[a..b] - [c..d] = [a - d..b - c].$$

Fig. 14.20 Pseudo code
description of the ExprDom
algorithm

```
Algorithm ExprDom
  Input:
      AST : Tree;
  Output:
      ExprDomain : Interval;
  Method:
      simpleOperators = [+, -, *, /];
      Foreach node in PreOrderTraversal( AST ) do
          node.prevDom = node.domain;
          if not node.isLeaf() then
              if node.val in simpleOperators then
                  let [a..b] = node.child[1].dom;
                  let [c..d] = node.child[2].dom;
                  case node.val of
                      + : node.domain = [a+b..c+d];
                      - : node.domain = [min(a-b,c-d)..Max(a-b,c-d)];
                      * : node.domain = [min(ac, ad, bc, bd),
                                          max(ac, ad, bc, bd)];
                      / : node.domain = [min(a/c, a/d, b/c, b/d)..
                          max(a/c, a/d, b/c, b/d)];
                  end case
              endif
              if node.isFunctionCall then
              Lb = searchForMin(node);
              Ub = searchForMax(node);
              Node.dom = [Lb..Ub];
              endif
          endif not isLeaf
      end foreach
end Algorithm
```

[a..b] × [c..d] = [min(ac, ad, bc, bd)..max(ac, ad, bc, bd)].
[a..b]/[c..d] = [min(a/c, a/d, b/c, b/d)..max(a/c, a/d, b/c, b/d)].

The example presented in Fig. 14.20 shows two function calls. The source code for these functions is not available. Therefore, to estimate the domain of these black-box functions, two functions, searchForMin and searchForMax are invoked. These two functions attempt to find the minimum and maximum return value of the black box functions by executing the functions several times with different values, selected from the domains of their input parameters. In fact, the searchForMin and searchForMax functions apply a binary search in the input space of the black box functions by executing them several times with different input values. A pseudo-code description of the ExprDom algorithm is given in Fig. 14.21.

14.7.1   Dynamic Domain Reduction

Dynamic Domain Reduction (DDR) is a test data generation method that takes a condition of the form:

left_expression rolatioal_operator right_expression (e.g. x >= y*x).

Algorithm GetSplit (LeftDom, RightDom, SrchIndx)

precondition: LeftDom and RightDom are initialized appropriately and
                SrchIndx is one more than the last time GetSplit was called
                with these domains for this expression.

postcondition : split >= LeftDom.Bot && split >= RightDom.Bot &&
                split <= LeftDom.Top && RightDom.Top

input :   LeftDom: left expr's domain with Bot and Top values
                RightDom: right expr's domain with Bot and Top values

output :   split Float
                -- a value that divides a domain of values into two subdomains .

Begin
-- Compute the current search point .
```
    static int srchIndex = 1;
    static int srchStep = 1;
-- srchPt= (1/2,1/4.3/4,1/8,3/8,5/8,7/8, 1/16,3/16...)
    for(k = 1, j = 0; (k<srchStep**2) && (j<srchIndex); j++ ) k += 2;
    srchPt = (k – 2)/2**srchStep; srchIndex++;
    if(2**srchStep == srchIndex) {srchIndex = 1; srchStep++;}
-- Try to equally split the left and right expression's domains .
```

```
--1    Left :            Bottom |------------------------|Top
--     Right :        Bottom |----------------------------------|Top
    if (LeftDom.Bot >= RightDom.Bot && LeftDom.Top <= RightDom.Top)
    then split = (LeftDom.Top - LeftDom.Bot)*srchPt + LeftDom.Bot
--2    Left :            Bottom |----------------------------------|Top
--     Right :        Bottom |------------------------|Top
    else if (LeftDom.Bot <= RightDom.Bot && LeftDom.Top >= RightDom.Top)
    then split = (RightDom.Top - RightDom.Bot)*srchPt + RightDom.Bot
--3    Left :                Bottom |------------------------|Top
--     Right :        Bottom |------------------------|Top
    else if (LeftDom.Bot >= RightDom.Bot && LeftDom.Top >= RightDom.Top)
    then split = (RightDom.Top - LeftDom.Bot)*srchPt + LeftDom.Bot
--4    Left :        Bottom |------------------------|Top
--     Right :        Bottom |------------------------|Top
    else if LeftDom.Bot <= RightDom.Bot AND LeftDom.Top <= RightDom.Top
    then split = (LeftDom.Top - RightDom.Bot)*srchPt + RightDom.Bot
    return split
end GetSplit
```

Fig. 14.21 An algorithm to find a split point to divide the domain

and invokes a function called GetSplit to determine the subdomains satisfying the condition [3]. For instance, consider the condition x >= y*x. The domains of the variables x and y are reduced in such a way that:

(a)   the new subdomains satisfy the condition, and
(b)   the size of the two domains is balanced.

With this modification, the proper subdomain for x and y is obtained. All values in the resulting subdomains will undoubtedly satisfy the condition x >= y, but these values may not satisfy the conditions after the condition x >= y in path constraint.

In such cases, the search process backtracks by invoking GetSplit to select another split point. This algorithm is critical to the search process and test case selection.

For example, if LeftDom = $[-20, 20]$, RightDom = $[-40, 30]$, and GetSplit then the first case, labelled 1, is satisfied. If it is the first tme that GetSplit is invoked:

srchIndex = srchStep = 1andsrchPtis1/2

SplitPoint = (LeftDom.Top $-$ LeftDom.Bot) $*$ srchPt $+$ LeftDom.Bot

$\qquad$ = $(20 - (-20))/2 + (-20) = 0.$

$\qquad$ => LeftDom = $[-20, -1]$

$\qquad\qquad$ RightDom = $[0, 30]$

Ontheotherhand, ifSrchIndxis2, andscrhStepis2 :

$\qquad$ = SrchPt = $3/8,$

SplitPoint = $(20 - (-20)) * (3/8) + (-20) = -5.$

$\qquad\qquad$ => LeftDom = $[-20, -6]$andRightDom = $[-5, 30]$

The DDR procedure starts with the first condition as the relational expression, "leftExp R rightExp," in the path constraint. It invokes GetSplit to find a split to split the leftExp, and rightExp domains so that the relation R holds. In the next step an algorithm called Update, shown in Fig. 14.10, is invoked to compute the value of the variables in leftDom and rightDom, based on their domains. The algorithm proceeds with the next condition. If it can not find subdomains of the input domain satisfying the following condition. It backtracks to the previous condition (decision point) and selects a new split point. Table 14.1 illustrates the main steps of the DDR algorithm to detect sub-domains, satisfying the path constraint (x >= y) && (3*y <= 10−x).

As shown in Table 14.1, DDR starts with the first ondition, x > y, where the domain of both the input variables, x, and y, is $[1, 64]$. DDR invokes the GetSplit function to select the next split point. GetSplit chooses the split point to be 32, leaving the domain for LeftDom to $[32, 64]$ and RightDom to $[0, 32]$. In this case, the domain of x and y will be restricted to $[33, 64]$ and $[1, 32]$, respectively. As shown in the first row of Table 14.1, the sub-domains $[33, 64]$ and $[1, 32]$ for respectively x and y cover the branch condition, "x > y." Here, the first shortcoming of the GetSplit algorithm is revealed. Although there are acceptable data in some intervals, such as $[16, 32]$ for x and $[1, 15]$ for y, the intervals are removed. As shown in row number 2 of the Table, the algorithm proceeds with the second constraint, 3 * y <= 10 − x. Given that, at this stage, the sub-domains for the input variables x and y are $[33, 64]$ and $[1, 32]$, respectively, the LeftDom and RightDom for "3 * y <= 10 − x" will be calculated as follows:

$$\text{LeftDom} = 3 * [1, 32] = [, 96]$$
$$\text{RightDom} = 10 - [33, 64] = [-54, -23]$$

Since there are now intersections between the two sub-domains, the algorithm backtracks to the first condition, "x > y. "This time, the search and split points are

Table 14.1 Step-by-step execution of the GetSplit function in the DDR approach

#	Condition	LeftDom	RightDom	Search point	Split point	X	Y
1	x>=y	<1..64>	<1..64>	Sp1=1/2	32	<33..64>	<1..32>
2	3*y<=10-x	<3..96>	<-54,-23>	backtrack to x>=y			
3	x>=y	<1..64>	<1..64>	Sp2=1/4	16	<17..64>	<1..16>
4	3*y<=10-x	<3..48>	<-54,-7>	backtrack to x>=y			
5	x>=y	<1..64>	<1..64>	Sp3=3/4	48	<49,64>	<1,48>
6	3*y<=10-x	<3..144>	<-54,-39>	backtrack to x>=y			
7	x>=y	<1..64>	<1..64>	Sp4=1/8	8	<9,64>	<1,8>
8	3*y<=10-x	<3..24>	<-54,1>	backtrack to x>=y			
9	x>=y	<1..64>	<1..64>	Sp5=3/8	24	<25,64>	<1,24>
10	3*y<=10-x	<3..72>	<-54,-15>	backtrack to x>=y			
11	x>=y	<1..64>	<1..64>	Sp6=5/8	40	<41,64>	<1,40>
12	3*y<=10-x	<3..120>	<-54,-31>	backtrack to x>=y			
13	x>=y	<1..64>	<1..64>	Sp7=7/8	56	<57,64>	<1,56>
14	3*y<=10-x	<3..168>	<-54,-47>	backtrack to x>=y			
15	x>=y	<1..64>	<1..64>	Sp8=1/16	56	<5,64>	<1,4>
16	3*y<=10-x	<3..12>	<-54,5>	Sp1=1/2	4	<6,6>	<1,2>

considered to be ¼ and 16, respectively, which leads to the intervals [17, 64] and [1, 9] for the variables x and y, respectively. Again, the condition $3 * y <= 10 - x$ is considered. Given the sub-domains for the variables x and y, LeftDom and RightDom will be equal to:

$$LeftDom = 3* \ < 1..16 > \ = \ < 3..48 >$$
$$RightDom = 10 - \ < 17..64 > \ = \ < -54.. - 7 >$$

If the domains LeftDom and RightDom of a given path constraint, represented as a relational expression, do not overlap, then the domains either completely satisfy or violate the constraint. On the other hand, overlapping domains contain sub-domains that satisfy the constraint.

It is observed that a poor selection of the split point may restrict the domains such that a later condition cannot be satisfied. In such cases, the algorithm backtracks and selects a different split point. The difficulty is that it does not make any consideration for its former selection and reexamines regions that are already examined.

In summary, there are three significant deficiencies concerning the use of DDR, as follows:

1. Getting trapped in backtracking loops.
2. Repeated unnecessary reexamining of some parts of the domains.

3. GetSplit should consider the intersections of the LefDom and RightDom. The decision about the rest is straightforward.

I have introduced a new method called RDDR [26] to resolve the difficulties with DDR. The GetSplit function reduces the domains of the expressions, lefExp, and rightExp, so that the relation R in "leftExp R rightExp" holds. After changing the domains of leftExp and rightExp, the domains of variables in leftExp and rightExp should be modified accordingly. The function Update, described in the next section, propagates the changes in the domain of an expression to the variables involved in the expression.

14.7.2 Update Function

The update function propagates any changes to an expression's domain back to the variables involved in the expression. The algorithm is invoked recursively to balance the changes to the variables' domains. For instance, consider the expression $A + B$, where the domain of A and B are [0,20] and [10,50], respectively. The ExprDom algorithm, in Fig. 14.20, computes the domain of $A + B$ as follows:

$(A + B).dom = [A.Bot + B.Bot, A.Top + B.Top] = [0 + 10, 20 + 50] = [10,70]$

Now, assume the domain of $A + B$ is reduced from [10,70] to [24,36]. The Update algorithm adjusts the domains of A and B as follows:

$$A = [A.Bot/2, \ A.Top/2] = [24/2, 36/2] = [12, 16]$$
$$B = [B.Bot/2, \ B.Top/2] = [24/2, 36/2] = [12, 16]$$

Our modified version of the Update algorithm is shown in Fig. 14.22. Jeff Offutt originally wrote the algorithm [3].

The update function relies on a recursive method to extract the domain of the constituent variables from an expression domain. Unlike the update function proposed by Jeff Offutt, our update function supports the expressions, including constants. Moreover, it does not need a backtracking loop to revise the split point of the input domain, which may result in reexamining of some regions. As an example, consider the function test in Fig. 14.23.

Table 14.2 illustrates the main steps of the DDR algorithm to detect sub-domains, satisfying the path constraint (x > y) && (x > y*y) && (3*y < 20-x). As shown in Table 14.2, the domain of x and y in the first constraint, x > y, is initially [0,64]. DDR reduces the domains LeftDom] and for RightDom to be [0,64]. In this case, the domain of y and x will be restricted to [0,32] and [33,64], respectively. The second row of the table shows domains of x and y*y as [33,64] and 0,1024], respectively. The Update function computes the domain of y as [3, 16] based on the domain of y*y, i.e., [0,1024].

Algorithm: Update

Input: Expr: An expression

Bot: A **bottom value of Expr's domain**

Top: A **top value of Expr's domain**

Output: The domain of expression constituent variables

```
1.   L = GetLeftExpression(Expr)
2.   R = GetRightExpression(Expr)
3.   op = GetArithmeticOperator(Expr)
4.   switch (op)
5.   {
6.     case "+":
7.       if(L is a numeric constant)
8.         leftDom.Bot = leftDom.Top = L;
9.         rightDom.Bot = Bot - L;
10.        rightDom.Top = Top - L;
11.      else if(R is a numeric constant)
12.        rightDom.Bot = rightDom.Top = R;
13.        leftDom.Bot = Bot - R;
14.        leftDom.Top = Top - R;
15.      else
16.        leftDom.Bot = rightDom.Bot = Bot / 2;
17.        leftDom.Top = rightDom.Top = Top / 2;
18.    case "-":
19.      if(L is a numeric constant)
20.        leftDom.Bot = leftDom.Top = L;
21.        rightDom.Bot = L - Top;
22.        rightDom.Top = L - Bot;
23.      else if(R is a numeric constant)
24.        rightDom.Bot = rightDom.Top = R;
25.        leftDom.Bot = Bot + R;
26.        leftDom.Top = Top + R;
27.      else
28.        leftDom.Bot = Top + m * (Top - Bot) / 2;
29.        leftDom.Top = (Top - Bot) / 2 + Top + m * (Top - Bot) / 2;
30.        rightDom.Bot = (Top - Bot) / 2 + m * (Top - Bot) / 2;
31.        rightDom.Top = Top - Bot + m * (Top - Bot) / 2;
32.    case "*":
33.      if(L is a numeric constant)
34.        leftDom.Bot = leftDom.Top = L;
35.        rightDom.Bot = Bot / L;
36.        rightDom.Top = Top / L;
37.      else if(R is a numeric constant)
```

Fig. 14.22 Using an expression's domain to adjust its variables' domain

```
38.       rightDom.Bot = rightDom.Top = R;
39.       leftDom.Bot = Bot / R;
40.       leftDom.Top = Top / R;
41.    else
42.       if(Bot >= 0 && Top >= 0)
43.          leftDom.Bot = rightDom.Bot = Sqrt(Top);
44.          leftDom.Top = rightDom.Top = Sqrt(Bot);
45.       else if(Bot < 0 && Top < 0)
46.          leftDom.Bot = 1;
47.          leftDom.Top = Abs(Top);
48.          rightDom.Bot = Bot / Abs(Top);
49.          rightDom.Top = -1;
50.       else if(Bot < 0 && Top >= 0)
51.          leftDom.Bot = - Sqrt(Abs(Bot));
52.          leftDom.Top =
53.               Min(Floor(Sqrt(Abs(Bot))), Floor(Top / Sqrt(Abs(Bot))));
54.          rightDom.Bot =
55.               - Min(Floor(Sqrt(Abs(Bot))), Floor(Top / Sqrt(Abs(Bot))));
56.          rightDom.Top = Sqrt(Abs(Bot));
57.    case "/":
58.       if(L is a numeric constant)
59.          leftDom.Bot = leftDom.Top = L;
60.          rightDom.Bot = L / Bot;
61.          rightDom.Top = L / Top;
62.       else if(R is a numeric constant)
63.          rightDom.Bot = rightDom.Top = R;
64.          leftDom.Bot = R * Bot;
65.          leftDom.Top = R * Top;
66.       else
67.          leftDom.Bot = (Top + 1) / (Bot + 1) * Pow(Bot, n + 1) * Pow(Top, n);
68.          leftDom.Top = Pow(Bot, n) * Pow(Top, n + 1);
69.          rightDom.Bot = Pow(Bot, n) * Pow(Top, n);
70.          rightDom.Top = (Top + 1) / (Bot + 1) *
71.                         Pow(Bot, n) * Pow(Top, n + 1);
72. }
73. if(leftDom.Bot > leftDom.Top)  Sawp(leftDom.Bot, leftDom.Top);
74. if(rightDom.Bot > rightDom.Top)  Sawp(rightDom.Bot, rightDom.Top);
75. Update(L, leftDom.Bot, leftDom.Top);
76. Update(R, rightDom.Bot, rightDom.Top);
```

Fig. 14.22 (continued)

Fig. 14.23 A function with
three decision-making points

```
void test1(int x, int y)
{
1.       if (x>y)
2.          if (x>y*y)
3.             if (3*y<20-x)
4.                printf("solved");
}
```

Table 14.2 Step-by-step execution of the GetSplit function

	K	Predicate	Input domain X	Y	Left dom	Right Dom	Split point	Srch indx
1	k1=1	x>y	[0,64]	[0,64]	[0,64]	[0,64]	32	½
2	k2=1	x>y*y	(32,64]	[0,32]	(32,64]	[0,1024]	48	½
3	k3=1	3*y<20-x	[48,64]	[0,6]	(3,18]	[-44,-28]	Backtrack	
4	k2=2	x>y*y	(32,64]	[0,32]	(32,64]	[0,1024]	40	¼
5	k3=1	3*y<20-x	[40,64]	[0,6]	(3,18]	[-44,-20]	Backtrack	
6	k2=3	x>y*y	(32,64]	[0,32]	(32,64]	[0,1024]	56	¾
7	k3=1	3*y<20-x	[56,64]	[0,7]	(3,21]	[-44,36]	Backtrack	
8	k2=4	x>y*y	(32,64]	[0,32]	(32,64]	[0,1024]	36	1/8
9	k3=1	3*y<20-x	(36,64]	[0,6]	(3,18]	[-44,-16]	Backtrack	
10	k2=5	x>y*y	(32,64]	[0,32]	(32,64]	[0,1024]	44	3/8
11	k3=1	3*y<20-x	[44,64]	[0,6]	(33,64]	[-44,-24]	Backtrack	
12	k2=6	x>y*y	(32,64]	[0,32]	(32,64]	[0,1024]	52	5/8
13	k3=1	3*y<20-x	[52,64]	[0,7]	(3,21]	[-44,-32]	Backtrack	
14	k2=7	x>y*y	(32,64]	[0,32]	(32,64]	[0,1024]	60	7/8
…	…	…	…	…	…	…	…	…

14.8   Why Domain Coverage

As described in chapter 10, the test result may be correct despite running the faulty statement. Such test cases are coincidentally correct. For instance, consider the following if statement:

if (i > 0.1000) j++; // should be: i > 0.1001.

The fault reveals only when 0.1000 < i <= 0.1001. In such cases, the tester is better off having the input domain covering an execution path rather than a single test data. One may argue that we already have enough to test each execution path even once; there is no room for more test data. Our job as test data generator is to ensure adequate test data. It is up to the tester to use as much as they can. Indeed, test data generation is one task, and test data selection is another. Test data generation requires providing a test suite covering the system behavior. However, depending on what operational domain the system under test belongs to, different adequacy criteria could be used for test data selection. For example, selecting a set of test data covering the behavioral domain is essential for safety–critical systems, where one failure may result in uncompensable consequences.

Secondly, fault proneness provides the tester with some prior knowledge about the regions of code to be tested relatively more cautiously. For example, in the case of cyber-physical systems, the tester can prioritize the paths with relatively more I/O interactions with the physical environment. The underlying rationale is that the I/O

Fig. 14.24 The fault in line 3 reveals only when $30 \leq x + y < 40$

```
void foo(int x, int y) {
1.  w=0;
2.  if (x>y)
3.      if (x+y<40)   //correct: if (x+y<30)
4.          w=1;
5.      else  w=2;
6.  else  w=3;
7.  if (x>2*y)
8.      w=3;
9.  else  w=4;
10. print(w);}
```

intensive paths are more likely to manifest faulty behavior. By having the feasible domain for the test paths, the tester can apply test selection methods such as fault-based testing to prepare a reliable test data set. Besides, some researchers propose to statically rank test paths using the program's abstract syntax tree under test [30].

Thirdly, having more than one test data covering the faulty execution path with different results helps the debugger guess where the suspicious region could be. On the other hand, testing a cyber-physical system is often based on a simulator. As simulation-based testing provides some representations of the system behavior, it is possible to guess the regions of the code that contribute to the undesirable behavior of the system. Therefore, the tester could prioritize those regions for testing. For example, consider a self-driving car system. Once the ego vehicle collides with a pedestrian crossing the road, the code segments corresponding to the automated emergency braking feature could be the suspicious regions. Moreover, in the case of a new software version, the newly modified or added regions of code and the regions of code depending on these regions are relatively more fault-prone.

Putting aside the fault proneness data, we also agree that testers do not have any prior knowledge about which paths may be faulty. However, when the tests fail, the debugging starts with prior knowledge of which execution paths can be faulty. Therefore, after detecting faulty execution paths, test data should be generated to further restrict the suspicious regions of code to branches and statements, including the root cause of the failure. For example, Fig. 14.24 represents a program that receives two integer variables as input parameters. A fault exists in line 3 of the program; it should be $(x + y < 30)$.

Suppose the initial domain of input variables x and y are in the range [0,64]. In Fig. 14.25, the small triangle includes coincidentally correct data points, (x,y), where $(x > y)$ && $(x + y < 30)$. The failure region is the red strip between the large and small triangles. Running the program with any value from this region causes the program to fail. Although selecting any value from the small triangle executes the faulty statement, it does not reveal the fault. This kind of data is known as the coincidentally correct test case Fig. 14.26.

Assume the initial domain of the variables x and y is divided into two subdomains. This division creates four distinct regions, shown in Fig. 14.25.

For each region, one test data is selected randomly and given as input to the program under test. Table 14.2 shows each test data's execution result and statement

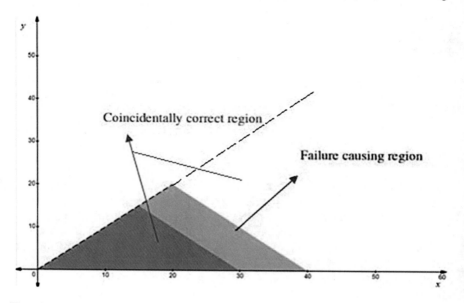

Fig. 14.25 Coincidentally correct regions

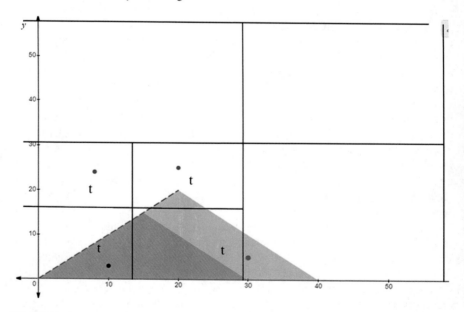

Fig. 14.26 the region including t4 is further partitioned (iteration 2)

Table 14.3 The running example diagnosis matrix (iteration 1)

Test data	x	y	S1	S2	S3	S4	S5	S6	S7	S8	S9	S10	Result
t1	5	50	•	•				•	•	t	•	•	Pass
t2	58	38	•	•	•		•		•		•	•	Pass
t3	51	4	•	•	•		•		•	•		•	Pass
t4	22	14	•	•	•	•			•		•	•	Fail
Suspiciousness scores are computed using the Ochiai metric — susp			0.5	0.5	0.58	1	0	0	0.5	0	0.58	0.5	
rank			4	5	2	1	8	9	6	10	3	7	

coverage. In the table, the bullet symbol in the intersection of the ith row and the jth column indicates the coverage of statement S_i by the test case t_j. Moreover, the last column represents whether the execution of a specific test data failed or passed. Ochiai formula is used to compute the suspiciousness score for the statements S1 to S10.

As shown in Table 14.3, the statements are ranked based on their suspiciousness scores as follows:

{S4, S3, S9, S1, S2, S7, S10, S5, S6, S8}

S4 is the most suspicious statement; hence, the region of the input domain covering S4 is further partitioned to locate the faulty statement. The result shows four new test data, t5, t6, t7, and t8, are generated. The Ochiai formula is used to compute the suspiciousness. Table 14.4 shows the coverage and test results for the eight test data, t1 to t8.

As shown in Table 14.4, the statements are ranked, for a second time, based on their suspiciousness computed by Ochiai, as follows:

{s4, s3, s1, s2, s7, s8, s10, s5, s6, s9}.

Since s4 is the most suspicious statement, one of the regions enclosing t6 or t8 is further partitioned. As shown in Fig. 14.27, the region containing t6 is partitioned, resulting in four more test data being created.

In this step, four new test data (t9, t10, t11, and t12) are selected from the newly created regions. Table 14.5 shows the test coverage and results. As shown in Table 14.5, the statements are ranked based on their suspiciousness scores as follows:

{s4, s3, s1, s2, s7, s10, s8, s9, s5, s6}.

The suspicious statements will be accurately revealed by repeating the process on the reduced domains.

Table 14.4 The diagnosis matrix for the running example (iteration 2)

tc	x	y	S1	S2	S3	S4	S5	S6	S7	S8	S9	S10	Execution Result
t1	5	50	●	●				●	●		●	●	Passed
t2	58	38	●	●	●		●		●		●	●	Passed
t3	51	4	●	●	●		●		●	●		●	Passed
t4	22	14	●	●	●	●`			●		●	●	Failed
t5	8	24	●	●				●	●		●	●	Passed
t6	10	3	●	●	●	●			●	●		●	Passed
t7	20	25	●	●				●	●		●	●	Passed
t8	30	5	●	●	●	●			●	●		●	Failed
Suspiciousness scores are computed using the Ochiai metric		susp	0.5	0.5	0.63	0.82	0	0	0.5	0.4	0	0.13	
		rank	3	4	2	1	8	9	5	6	10	7	

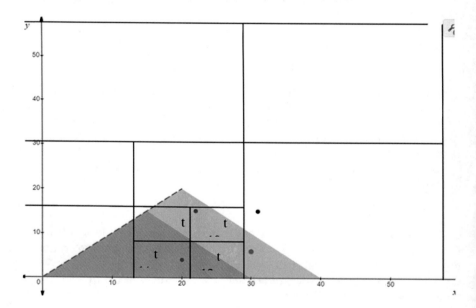

Fig. 14.27 The region, including t6, is further partitioned (iteration 3)

Table 14.5 The diagnosis matrix for the running example (iteration 2)

tc	x	y	S1	S2	S3	S4	S5	S6	S7	S8	S9	S10	Execution Result
t1	5	50	•	•				•	•		•	•	Passed
t2	58	38	•	•	•		•		•		•	•	Passed
t3	51	4	•	•	•		•		•	•		•	Passed
t4	22	14	•	•	•	•			•		•	•	Failed
t5	8	24	•	•				•	•		•	•	Passed
t6	10	3	•	•	•	•			•	•		•	Passed
t7	20	25	•	•				•	•		•	•	Passed
t8	30	5	•	•	•	•			•	•		•	Failed
t9	22	15	•	•	•	•			•	•		•	Failed
t10	31	15	•	•	•		•		•		•	•	Passed
t11	20	4	•	•	•	•			•	•		•	Passed
t12	30	6	•	•	•	•			•	•		•	Failed
Suspiciousness scores are computed using the Ochiai metric		susp	0.58	0.58	0.67	0.81	0	0	0.58	0.61	0.20	0.58	
		rank	3	4	2	1	9	10	5	7	8	6	

14.9 Summary

Some latent faults are not revealed unless the program under test is executed with a specific combination of input data. A faulty statement may be executed several times without revealing its fault. To this end, this chapter suggests providing the tester with subdomains of the input domain rather than test data. In this way, considering the subdomains for executing a particular target, the tester may generate as much test data as required.

Test data adequacy is achievable by providing the testing process with domains of input variables rather than test data. Test data generation techniques should be improved to detect the domains of input variables, covering a specific part of the program under test. In this respect, all the existing coverage criteria, such as statement, branch, path, mutation, control flow graph edge, and control flow graph node, could be extended to domain coverage criteria. For instance, the branch domain coverage criterion evaluates the subdomain, which is supposed to cover a certain program branch.

In order to determine whether a constraint is satisfiable, the SCP algorithm can be used. This algorithm compares the domains of the expressions on two sides of the relational expression, representing a constraint. If the domains overlap, then the constraint is definitely satisfiable.

Test input generation is an essential task in automated testing. Most existing techniques generate a single input for each program's execution path, which may not be enough in specific scenarios. Given a path condition and the initial domains of the input variables, domain reduction techniques gradually reduce the domains so that all the inputs constructed using values from the resultant domains are valid for triggering the target execution path. The Euler/Venn reasoning system is applied to decide how the input domain should be reduced step by step and organize the reduction steps into a DAG so that the whole process is better tracked.

14.10   Exercises

1. Complete the constraint propagation algorithm in Sect. 14.6 to consider constraints as logical expressions rather than simply relational expressions. Use the syntax for logical expressions presented in Fig. 14.16.
2. Black box function is an unresolved symbolic and dynamic-symbolic testing problem. Describe how the SCP algorithm, presented in Fig. 14.18, has coped with this problem.
3. Write an algorithm to accept a subdomain covering a suspicious fault or critical region of code, and generate test data to examine the region. You should bear in mind that failure-causing inputs, indeed, form contiguous regions. Therefore, common sense tells us that test cases should be distributed evenly across the subdomain to improve the chance of hitting the failure sub-regions.
4. Follow the DomainCoverage algorithm, presented in Fig. 14.13, step by step to generate test data for the following path constraint:

$$PC : y^2 < x * 0.5 \&\& y <= \sin(20 * x) \&\& y >= x * 0.6$$

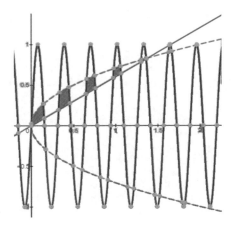

5. Complete the DDR process in Table 14.2 to calculate the domain of inputs, satisfying the path constraint (x > y) && (x > y*y) && (3*y < 20-x).
6. Use the IP-PRT algorithm instead of the GetSplit to calculate leftDom and rightDom for the path constraint (x > y) && (x > y*y) && (3*y < 20-x). Compare the results with the domains obtained when using the GetSplit function.
5. Use the result of exercise 1 and the IP-PRT algorithm to develop your own constraint solver and compare it with the Z3 solver.
6. To understand the domain coverage and search-based testing practices and techniques, it helps to look at the critical scenarios in CARLA simulator. The simulator is intended for self-driving cars. Here is the latest source code release: https://github.com/carla-simulator/carla. Follow the ReadMe file and install the simulator. Then use the python API called Senario_Runner and automatically generate your own scenario or try one of the predefined scenarios that led to the crash. At first, instrument the simulator code and find the critical region leading to the crash. Then apply the CodeCoverage or IP-PRT algorithm presented in this chapter and detect the domain of input variables covering the critical region. Now, apply a search-based method to look for other scenarios that result in a crash. To do so, search for the execution paths, each having the highest overlap with the original path.

References

1. Parsa, S.: A reasoning-based approach to dynamic domain reduction in test data generation. Int. J. Softw. Tools Technol. Transf. (STTT) Arch. **21**(3), 351–364 (2019)
2. Gotlieb, A., Petit, M.: A uniform random test data generator for path testing. J. Syst. Softw. **83**, 2618–2626 (2010)

3. Offutt, A. J.: Automatic Test Data Generation. Ph.D. thesis. Georgia Institute of Technology, Atlanta, GA, USA (1988)
4. Offutt, A.J., Jin, Z., Pan,J.: The dynamic domain reduction approach for test data generation: design and algorithms. Technical Report ISSE-TR-94–110. George Mason University, Fairfax, Virginia (1994)
5. Offutt, A.J., Jin, Z., Pan, J.: The dynamic domain reduction procedure for test data generation. Softw. Pract. Exp. **29**(2):167–193 (1999)
6. Parsa, S.: Path-oriented random testing through iterative partitioning (IP-PRT). Turk. J. Electr. Eng. & Comput. Sci. (2018)
7. Parsa, S.: Enhancing path-oriented test data generation using adaptive random testing techniques. In: 2015 2nd International Conference on Knowledge-Based Engineering and Innovation (KBEI), pp. 510–513 (2015)
8. Nicolini, E.: Combined decision procedures for constraint satisfiability. Ph.D. thesis, Mathematics department, University of Milan (2006)
9. Swoboda, N.: Implementing Euler/Venn reasoning systems. In: Diagrammatic representation and reasoning, pp. 371–386. Springer, London (2002)
10. Schmitt, M., Orsoy, A.: Automatic Test Generation Based on Formal Specifications Practical Procedures for E±cient State Space Exploration and Improved Representation of Test Cases. Ph.D. thesis (2003)
11. Dijkstra, E.W.: The humble programmer. ACM Turing Lectures (1972)
12. Weyuker, E. J., Jeng, B.: Analyzing Partition Testing Strategies. IEEE Trans. Softw. Eng. **17**(7) (1991)
13. Rapps, S., Weyuker, E.J.: Data flow analysis techniques for program test data selection. In: Proc. 6th Int. Conf: Software Eng., pp. 272–278 (1982)
14. Rapps, S., Weyuker. E.J.: Selecting software test data using data flow information. IEEE Trans. Software Eng., **SE-II**, 367–375 (1985)
15. Stapleton, G.: A survey of reasoning systems based on Euler diagrams. Electron. Notes Theor. Comput. Sci. **134**, 127–151 (2005)
16. Schmitt, M., Orsoy, A.: Automatic Test Generation Based on Formal Specifications Practical Procedures for Efficient State Space Exploration and Improved Representation of Test Cases, Ph.D. thesis (2003)
17. Offutt, A.J., Lee, A., Rothermel, G., Untch, R.H., Zapf, C.: An Experimental determination of sufficient mutant operators. ACM Trans. Softw. Eng. Methodol. **5**(2), 99–118 (1996)
18. Weyuker, E.J., Jeng, B.: Analyzing partition testing strategies. Trans. Softw. Eng. **17**, 703–711 (1991)
19. Schmitt, M.: Automatic Test Generation Based on Formal Specifications Practical Procedures for Efficient State Space Exploration and Improved Representation of Test Cases, Ph.D. thesis, Georg August University, of Gottingen (2003)
20. Bartak, R.: Constraint programming: in pursuit of the Holy Grail. ACM Comput. Surv. **28A**(4)
21. Tsang, E.: Foundations of Constraint Satisfaction. Department of Computer Science. University of Essex, Colchester, Essex, UK (1996)
22. ILOG: ILOG Solver 4.4 – Reference Manual. ILOG S.A., France (1999a)
23. ILOG. ILOG Solver 4.4 – User's Manual. ILOG S.A., France (1999b)
24. Kubica, B. J.: Interval Methods for Solving Nonlinear Constraint Satisfaction, Optimization and Similar Problems From Inequalities Systems to Game Solutions. Springer (2019)
25. Hansen, E., Walster, W.: Global Optimization Using Interval Analysis. Marcel Dekker, NewYork (2004)
26. Parsa, S.: Improving dynamic domain reduction test data generation method by Euler/Venn reasoning system. Softw. Qual. J. **28**(2), 823–851 (2020)
27. Parsa, S.: A comprehensive framework for automatically generating domain-oriented test suite. Inf. Softw. Technol. (2022)

Printed in the United States
by Baker & Taylor Publisher Services